Human Success

Human Success

Evolutionary Origins and Ethical Implications

Edited by

HUGH DESMOND AND GRANT RAMSEY

OXFORD
UNIVERSITY PRESS

Oxford University Press is a department of the University of Oxford. It furthers the University's objective of excellence in research, scholarship, and education by publishing worldwide. Oxford is a registered trade mark of Oxford University Press in the UK and certain other countries.

Published in the United States of America by Oxford University Press
198 Madison Avenue, New York, NY 10016, United States of America.

Library of Congress Cataloging-in-Publication Data
Names: Desmond, Hugh (Philosopher), editor. | Ramsey, Grant, 1972– editor.
Title: Human success : evolutionary origins and ethical implications /
[edited] Hugh Desmond, Grant Ramsey.
Description: New York, NY : Oxford University Press, [2023] |
Includes bibliographical references.
Identifiers: LCCN 2022049824 (print) | LCCN 2022049825 (ebook) |
ISBN 9780190096168 (hardback) | ISBN 9780190096182 (epub)
Subjects: LCSH: Human evolution—Philosophy.
Classification: LCC GN281.4 .H857 2023 (print) | LCC GN281.4 (ebook) |
DDC 599.93/8—dc23/eng/20230103
LC record available at https://lccn.loc.gov/2022049824
LC ebook record available at https://lccn.loc.gov/2022049825

DOI: 10.1093/oso/9780190096168.001.0001

Printed by Integrated Books International, United States of America

Contents

PART III: HUMAN SUCCESS IN THE ANTHROPOCENE

Contributors

Susan C. Antón is the Silver professor of Anthropology, Center for the Study of Human Origins, New York University and studies the anatomy and evolution of early members of the genus *Homo*. Her pioneering studies of cranial variation in *Homo erectus, Homo neanderthalensis*, and other *Homo* species explore the role of (genetic) isolation and local environment in shaping cranial form. Her innovative studies of the dispersal dynamics of early *Homo* species show that eclectic diets, high mobility, behavioral flexibility, and cohesive social organization contributed to the ability of early humans to move fast relative to other mammals. She is past editor of the Journal of Human Evolution, elected member of the American Academy of Arts and Sciences, and former president of the American Association of Physical (now Biological) Anthropologists. A former Ford Diversity Fellow, she currently serves as Treasurer of the Society of Senior Ford Fellows.

Allen Buchanan was the James B. Duke Distinguished Professor of Philosophy at Duke University and is currently a professor of philosophy at the University of Arizona. He is the author of *Our Moral Fate: Evolution and the Escape from Tribalism* (2020), *The Evolution of Moral Progress: A Biocultural Account* (2018, with Rachell Powell), *Beyond Humanity: The Ethics of Biomedical Enhancement* (2013), *The Heart of Human Rights* (2013), and eleven other books.

Kathryn Demps is an associate professor in anthropology at Boise State University. She specializes in the study of the evolution of human behavior, focusing on human behavioral ecology and cultural evolution. Her work has explored how humans are flexible decision-makers, using social and ecological contexts to exploit their environment. This has included contexts as diverse as collecting wild honey in south India, outdoor recreation in southwest Idaho, and separating household consumption and production to participate in marketplace exchange.

Hugh Desmond is a research fellow at the Leibniz University of Hannover and assistant professor at the University of Antwerp. He received his PhD in philosophy from the KU Leuven and has held research and visiting positions at the Paris I-Sorbonne, KU Leuven, Princeton University, New York University, and the Hastings Center. He has a strongly interdisciplinary background, with degrees in physics and mathematics (KU Leuven) as well as music performance (Royal Conservatory of Brussels; Conservatory of Paris). His work centers on the philosophy and ethics of science and technology, with particular emphasis on biology. In the latter area, most of his work has centered on the interplay between agency and natural selection in evolutionary theory, and reflecting about consequences for evolutionary progress, human nature, and human enhancement.

Agustín Fuentes is an anthropologist at Princeton University whose research focuses on the biosocial, delving into the entanglement of biological systems with the social and cultural lives of humans, our ancestors, and a few of the other animals with whom humanity shares close relations. From chasing monkeys in jungles and cities and exploring the lives of our evolutionary ancestors to examining human health, behavior, and diversity across the globe, he is interested in both the big questions and the small details of what makes humans and our close relations tick. Earning his BA/BS in anthropology and zoology and his MA and PhD in anthropology from UC Berkeley, he has conducted research across four continents, multiple species, and two million years of human history. His current projects include exploring cooperation, creativity, and belief in human evolution, multispecies anthropologies, evolutionary theory and processes, and engaging race and racism. Fuentes is an active public scientist, a well-known blogger, lecturer, and tweeter and an explorer for National Geographic. He was recently awarded the Inaugural Communication & Outreach Award from the American Association of Physical Anthropologists and the President's Award from the American Anthropological Association and was elected to the American Academy of Arts and Sciences.

Matt Grove is a reader in evolutionary anthropology in the Department of Archaeology, Classics and Egyptology at the University of Liverpool. His interests include the Middle Stone Age of Eastern Africa, the effects of paleoclimatic change and variability on hominin evolution and dispersal, the evolution of hominin life history, and the evolution of hominin cognition and culture. He works on developing new quantitative and computational methods for studying the archaeological record. His research focuses on the impacts of climatic change and variability on human evolution, with a particular emphasis on the evolution of behavioral plasticity and its relationship to the archaeological and fossil records of *Homo sapiens* dispersal.

Marion Hourdequin teaches in the Department of Philosophy at Colorado College and specializes in environmental philosophy. Her research and teaching interests also include ethics, comparative philosophy, animal studies, and philosophy of science. Her current research focuses on climate ethics, climate justice, and the social and ethical dimensions of ecological restoration. She is the author of *Environmental Ethics: From Theory to Practice* (2015) and the editor, with David Havlick, of *Restoring Layered Landscapes* (2015). She serves as an associate editor for the journal *Environmental Values*, is on the editorial board of *Environmental Ethics*, and is President of the International Society for Environmental Ethics.

Daniel W. McShea teaches in the Department of Biology and Department of Philosophy at Duke University. His earlier work was in the field of paleobiology, where his focus was on large-scale trends in the history of life, especially documenting and investigating the causes of the (putative) trend in the complexity of organisms. A significant part of this work involved operationalizing concepts such as complexity and hierarchy, as well as clarifying conceptual issues related to trends at larger scales. More recently, he has been interested in teleology, in developing a theory of goal-directedness that explains how all goal-directed systems work, from organismal development and human wanting to the process of natural selection itself. He has three published books: *The Missing Two-Thirds of Evolutionary Theory* (2020) and *Biology's First Law* (2010, both with Robert Brandon) and *The Philosophy of Biology: A Contemporary Introduction* (2007, with Alex Rosenberg).

Richard Potts is a paleoanthropologist and has been the director of the Smithsonian Institution National Museum of Natural History's Human Origins Program since 1985. He is the curator of the Hall of Human Origins at the Smithsonian, where he holds the Peter Buck Chair in Human Origins. In partnership with the National Museums of Kenya, he leads ongoing excavations at field sites in southern and western Kenya. After receiving his PhD in biological anthropology at Harvard University in 1982, he taught at Yale before joining the Smithsonian in 1985. His research focuses on Earth's environmental dynamics and the processes that have led to human evolutionary adaptations. Recent contributions focus on how environmental instability has influenced human evolution, an idea that has stimulated new studies in earth sciences, paleontology, and computational biology. He is also the curator of the exhibit *Exploring Human Origins*, which has been traveling across the United States since 2015. Among many publications for scientific audiences and educators, he is the author of the exhibit companion book *What Does It Mean to Be Human?* (2010).

Rachell Powell (formerly Russell) is an associate professor in the Department of Philosophy at Boston University. Holding advanced degrees in philosophy, evolutionary biology, and law, she has published in areas ranging from the philosophy of biology and bioethics to political and legal philosophy. Her recent monographs include *Contingency and Convergence: Toward a Cosmic Biology of Body and Mind* (2020) and *The Evolution of Moral Progress: A Biocultural Theory* (2018, with Allen Buchanan). She is currently working on two books: one on the evolution of normativity, and the other on genetic engineering and the future of human nature. She has held fellowships at the National Humanities Center, the American Council of Learned Societies, and the Konrad Lorenz Institute for Evolution and Cognition Research, and she has been a principal investigator on grants from the Templeton World Charity Foundation, National Endowment for the Humanities, the John Templeton Foundation, and the Mellon Foundation. She served from 2011 to 2016 as associate editor for the *Journal of Medical Ethics*.

Grant Ramsey is a BOFZAP Research Professor in the Institute of Philosophy at KU Leuven, Belgium. His work centers on the philosophical problems at the foundation of evolutionary biology. He has published widely in this area as well as in the philosophy of animal behavior, human nature, and the moral emotions. He has published three books: *Chance in Evolution* (2016, with Charles Pence), *The Dynamics of Science* (2022, with Andreas De Block), and *Human Nature* (2023).

Peter J. Richerson is a Distinguished Professor Emeritus at the Department of Environmental Science and Policy, UC Davis. He is best known for his seminal work on cultural evolution. Much of this work is summarized in his 2005 book with Robert Boyd, *Not by Genes Alone: How Culture Transformed Human Evolution*. He continues to publish in this field, coediting a 2013 book, *Cultural Evolution: Society, Technology, Language, and Religion*, with Morten Christiansen. He was trained as a limnologist and has worked at Lake Tahoe, Lake Titicaca, and Clear Lake. He is past president of the Evolutionary Anthropology Society, the Human Behavior and Evolution Society, the Society for Human Ecology, and the Society for Cultural Evolution.

Bernd Rosslenbroich is an evolutionary biologist at Witten/Herdecke University, Germany. He is interested in the study of major transitions in animal and human evolution. The approach of his team is to analyze morphological and physiological patterns and trends and then to learn about underlying processes. In addition he is interested in organismic concepts of biology which take due account of the specific characteristics of the organism.

Geerat J. Vermeij is a Distinguished Professor of Marine Ecology and Paleoecology at the University of California, Davis. He is probably best known for his work chronicling the arms race among long-extinct mollusks and their predators. By examining and analyzing fossils for evidence of interspecies competition and predation, Vermeij has prompted the field of paleobiology to acknowledge the profound influences creatures have on fashioning each other's evolutionary fates. Vermeij has more than 280 publications, including six published books as well as papers in such leading journals as *Paleobiology, Science, American Naturalist*, and the *Philosophical Transactions of the Royal Society B*. He served as editor for *Evolution*; received a MacArthur Fellowship, the "genius" grant, in 1992; and in 2000 was awarded the Daniel Giraud Elliot Medal from the National Academy of Sciences. He was elected to the American Academy of Arts and Sciences in 2021 and the National Academy of Sciences in 2022.

Colin Waters is an honorary professor at the University of Leicester and chair (and former secretary) of the Anthropocene Working Group (AWG) of the Subcommission on Quaternary Stratigraphy, the body investigating the Anthropocene as a potential geological time unit. He was lead author of the high-impact 2016 *Science* paper on the Anthropocene and has had a central role in coordinating AWG activities in seeking formalization of the Anthropocene as a stratigraphic unit. He was formerly principal scientific officer with the British Geological Survey, with a 29-year career in which he developed particular expertise in the geology of British strata of the Carboniferous age (approximately 300 to 360 million years ago). Such strata have historic significance as a major source of minerals, including coal, but are also associated with much of the industrialization and urbanization of parts of the United Kingdom, which led to his interest in anthropogenic processes as a component of contemporary geology.

Mark Williams is a professor of paleobiology at Leicester University. He is a member (and former secretary) of the Anthropocene Working Group. His work is predicated on understanding major innovations in the biosphere over geological time. The co-author, with Jan Zalasiewicz, of the popular science books *The Goldilocks Planet* (2012), *Ocean Worlds* (2014), *Skeletons* (2018), and *The Cosmic Oasis* (2022). He co-authored the book *The Anthropocene: A Multidisciplinary Approach* (2020, with Jan and Julia Thomas).

Jan Zalasiewicz is emeritus professor of geology (paleobiology) at the University of Leicester. He is a member, formerly the chair, of the Anthropocene Working Group of the International Commission on Stratigraphy and the current chair of the Subcommission on Quaternary Stratigraphy. The author of many articles and books, he recently coedited *The Anthropocene as a Geological Time Unit* (2019). In 2008 he and his colleagues put forth the first proposal to adopt the Anthropocene Epoch as a formal geological interval.

1
Introduction
The Manifold Challenges to Understanding Human Success

Hugh Desmond and Grant Ramsey

1. Introduction

We often speak of our own species as being remarkably—or even uniquely—successful. As one would expect, such talk of "human success" can most often be found in popular science communications, such as newspapers or magazines, where the human species may be termed an "evolutionary success story" (Tarlach 2018) or, in a similar vein, "the dominant species on our planet" (Stringer 2019). The word "success" clearly speaks to the imagination. While "success" is probably not viewed as sufficiently precise to be of use in academic contexts, this way of speaking about humans does sometimes appear in academic and semi-academic contexts as well, usually in passing and as a starting point for scientific inquiry and explanation. For instance, some take the success of humans compared to other primates and large mammals as an explanandum for further inquiry (Sterelny 2012). Others see humans as a comparatively successful species tout court (Henrich 2016). In yet other places, humans are simply introduced as "undoubtedly the most successful species on Earth" (Smith 2020, 1).

Drawing attention to the language of "success talk" in this way will inevitably give the impression that this volume is merely skeptical of such language. This is not the case. However, the word "success" has multiple layers of meaning, and the mere fact that scientists turn to that word as a hook for a broader audience indicates that we are all, at some level, aware of these layers. Saying that the human species is a "success" draws the attention in ways that neutral, factual descriptions (e.g., "humans are bipedal, large-brained primates") do not.

Hugh Desmond and Grant Ramsey, *Introduction* In: *Human Success*. Edited by: Hugh Desmond and Grant Ramsey, Oxford University Press. © Oxford University Press 2023. DOI: 10.1093/oso/9780190096168.003.0001

At first sight, "success" could seem to be simply a neutral shorthand for some type of descriptive species-level metric, including geographic range; ecological dominance (as measured, for instance, by the consumption of biomass production in ecosystems); or population size. In this narrowly circumscribed sense, attributing success to humans is no different from the way success is attributed to an ant or bird species spreading around the globe or to some now extinct species that once roamed the Earth. According to such an analysis, there is not one single metric of species success, but many different metrics, and different species can be successful in different ways. It may be difficult to think of another large animal that comes close to *Homo sapiens* in population size or ecological dominance (except perhaps for the domesticated animals that humans breed for consumption), but what about species whose worlds take place at much smaller spatial scales and who are more or less oblivious to the evolution of the human species? Tardigrades—the phylum of minuscule eight-legged creatures known as "water bears"—come to mind as a very successful species in terms of population size or geographic range but that do not compete directly with the human species, being about four orders of magnitude smaller in body size.

If such analyses exhausted the meaning of "success," we could stop here and there would be no need for a volume on the subject. However, the concept of success connotes far more than the growth or persistence of species. As a usage of "success" that may seem at first entirely unrelated, think of how ethicists are traditionally concerned with the question *What is the good life?* For Aristotelian and neo-Aristotelian thinkers, who ascribe a purpose or teleology to human life, a life can be a "success" when that purpose is realized. For some neo-Aristotelians, human success involves the realization of intrinsic human capacities, especially cognitive capacities (e.g., Nussbaum 2000). Such a concept of human success depends on a concept that is perhaps even more controversial: human nature.

Of course, today many biologists, evolutionary anthropologists, and philosophers of biology are skeptical of the existence of human nature, or of the usefulness of the concept (see Hannon and Lewens 2018). However, human success need not be understood in relation to human nature. For a different understanding of human success, think of the expression "He was driven by a desire for success." The expression does not conjure up the image of a Socratic figure endeavoring to live a life of rational virtue. Instead, it calls to mind an individual who sacrifices friends, family, or even personal well-being in the pursuit of wealth, social status, or professional or political success. Success here connotes having power rather than the good, and illustrates how success has an ambiguous ethical status.

It is tempting to think that this ambiguity helps explain our attraction to the word "success": it seems like a good fit for a globalized, competitive era. Like a burger restaurant hitting on a successful formula and turning into a franchise invading markets worldwide, the human species seems to have hit on a pattern of behavior that has allowed it to invade ecosystems radically different from the ones primates typically inhabit. Or, like the successful influencer, the human species has "gone viral." In the process, humans have spread over the Earth and settled in regions as diverse as deserts, arctic tundra, and tropical rainforests. In so doing, the human population has risen many orders of magnitude above that of other primate species.

These comparisons only underline the darker side of human success. In the analogy with the burger franchise, invading new markets means that mom-and-pop restaurants are put out of business, straining the conditions of employment as well as the involvement of local communities. Similarly, the success of the human species has disrupted ecosystems and caused biodiversity to decrease. One could compare the way in which the human species exploits and manages nonhuman species with the way colonial powers both exploited other societies and governed them according to their own cultural ideas and norms.

These overtones evoked by the word "success" illustrate how human success is not merely a causal-descriptive concept that can be neatly captured by empirical metrics. And herein lies the danger: in reflecting on the success of the human species, we need to carefully consider whether we are just exporting our anthropocentric conceptions about what it is to be successful, either in terms of ethics or power. This dimension of success was anticipated by E. O. Wilson (1990, 19) in a passing remark in his overview of the ecology and evolution of social insects:

> "Success" and "dominance," laden with the history of Western culture, are risky words to use in biology. But they are nonetheless used all the time and sometimes interchangeably. We intuitively feel them to be necessary words in the biological vocabulary. Indeed, some kinds of organisms are successful and dominant. There is real biology in the distinction, and neologisms that we might invent would not be welcomed.

In a sense, the present volume aims to unpack the elements present in this quote. We agree that there is real biology in the term "success" and at the same time agree that the social and political connotations cannot be ignored. The term sits uneasily at the interface between evolutionary biology and ethics and raises many of the same types of issues—albeit in a more contemporary guise—that the concept of evolutionary progress did for many 20th-century evolutionary thinkers.

2. Success and Progress

The history of biological thought is replete with concepts that at first seem to describe a causal state of affairs but that, once scratched, reveal an underlying ethical-normative layer. Aristotelian biology would, of course, be the most obvious example, where the causal and the ethical are fused: the ethical good for humans, *eudaimonia,* is the fulfillment (*telos*) of human potentialities in a way that is structurally identical to how the oak tree represents the fulfillment of the acorn. However, concepts that muddle the fact-value distinction are not strange to the Aristotelian framework. Only in a post-Darwinian intellectual landscape, where the main mechanism of adaptive change (natural selection) seems to have nothing to do with what is ethically good, do they become puzzling phenomena.

Historically, the concept of evolutionary progress is one of the most important examples of such a muddling of facts and values, causes and reasons. On the one hand, progress seems to describe—and even causally explain—a state of affairs (namely, large-scale trends). On the other, it also judges the later states of affairs to be somehow "better" than the earlier ones. As intellectual heirs to the Enlightenment, it was difficult for 19th- and 20th-century naturalists to resist viewing the history of life and human evolutionary history in analogous terms.

Darwin's own views on progress are notoriously ambiguous (cf. Ruse 1996, 136–177; Richards 1988). Even though he made liberal use of the terms "higher" and "lower" in his references to taxa, he also worried about whether their use implied using humans as the standard of comparison. For example, in an 1854 letter to J. D. Hooker, he reflects on the concept:

> With respect to "highness" & "lowness," my ideas are only eclectic & not very clear. It appears to me that an unavoidable wish to compare all animals with men, as supreme, causes some confusion; & I think that nothing besides some such vague comparison is intended, or perhaps is even possible,

> when the question is whether two kingdoms such as the articulata or mollusca are the highest. (Darwin 1854)

In a similar vein, Darwin initially preferred to avoid the term "evolution" in favor of "transmutation" or "descent with modification," since "evolution" (from the Latin *e-volvere:* unrolling or unfolding) had the preformationist connotations of the unrolling of a predetermined plan, usually from a simple beginning to a more complex final state (Bowler 1975).

Reflecting the spirit of his times, Darwin (1871) was more overtly progressionist in his *Descent of Man,* for instance with his distinction between "savages" and "civilized man." Around the same time, Galton (1869) was using the theory of natural selection in order to draw eugenic lessons about the future of the human species. The ethical and political implications of the theory of natural selection seemed straightforward for 19th-century philosophers and naturalists: the lower socioeconomic classes were outreproducing the higher ones, and this distortion of natural selection seemed to imply the "degradation" of the human species. The eugenicist response was that various forms of artificial selection were needed to counteract this development; not only would this prevent the degradation of the human species, but through judicious interventions on reproduction patterns, this artificial selection could help the human species (and indeed all of evolution) enter a new era of progress. In this way, the eugenicist concept of evolutionary progress drew on the lawlike normativity of the theory of natural selection in order to make judgments on how to control human evolution and steer it toward more ethically desirable states of affairs (for an example, see Huxley 1936).

How should the concept of evolutionary success be situated vis-à-vis that of progress? First, the concept of success does not seem to have the same baggage as progress does, and not merely because it has played a less prominent role in the history of biology. It seems to be more straightforward to present "success" as ethically neutral, precisely because it can be operationalized with precise metrics (population size, geographic range, etc.). By contrast, progress depends on the operationalization of "higher" and "lower," which in turn depends on concepts such as complexity that are not operationalizable, or only with considerable simplification (e.g., where complexity is measured in terms of the number of part types: McShea and Brandon 2010). This is part of what is attractive about talking about success: it seems to involve neutral observations based on clearly definable metrics.

At a conceptual level, success and progress may indeed be distinct. Holding that one species is more successful than another does not imply that the former represents "progress" over the latter. If the human species is deemed a remarkable evolutionary success, this does not necessarily imply that it represents a pinnacle of progress, any more than it should imply that opossums are the pinnacle of marsupial evolution. Nonetheless, success and progress are not wholly independent either. Success-talk recalls historical terminology that we now tend to reject. Think of how, in older textbooks (Matthew 1928; Colbert 1955, 1965), ecologically dominant taxonomic groups would be used to categorize geological epochs. Thus, paleontologists spoke of the "Age of Reptiles" to refer to the Mesozoic, and the "Age of Mammals" to refer to the Cenozoic. (Today, curiously, only the "Age of Man" phrase remains, under the guise of the "Anthropocene.")

This language can be seen as progressivist in the following sense: if one taxon wanes in dominance, only to be replaced by another dominant taxon, it is a psychologically small (though not logically valid) step to conclude that the later taxon is "more dominant" than the earlier taxon. This inference of course presupposes that the success of various species or other taxa can be compared despite living in different epochs and in different circumstances—a presupposition that, at least in general, seems false, even though it may be true under specific circumstances. In fact, when Wilson (1992, 187ff.) used the same phrasing ("Age of . . .") decades later, it was in connection to a progressivist view of evolutionary history.

3. Success and Human Uniqueness

A second concept that seems both to reflect neutral facts and to entail considerable value judgments is the concept of *human uniqueness*. Here also Aristotle led the way with a first categorization of humans as the *zoon logon*, the "rational animal." The capacity of reason sets humans apart from animals, and much effort in identifying the capacities that set *Homo sapiens* apart—whether it be toolmaking, language, or culture—could be viewed as variations on the Aristotelian theme.

It would be difficult to understand our fascination with human uniqueness if it were a merely causal-descriptive concept. Such a concept would appear to be almost scientifically trivial. Every species has a unique set of traits (or, more precisely, a unique range of traits), so why care about discovering what precisely makes humans unique? Matt Grove in this volume starts out with precisely this observation, citing Foley (1987).

When phrased in these terms, it is indeed tempting to dismiss our concerns with human uniqueness as an antiquated form of anthropocentric navel-gazing. However, from a broader philosophical perspective, the concept of human uniqueness plays some crucial normative functions. For instance, without some concept of what sets humans apart from nonhuman animals, the distinction between human and animal rights disappears. Even Peter Singer, the original crusader against speciesism—discrimination on the basis of species instead of on the basis of what matters morally, namely the capacity to suffer and the possession of preferences—held that, in the choice between saving the life of a (healthy, fully developed) human and that of a (healthy, fully developed) nonhuman animal, preference should still be given to the former. When faced with a situation that requires a choice between promoting human welfare and promoting nonhuman animal welfare, even proponents of animal welfare choose the first in general. (Exceptions for Singer, notoriously, can be made when the human in question is disabled or afflicted with neurogenerative diseases.)

In political philosophy also, concepts of moral personality or moral rights are typically grounded on some concept of the unique capacities of humans. In Rawls's veil of ignorance, individuals are stripped of all traits (personality, socioeconomic class, etc.) except for their reason. The resulting concept of justice has been criticized as discriminatory to those without the requisite reasoning capacities (Sen 1979). However, one can argue that all sentient or agential beings—rational or not—possess moral standing without collapsing the difference in moral status between humans and nonhuman animals. We need not delve further into this issue, but such considerations illustrate how concepts of human uniqueness are not just quaint forms of anthropocentric confabulations: common moral judgments and legal practices currently presuppose it.

As with progress and success, some conceptions of human uniqueness have a darker side. Today we would likely not be entirely comfortable in following Aristotle to the conclusion that the goal of human life is the development of reason, and that this goal is what grounds the superiority of human beings. This line of reasoning introduces too large a gap between the weight given to human needs and the weight given to nonhuman needs, and can ultimately be used to justify the exploitation of nonhumans or of humans lacking a rational capacity.

From this perspective, the concept of success can be used to modify the concept of uniqueness to help to avoid some of those darker connotations. Instead of asking what makes humans superior, one can ask: What is the unique formula that lies behind the success of *Homo sapiens*? Indeed, a cursory inspection of the properties that are often associated with human uniqueness—problem-solving, toolmaking, language, and cooperation come to mind—shows a striking overlap with some of the properties that are often understood to be causally responsible for human evolutionary success. It suggests a narrative of human evolution where a certain ensemble of unique traits (cooperation, cumulative culture) led to ecological dominance. From this perspective, these traits are not significant in themselves, but only because they provided a way to success.

Whether thinking of the human evolutionary trajectory in terms of success instead of uniqueness or progress can help avoid some of the pitfalls of the latter, or whether success is like the attractive new acquaintance whose character flaws still have to become apparent, is, of course, not a question this volume will be able to answer. However, in problematizing evolutionary success in a systematic, transdisciplinary way, hopefully some of the debate about success can be expedited to come to a balanced judgment.

4. Overview of the Volume

The first part of the volume tackles the question of what human success—or species success, more generally—can mean. This helps to orient the reader, to get a sense of the landscape. However, the chapters in this part do not constitute the last word on the subject, since the second and third parts of the volume explore, respectively, what caused human success and what the ethical implications are of human success. These parts help to further reflect on and critique our understanding of human success.

In the first chapter, Daniel McShea provides a bird's-eye view of some of the normative dimensions of success and considers how success could reflect the type of normativity of a natural law, but also ethical normativity. He draws on Robert McShea's distinction between six value standards and discusses to what extent each is applicable to evolutionary success. He makes no firm conclusions about how success should be understood—and even gives reasons to doubt that any firm conclusion can ever be drawn—but does strike a cautionary note about the work that needs to occur to make evolutionary success a "scientifically wholesome" concept.

The next chapter, by Bernd Rosslenbroich, shows how key factors in success—such as freedom and autonomy—have antecedents in the very structural organization of organisms. Rosslenbroich argues that viewing human evolution in terms of changes in biological autonomy can help undermine false dichotomies between biology and culture, fact and value.

While McShea and Rosslenbroich point to work that would need to occur to place evolutionary success on a scientific foundation, in the last chapter of the first part Marion Hourdequin explicitly argues against reducing success to a scientific concept and holds that it is paramount to maintain a pluralism of success concepts that reflect a pluralism of goals. She illustrates this by discussing how different notions of success are at play in the Anthropocene, and thus her chapter anticipates later chapters devoted to climate change and environmental ethics.

In the four chapters of the second part, we get four perspectives on the causal antecedents of human success: from a paleobiologist, two physical anthropologists, and two biocultural anthropologists. Geerat Vermeij draws on his work on evolutionary escalation and paleoecology to provide a zoomed-out picture of human evolution in terms of resource consumption, competition, cooperation, colonization, adaptation, and evolutionary feedback loops. It is a thought-provoking read for those interested in the big picture. Vermeij also discusses the roles of predictability and contingency in human evolution.

Susan Antón's chapter focuses on a shorter time span of human evolution—that of *Homo*—and, in particular, examines the evolutionary fate of *Homo erectus*. Like *H. sapiens, H. erectus* dispersed across a vast territory, far beyond any of its hominin or primate ancestors. In accounting for this broad ecological tolerance, Antón reviews and examines the evidence for developmental plasticity in *H. erectus*, that is, its capacity to develop differently (and adaptively) in different environments. There are clear, large-feature differences with *H. sapiens* (the latter has much shorter intervals between offspring and achieved a much higher population density compared to *H. erectus*), but the fact remains that *H. erectus* was quite successful by our own standards, and yet is now extinct. By some measures, it was more successful, having persisted for two million years—something that cannot yet be said of our own species.

Matt Grove concentrates on one crucial feature of human evolution: how increasing brain sizes (encephalization) went hand in hand with life history changes (especially a longer life expectancy and a longer youth). This implied a vast range of further changes, for instance how human infants required increasing amounts of nurturing and parental (and especially maternal) investment. Grove's hypothesis is that encephalization slowed our ancestors' reproductive rate and that this slowed the rate of biological evolution; given heterogeneity in the environment, this provided the conditions in which the capacity for culture would be adaptive. In short, the human capacity for culture—a core part of the story of human success—did not evolve as a direct consequence of larger brain size but rather because increasingly large brains meant hominins did not reproduce as rapidly and thus could not adapt to changing environments through genetic evolution alone.

Kathryn Demps and Peter J. Richerson, in the final chapter of Part II, also examine the rise of culture in human evolution and its role in human success. They use a gene-culture coevolution framework and focus on human evolution over approximately the past 50,000 years. They cover some of the same ground on encephalization as did Grove, and review how, once human populations grew large enough, and perhaps also in response to intense climate variation, the rate of cultural adaptation increased significantly around 50,000 years ago. This rate increased after the last ice age, intensified by a feedback loop between increased exploitation of natural resources, increased human population size, and increased technological and social innovation. Their narrative ends at the Anthropocene, and given the explosive growth of the human species in what amounts to less than the blink of an eye in evolutionary terms, they are clearly cautious about whether *H. sapiens* should already be deemed an evolutionary success.

The implied threat that the Anthropocene is actually the start of an ecological collapse brings us to the final part of the volume, which concerns the ethical implications of success. Geologists Jan Zalasiewicz, Mark Williams, and Colin Waters review the evidence for the claim that the human global impact marks the beginning of a novel geological epoch. They give us an accessible and informative primer on the subject matter and proceed with a big-picture discussion of the coevolutionary processes occurring at a global scale—and they advance the notion of "technosphere" in analogy to the biosphere, lithosphere, and atmosphere to help make sense of these processes.

Agustín Fuentes discusses what human success should look like in the Anthropocene from the perspective of a primatologist and biocultural anthropologist. He draws on the conceptual resources of recent developments in evolutionary biology (under the banner of the extended evolutionary synthesis) in order to highlight the epigenetic, behavioral, and cultural channels of inheritance beyond the genetic channel. Fuentes uses this to ground a concept of human success that goes beyond population size or geographic expansion: level of equity. In other words, the human species cannot count itself "successful" unless its individuals receive opportunities for flourishing and expressing their capabilities.

Richard Potts sketches a vision for future human success from the perspective of a physical anthropologist who has studied how climates have warmed up and cooled down many times since the dawn of the genus *Homo*. We seem now to be faced with worrying and unprecedented change. Nonetheless, drawing on previous work, Potts reminds us that adaptation to environmental uncertainty is a recurring theme in the story of human evolution—the motor that has driven many key changes in the past. What precise forms this adaptability will take is difficult to say, but since what is required is an unprecedented level of international cooperation, it will likely come in the guise of novel organizational networks as well as novel cooperative moral frameworks. What may be happening is unprecedented, but then again, "unprecedented" is business as usual for *Homo sapiens*.

The subject of human flourishing is central to Allen Buchanan and Rachell Powell's chapter, where they draw on their work on the evolution of morality to propose a view in which "human success" should be measured in terms of inclusivist morality. In other words, human success involves expanding the circle of moral concern far beyond what could be considered "adaptive" in terms of reproductive success alone. Concepts such as human rights or animal well-being illustrate how humans have created conditions allowing for flourishing and, in the process, leveraging the plasticity of our (moral) cognition to resolve issues far beyond the problems of social cooperation our moral psychology originally was selected for.

Implicit in Buchanan and Powell's emphasis on cognitive plasticity is a criticism of the idea that technological enhancement should play an outsized role in engineering human (moral) success. The idea that we can and should control future human evolution through physical interventions on human genotypes and phenotypes is an old one—going back to eugenics—but remains quite alive today. After all, could we not solve all the problems associated with the Anthropocene—ranging from our tendency to overexploit natural resources to our exclusivist or tribal behavioral patterns—by technologically enhancing our brain in some fashion? This idea is explicitly targeted by Hugh Desmond in the last chapter of the volume, where he lays out the evolutionary rationale often given for "techno-libertarian success" and concludes that it is poorly supported by what we know about human cultural evolution. Instead, concepts of future human evolutionary success should explicitly integrate community-level (and not just species-level or individual-level) metrics.

5. Whither Human Success?

Even though an edited volume should raise more questions than it answers and present more divergence than convergence in views, some common themes are woven throughout the chapters. The first is the frailty of human success: even in merely eco-evolutionary terms, current success is no guarantee of future success. The second is the incompleteness of this concept of success, even from a purely causal-evolutionary perspective: there has been much more to human evolution than maximizing population size or resource extraction, and the reasons why—our cooperation, our adaptability, our cognitive plasticity, our communal culture, and so on—are crucial to a more satisfactory notion of human success. Any hard distinction between biology and culture, genes and values, is hard to justify in the case of human success.

One worry worth highlighting is Daniel McShea's early warning that success may be yet another "obnoxious" and "untestable" notion (as S. J. Gould judged the concept "progress" to be: see discussion in Desmond 2021) that is best either avoided or neutered into a more acceptable concept. In this sense, success may indeed be considered the offspring of progress—even though that would not be a lineage to be proud of. However, in later chapters it becomes clear that other concepts of human success seem possible—ones that are inclusive, community-oriented, or focused on flourishing—that are at least not *obviously* obnoxious. This would be a different kind of heir, one inheriting the value-ladenness of the parent concept, but without some of its baggage.

However, even in the latter case, questions remain. Do these involve an unacceptable imposition of contingent value standards onto interpretations of the human evolutionary sciences? A number of contributions try to square success with pluralism, even though speaking of success implies the possibility of failure, and thus that some social configurations or ways of living would need to be judged to be such. What rational response, beyond appealing to one's own moral intuitions, could be given to one disagreeing with such a concept of success? Must one flesh out inclusive, pluralistic concepts of success in terms of power or exploitation? The concept of human success opens up new opportunities to revisit old questions about facts and values, evolution and ethics, and about how human evolutionary history should be understood.

References

Bowler, P. J. 1975. "The Changing Meaning of 'Evolution.'" *Journal of the History of Ideas* 36 (1): 95–115. https://doi.org/10.2307/2709013.

Colbert, E. H. 1955. *Evolution of the Vertebrates: A History of the Backboned Animals through Time*. New York: Wiley.

Colbert, E. H. 1965. *The Age of Reptiles*. New York: W. W. & Norton.

Darwin, C. 1854. "Letter No. 1573." Darwin Correspondence Project. Accessed December 13, 2021. https://www.darwinproject.ac.uk/letter/?docId=letters/DCP-LETT-1573.xml.

Darwin, C. 1871. *The Descent of Man*. New York: D. Appleton.

Desmond, H. 2021. "The Selectionist Rationale for Evolutionary Progress." *Biology & Philosophy* 36 (3): 32. https://doi.org/10.1007/s10539-021-09806-1.

Foley, R. 1987. *Another Unique Species: Patterns in Human Evolutionary Ecology*. Harlow: Longman Scientific & Technical.

Galton, F. 1869. *Hereditary Genius: An Inquiry into Its Laws and Consequences*. London: Macmillan.

Hannon, E., and T. Lewens. 2018. *Why We Disagree about Human Nature*. Oxford: Oxford University Press.

Henrich, J. 2016. *The Secret of Our Success: How Culture Is Driving Human Evolution, Domesticating Our Species, and Making Us Smarter*. Princeton, NJ: Princeton University Press.

Huxley, J. S. 1936. "Natural Selection and Evolutionary Progress." *Nature* 138: 603–605. https://doi.org/10.1038/138603a0.

Matthew, W. D. 1928. *Outline and General Principles of the History of Life*. Berkeley: University of California Press.

McShea, D. W., and R. N. Brandon. 2010. *Biology's First Law: The Tendency for Diversity and Complexity to Increase in Evolutionary Systems*. Chicago: University of Chicago Press.

Nussbaum, M. C. 2000. *Women and Human Development: The Capabilities Approach*. Cambridge: Cambridge University Press.

Richards, R. J. 1988. "The Moral Foundations of the Idea of Evolutionary Progress: Darwin, Spencer, and the Neo-Darwinians." In *Evolutionary Progress*, edited by M. H. Nitecki, 129–148. Chicago: University of Chicago Press.

Ruse, M. 1996. *Monad to Man: The Concept of Progress in Evolutionary Biology*. Cambridge, MA: Harvard University Press.

Sen, A. 1979. "Equality of What." Tanner Lecture on Human Values. May 22. http://www.ophi.org.uk/wp-content/uploads/Sen-1979_Equality-of-What.pdf.

Smith, D. 2020. "Cultural Group Selection and Human Cooperation: A Conceptual and Empirical Review." *Evolutionary Human Sciences* 2: e2. https://doi.org/10.1017/ehs.2020.2.

Sterelny, K. 2012. *The Evolved Apprentice*. Cambridge, MA: MIT Press.

Stringer, C. 2019. "Meet the Relatives: The New Human Story." *Financial Times*, July 26. https://www.ft.com/content/6fc26e8c-ada8-11e9-8030-530adfa879c2.

Tarlach, G. 2018. "The Generalist Specialist: Why Homo Sapiens Succeeded." *Discover Magazine*, July 30. https://www.discovermagazine.com/planet-earth/the-generalist-specialist-why-homo-sapiens-succeeded.

Wilson, E. O. 1990. *Success and Dominance in Ecosystems: The Case of Social Insects*. Oldendorf: Ecology Institute.

Wilson, E. O. 1992. *The Diversity of Life*. Cambridge, MA: Harvard University Press.

PART I
WHAT IS EVOLUTIONARY SUCCESS?

2

Evolutionary Success

Standards of Value

Daniel W. McShea

1. Introduction

"Success" is a value term. Values imply a standard. So in assessing the degree to which humans have been successful, what standard of value shall we choose? What standard of success shall we apply?

For much of the 20th century, the most common answer would have been that all values are cultural, so the only standard of success available is a cultural one (Herskovits 1972; Mead 1928). The trouble with that answer here is that different cultures will have different standards of success. For any specific standard of success that might be proposed—for instance, population size, ecological dominance, tool use or technology—it is easy to imagine a plausible culture that values the opposite. We might value human ecological dominance and call that success, but some other culture might value ecological unobtrusiveness and call dominance failure. For a scientific investigation of success, that sort of cultural relativity is unsatisfying.

Happily, there are other value standards. In fact, as I shall argue later, there are five, in addition to the cultural standard: God, nature, reason, the autonomous individual, and human nature. These standards can be thought of as "value bases." Each is a possible answer to the question of what we mean by success, of where to turn to evaluate success or failure. If we adopt God as our value basis, then we turn to God to discover the degree to which an entity or system is successful. God's preferences are what count. "X is successful" simply means "God prefers X." The next section discusses this six-option value-basis scheme and argues that only one, the human nature basis, may be suitable for present purposes, although applying it raises some difficult issues. Another value basis, the nature basis, could be modified to make it

Daniel W. McShea, *Evolutionary Success* In: *Human Success*. Edited by: Hugh Desmond and Grant Ramsey, Oxford University Press. © Oxford University Press 2023. DOI: 10.1093/oso/9780190096168.003.0002 useful, although, as will be seen, even suitably modified it too raises serious problems.

It might seem that for evolutionary success, no explicit value basis is needed because the standard of success is implicit in the evolutionary process. The thought is that natural selection is largely about competition, competition produces winners, and the winners survive and reproduce, generating offspring that are present in subsequent generations. Success is about winning. In other words, a standard of success is implicit in the process, just as it is implicit in competitive sports. For example, one could argue that in tennis, success is winning games and tournaments. Winning is implicit in the game in that it is built into the rules, the psychology, and the organization of the sport. The word "success" appears from time to time in the evolutionary literature, usually in passing, and when it does, success is usually meant to be understood in this sense, with no discussion of values seeming to be necessary. Success is just what the process favors.

There are several problems with this line of thought. First, if our intent is to talk in a value-neutral way about evolution, why use such a value-laden word like "success," a word with such positive connotations? We could instead use an unambiguously value-neutral word like "trend." That would be fine, but I sense that the choice of "success" in this context is not casual, that the intent of those discussing it is to put a positive spin on certain kinds of trend. (The same is true of the word "progress" in discussions of evolutionary progress.) As I shall discuss, one way to accommodate this problem is to acknowledge a value component in "success" but treat it as metaphorical. Success is what the evolutionary process "values."

Second, however we maneuver to avoid the notion of value in a discussion of evolutionary success, values are always present. Imagine a species that survives, reproduces, and even expands on account of its profligate use of resources and its ability to destroy its competition, thereby depleting its environment and reducing diversity. Even if it finds a way to survive and to continue to expand, would we be happy to call it successful nonetheless? We probably would not, or at least many would feel uneasy doing so. The reason has to do with our values, values that are independent of the process that favored this species. No matter how the issue is framed, even if we claim to have no standard beyond what the evolutionary process favors, the word "success" drags true values into the discussion willy-nilly. The use of "value" in a metaphorical sense has the virtue that it reveals the conflict to be between standards that are, as I will argue, on the same footing, that are alternatives.

Which in turn allows us to say, for example, that a given species is successful by one standard and not by another.

Third, since I have raised the notion of selection favoring success, it needs to be pointed out that both in sports and in evolution, success is not always and only about winning. It is true that tennis star Roger Federer was considered successful because he won frequently. But that is not all there is to success in tennis. Federer is also polite and sportsmanlike. In calling him successful, we could mean that he excels in those areas. Another possible standard is the impact that a player has on the game, irrespective of how often they win. Venus Williams had a significant impact on the game, inspiring people of color to follow in her footsteps. And she continues to inspire, despite not winning quite as often in recent years. Thus, she could be said to be successful by a standard that has no necessary relationship to winning. Likewise, evolution is not about natural selection only. The process favors traits other than fitness. For example, in the absence of selection, individuals in populations tend to accumulate accidents and therefore to become different from each other. Diversity increases. Likewise, in the absence of selection, parts within individuals differentiate spontaneously. Complexity increases. This is the zero-force evolutionary law (ZFEL; McShea and Brandon 2010), which says that the evolutionary process has embedded within it, independent of selection, a tendency for diversity and complexity to increase. Thus the evolutionary process favors diversity and complexity, as well as fitness, and we will need to include these in our understanding of evolutionary success. Of course, doing so raises the question of what to do when conflicts arise. How shall we handle cases where one aspect of success increases at the expense of the other, say, a complex species that is extinction prone, perhaps on account of its complexity? Is such a species successful?

I will address these issues, but the discussion begins here with the various alternative ways to understand success. In what follows, I outline a value-basis scheme, covering all six of the possible standards of value that could be invoked for success. I shall argue that in talking about success, we necessarily invoke one of these value bases. As will be seen, there are good reasons not to invoke more than one—at least, not more than one at a time. Further, I argue that the scheme is complete. If we reject all six, there is nowhere else to go.

The value-basis scheme is a mechanism for producing clarity, but it is also a kind of safety technology. Like antilock brakes, it helps to prevent the skids—here, intellectual skids—to which we are all prone. Not uncommonly in evolutionary studies, the discussion of a trend in some feature of the evolutionary process starts with a well-articulated standard for assessing that feature, but then later, when attention flags, slips into using another, unstated standard. I have seen this in discussions of ostensibly non-value-related variables like complexity. One starts out with an understanding of complexity as, say, hierarchy, understood as the number of levels of parts within wholes in an organism, which when applied to the history of life reveals an increasing trend. And then the slip occurs. Seemingly out of nowhere, the claim is made that humans represent a new level of complexity because we have language. Perhaps language makes us complex in some sense, but it has no obvious connection to the stated standard, hierarchy. The same sort of slip occurs in discussions of evolutionary progress.

For evolutionary success, such a slip would occur if we start by adopting a standard of success like, say, population growth, applying it consistently over most of human history to demonstrate human success, but then conclude that later human efforts toward population control would, if effective, be further evidence of our success. In this hypothetical case, the skid is from a nature-based standard—population growth—to one based on human nature, our shared preference for plentiful resources and happiness. I am not challenging the use of any particular standard, nor any particular conclusion about human success, just pointing out the obvious: rigor and coherence require consistency, a determined and uncompromising value-basis consistency.

In what follows, I discuss the various alternative value bases and the reasons why all but two are not useful for present purposes. I then explore the two that could be useful, a human nature basis and a modified nature basis, pointing out some of the difficulties in applying the former and offering some lessons from the study of large-scale trends that could be useful in applying the latter. (As will be seen, Hourdequin [chapter 4, this volume] identifies essentially the same two value bases for success, in her terms, the ethical and the evolutionary approaches.)

2. Alternative Value Bases

A classification of standards, or value bases, has been developed by the political philosopher Robert McShea (1990). McShea identifies six bases, three of them human (the unique individual, culture, and human nature) and three humanly transcendent (God, reason, and nature) (see Figure 2.1). I start with

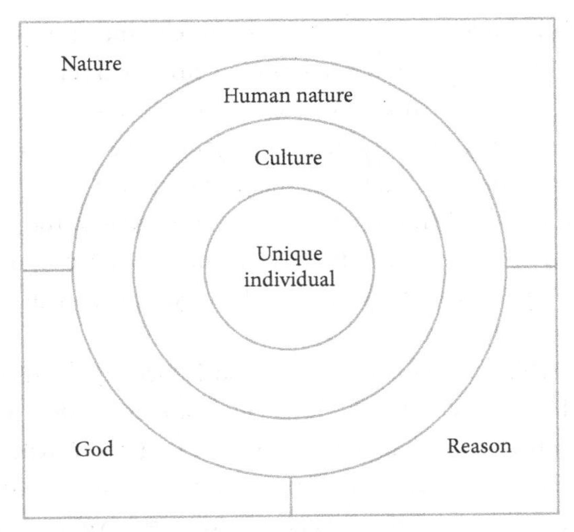

Figure 2.1 Six alternative value bases. Three are human (unique individual, culture, and human nature) and three are humanly transcendent (God, reason, nature).
Source: redrawn from R. McShea 1990.

the three humanly transcendent ones, dismissing them with only cursory arguments. I feel free to do this because I do not think anyone would advance and defend them in earnest in a discussion of evolutionary success. Mentioning them is important for completeness, however, to help make it clear that the value-basis classification is exhaustive and that there are no serious alternatives available beyond it.

God. A meaningful theoretical treatment of this value basis is beyond the scope of this chapter, and also beyond my competence. But I will point out a practical problem with it, one that strikes me as fatal. We cannot use God's preferences as a standard of success, because—putting aside the question of his existence—most of us have no direct way of knowing what God prefers. And reports from those claiming to have direct access to God are always open to doubt and to multiple interpretations. A decision that success is what God prefers will be followed inevitably by the question *What does God prefer?*, with no reliable answer forthcoming.

Reason. This value basis will not work either. Touted since the Enlightenment as *the* guidepost for secular thought, reason turns out not to offer any value guidance at all. As Hume demonstrated, no values follow logically from any set of facts (at least, none with any content to them; cf. Singer

2015). I am, and Hume was, using "reason" here in the narrow sense of strict logical entailment, the sense that has guided empirical science for centuries. Nothing about the better and the worse, nothing about what constitutes success and failure, is logically entailed by anything we know or anything we could learn about the world. It is said that Kant found a route to values based on reason, but it is doubtful that he was using reason in the modern sense, and certain that he is not using it in Hume's sense. In any case, it is hard to see how his moral conclusions (e.g., the good is a good will) will help us understand evolutionary success.

Nature. Nor will nature work as a value basis, because, as discussed, it is—we believe today—value-neutral. Things just happen, or do not, and the fact of their happening or not happening says nothing about whether that outcome is good or bad. Or at least, to put it more circumspectly than is necessary, if nature has values, then—as for God—we have no way of knowing what they are. Notice that a notion like "degree of naturalness" will not help here. Everything that happens is natural, and everything that happens is equally natural, leaving no room for distinctions among things or events, of some as being more natural than others. In any case, even if there were some principled way to make such distinctions, the fact-value gap, the logical disconnect identified by Hume between values and the natural, would remain (Buchanan and Powell, chapter 12 in this volume). It may be natural for me to care very little about the welfare of strangers, but that does not make it good for me to do so.

There might seem to be another way to salvage the nature basis: by combining it with reason. Perhaps we can apply reason to things or events in nature to get values. To this option Hume again delivers a one-punch knockout. Things and events in nature are, in Hume's language, "original existences" that have no "representative quality." A hiccup or a thunder roll is an event, or a "thing," if you like. It is an existence, not an idea or a representation of anything. And reason can be applied only to ideas or representations. Only these can have logical entailments. To say that some value standard is logically entailed by some thing or event is a category mistake (Hume [1740] 1978; Shaw 1998).

For those who accept Hume's view, this will end the discussion. For those who do not, there is a practical problem. If nature has values, there is no logic that demands that they also be our values. If nature is somehow discovered to value complexity, there is no reason that we cannot instead value simplicity. As a practical matter, nature's values would have no authority over us.

Still, there is more to say about the nature value basis, because while the nature-value disconnect is technically absolute, a connection can be made if we are willing to adjust our understanding of the word "value." Instead of asking nature what it values, which nature answers with a hollow silence, we can ask what nature "values" in the sense of *supports* or *promotes*. Not just winning or adaptation, as discussed earlier, but *everything* it supports or promotes, by whatever mechanism (with consequences alluded to earlier and discussed more later). We are now using the word "value" in a metaphorical sense, which saves it from the fact-value-gap objection. Thus, by this value standard, one could meaningfully propose that nature "values" entropy, meaning only that it consistently promotes the production of entropy. In this view, our mission in the study of evolutionary success is to discover what evolution tends to support or promote. I shall call this the nature "value" basis, with the quotes to remind us that the valuing is metaphorical.

There is an apt remark by Huxley in his 1936 paper on evolutionary progress: "Any purpose we find manifested in evolution is only an apparent purpose. If we wish to work towards a purpose for the future of man, we must formulate that purpose ourselves" (Huxley 1936, 604). Substituting "value" for "purpose," the remark nicely expresses the artificiality, the as-if quality, of the nature "value" basis, harmless provided we keep in mind that it is not the real thing. The remark also carries us forward to consideration of the three human value bases in the next three subsections.

The Unique Individual. We could adopt a value standard that says there is no standard of value higher than individual preferences and intuitions. In the present context, that would mean evolutionary success is what I say it is. Of course, that standard will be valid only for me. You will have your own standard, and it will be valid for you. If I say success is population growth, and you say success is improved efficiency of resource use, then I am right for me and you are right for you, and there is no way to settle the disagreement. So we have a problem, and it is the same problem raised earlier for a cultural basis, namely value relativism.

But for the unique individual value basis, relativism is not the worst problem. The worst is what might be called value instability. In Shakespeare's *The Taming of the Shrew*, Petruchio bullies his new wife, Katherina, threatening to abort their journey home unless she accedes to his absurd declarations, first insisting that what is obviously the sun is actually the moon, and then reversing himself and declaring it to be the sun. Katherina in frustration finally concedes, saying (in a heavily ironical tone), "Then, God be bless'd, it is the blessed sun. But sun it is not, when you say it is not. And the moon changes even as your mind." Success is population growth when you say it is population growth, and it is population control five minutes later when you say it is population control, and it changes even as your mind.

No one holds or would seriously defend the unique-individual position, although in my experience it is subtly present in a fair amount of casual value discussion.

Culture. Cultural values are less variable than individual values, but as discussed, the problem of relativism remains. We would like our study of evolutionary success to be objective, at least in the sense of cross-culturally stable. If it is not, we will be faced with some awkward problems. For example, suppose we adopt a standard of success based on some value that is widely accepted in our shared culture, say, success as technological sophistication. And suppose we then go on to conduct studies and develop an extended discourse around that notion, with technological sophistication understood in some operational way, perhaps as tool use. It will be embarrassing when someone from another culture comes along and says that technological sophistication, by their cultural standard, signifies failure. Their culture values simplicity. With a cultural value basis, we have no way to argue that they are wrong. Indeed, they are right, for them. And from their perspective, and from that of every single other culture that disagrees with us, our efforts will have been pointless. (The notion that culture can be separated out as distinct from other sources of value, such as human nature, is fraught in a number of ways, of course, and I address some of them in next section.)

3. The Human-Nature Value Basis

The human-nature value basis says that the good is what human beings approve of. A culturalist will object that we have few or no shared values, and will point to the sometimes huge differences among cultures in behavior, language, and conceptual categories, as well as in the stories and myths that underlie our understanding of the world. Defenders of the human nature basis will want to debate the extent of these differences but can cheerfully acknowledge that many of them are real, while pointing out that most have no bearing on the degree of *affective* similarity among people. For value judgments, the critical commonality across cultures is not in what people say, do, and think, but in their wants, preferences, and cares, or what Hume called the sentiments. (See Hume [1740] 1978; for modern treatments of the human-nature value basis, see McShea 1990; Ruse 1986; Midgley 1978.)

How can behavior, language, and so forth vary so much, while the sentiments vary so little? Affective reactions are situation-specific, keyed to the details of how the situations that evoke them are understood. And those understandings can vary enormously among cultures. A crowd stands to sing the national anthem, saluting the flag waving above them, with spirits soaring in the shared experience of a common heritage and purpose, while someone from a radically different culture looks on bewildered, puzzled by the lyrics in an alien language of course but also by all the attention directed to a bit of cloth curiously mounted on a stick. The affective responses differ, but that is because the understanding of the context differs. Shared experiences of a common heritage and purpose evoke roughly the same affective reaction in people everywhere; it is just that the symbols and contexts that evoke that reaction vary, and indeed are sometimes unique to a given culture. Likewise, there are characteristically human responses to loyalty and cheating, compliments and insults, empathy and threats; it is just that the words and situations that constitute loyalty, threats, and so on can differ enormously among cultures. The same is true over a wide swath of contexts, on matters related to children, friendship, kinship, cooperation, reciprocity, play, and social status. (See Brown 2004 for many, many more.) Corrected for differences in understanding, we have broadly similar sentiments, the human nature position claims.

Thus, here is how the human-nature value basis handles the question of evolutionary success. Success is what humans generally approve of. And a successful species, including our own, is one with properties and capacities that elicit a favorable response from our shared affective nature.

Apparent Difficulties. Using a human-nature value basis, a different kind of relativism comes into play. Values become species-relative. Following Hume, the human nature basis understands values as secondary qualities, like sound, color, and odor. And affective reactions differ among species just as the perception of sounds, colors, and odors does. Each species has what might be called an affective profile, the total stimulus-specific and situation-specific repertoire of sentiments. And just as each species has a unique profile of perceptual capabilities, each also has a unique affective profile. An orange smells different to a chimp than it does to a housecat. And lion feelings about other lions are different from tiger feelings about other tigers. (Lions' affective responses are highly social, while tigers are the epitome of antisociality.)

And to the extent that species' affective profiles are different from each other, values will be different. If black widow spiders could make value judgments and speak, they would disagree with us about how to treat mates, and there would be no principled way to settle the matter. In the present context of a discussion of the meaning of success, species relativity is certainly a weakness, but not much of one given that we are the only species that is presently able to participate in the conversation (so far as we know).

I have not yet mentioned the lively and growing literature on emotion and the long-running debate on whether there is a set of universal emotional responses shared among all people in all times and all places (Ekman 2009), or whether emotions are more plastic, varying with culture and personal experiences (Barrett 2017; Haidt 2007). The human nature view points out that emotions are strong, eruptive responses, and that the responses that underlie value judgments are often calmer, more properly called wants, preferences, cares, and so on. They are what Hume called the calm passions. Thus, the debate over emotion is only partly relevant in a discussion of values. In any case, whatever the very real differences in affective profile among cultures and among individuals, we know that strong similarities remain, as evidenced by, again, the characteristically human responses to loyalty and cheating, compliments and insults, empathy and threats, and so on.

The human nature basis is easily confused with emotivism. Emotivists argue that there is only one possible motivation to make a value statement, and that is the occurrence of a feeling, and from this they conclude that feelings and value judgments are trivial and of no consequence (Ayer 1950). The human nature basis agrees with the premise but not with the conclusion. To the extent that feelings are consistent within an individual over a lifetime, and to the extent that they are shared among all people, they offer a stable and enormously consequential value basis.

The Culture–Human Nature Separation. There are good reasons to worry about the conceptual separation between culture and human nature, indeed to worry that neither, as popularly understood in recent times, is quite real. Thankfully, while the debate about their reality has been entertaining (e.g., Pinker 2003; Sahlins 2008), I do not need to enter into it here. Here, culture versus human nature is my shorthand way of referring to the depth of entrenchment (Wimsatt 1986) or the embeddedness of the affective profile in a developmental system (Oyama 2000). With the phrase "human nature," I am referring to those aspects of the affective profile that are less variable than others, less easily modified, and more foundational in the development of human psychology. Variants arising in these tend to have more downstream developmental consequences (Wimsatt 1986). Sentiments arising from culture are those that are more variable and more easily modified, with fewer downstream consequences. Thus, the view of human nature intended here is roughly Hume's ([1740] 1978), a view that, as Lewens (2018, 16) nicely puts it, "enables us to make general claims about what members of *Homo sapiens* are typically like, and the ways in which they may be similar to, and different from, the typical members of other species." In my terms, it is a view that acknowledges the deeply entrenched similarities among the actual and potential developmental trajectories (*sensu* Ramsey 2018) of all human affective profiles, while allowing for more shallowly entrenched differences among cultures and deeper differences between humans and other species. Importantly, the human nature it points to has no essence, no precultural existence, and no necessary special role for genes in its etiology.

For political and moral purposes beyond this volume, quite a bit hinges on the claims of the human nature basis being true. In particular, it gives us a place to stand in judgment on the abhorrent practices of other cultures. Using a cultural value basis, if some culture practices ritual maiming of its members, we would have to concede that the practice is right, for them. In any case, my main purpose here is not to proclaim the superiority of the human nature view. Rather it is to make the more modest claim that human affective profiles are broadly similar, and to the extent that they are, agreement is possible on what constitutes success and failure.

Exactly What Is Our Shared Human Nature? It turns out to be very difficult to say exactly what the human affective profile is and therefore what constitutes success. Partly the problem is that we lack a good vocabulary for describing preferences, feelings, and affective responses generally. Partly it is that calm affective responses are just vague, at least as seen from the perspective of an introspective mind. Describing them is like describing the shape of a cloud as seen from the perspective of a window seat in a plane flying through it. And then of course, there are the differences in understanding among individuals and cultures, evoking different affective responses from our shared feeling profile. When it comes to seeking agreement on our shared values, all of these are troublesome enough. But more troublesome still is the incompleteness of our knowledge about ourselves. We seek happiness, but we often do not know what will make us happy. This will be especially problematic for our judgments about human success. It might seem reasonable to say that success is efficiency of resource use, because efficiency is a sensible way to sustain happiness over the long term for a large number of individuals. But the demand for efficiency may also lead to a restriction on the expression of human potential. It could cramp our style. If both sustainability and the unleashing of potential are values for us, what combination of the two is best? What combination will evoke a positive affective response in a human being? One could argue that we simply do not know ourselves well enough to answer questions like that.

But these are problems in application, not in theory. The human nature position is that some degree of value agreement is possible in principle, and—most important—the shared affective profile is where defenders of any particular standard of success need to target their argument. What this means in practice is that proposals for success need to be pitched in a way, and at a level of abstraction, that transcends individual and cultural differences. Perhaps success concerns health, not income? Or might success be the degree of satisfaction of basic human needs, not habitat expansion? It is not that income increase and habitat expansion are bad. It is that it would, I think, be hard to get substantial agreement on them across individuals and cultures.

Notice that the human-nature value basis does not line up perfectly with various claims that have been made for a universal morality, nor does it necessarily contradict the notion of moral pluralism (Buchanan and Powell, chapter 12 in this volume). The central claim of the human nature basis is that what is shared are the context-specific, individual-level, human sentiments. Morality, in contrast, might be understood to have more to do with the language, social norms, laws, and institutions that grow up around those sentiments. Great variability in the latter is consistent with much less variability in the former.

I shall have no more to say about the human nature basis. I am no more qualified than anyone else to articulate the contents of our shared affective profile or to judge which aspects of it will be useful in assessing our success. Here my point has been mainly that it offers a perfectly good—albeit species-specific—value basis, that we can meaningfully ask about the degree to which humans have been successful by the standard of our shared preferences.

4. Evolutionary Success and Value Bases

The Six-Part Scheme. I do not expect that the value-basis scheme I am using here will be uncontroversial. Nor do I claim that it is the only possible starting point for a discussion of values and of evolutionary success. It is especially appropriate here, I think, because it highlights the contrast between "values" in the metaphorical sense, the sense in which the evolutionary process can value something, and all other possible value bases. Also, it reveals the number of possible value bases to be fewer than might be expected, simplifying the search for an appropriate one for a discussion of success. And, happily, it reveals that there is a value basis, the human-nature value basis, for which general agreement on values is the expectation.

Identifying an Implicit Value Basis. In some cases, arguments about evolutionary success wear their value bases on their sleeves. Consider standards of success like a species' ability to expand its habitat, grow its population, and transform its environment. None of these mentions fitness or natural selection, but they are pretty clearly invoking a nature-based "value" standard. They are understood to be things that the process favors. For other standards, it is difficult to know. What value basis is implicitly invoked by an argument that success is the ability of a species to influence the course of its own evolution? Or an argument that success is complexity of social organization? Or that success is sustainability? The question is whether the person making the case is invoking personal values, or some set of cultural values, or values that most humans in most times and places would assent to. I have no special insight to offer here and am happy to offload to others the problem of situating particular existing arguments in the value scheme. My interest is more prospective: we can avoid the issue of retrospectively situating an argument if we simply start with a value basis. That is, in beginning an investigation into evolutionary success, we should begin by choosing a value basis.

The Incommensurateness of Value Bases. Imagine a species that wipes out its competition and is therefore successful by the nature standard, but in doing so threatens the planet's diversity, leaving us inclined to call it unsuccessful by the standard of human values. There is no contradiction here. Different standards of value can yield different judgments. Nor is there any way—even in principle—to settle the matter, no way to decide whether that species is actually, at bottom, successful or unsuccessful. The reason is that value bases are incommensurate. They are assessments by different standards, with no necessary relationship among them. They address success in different senses. Hourdequin (chapter 4, this volume) also treats her evolutionary and ethical value bases as independent of each other. And see Desmond (chapter 13, this volume) on the gap between population size and ecological dominance and the things "humans actually care about."

The same point can be framed as a warning against slippage. We cannot begin by arguing that humans have been successful because we have prospered by the standard of the evolutionary process (nature in the metaphorical sense), as evidenced perhaps by our increasing population size and resource use, and then support this claim with evidence that human happiness has risen (human nature basis). For one thing, there is no reason to think that evolutionary processes favor our happiness. They may sometimes, but they need not always, or even usually. But in any case, whether or not they do, the two value bases are non-overlapping alternatives, and as a matter of logic, success by one standard is not evidence for—or even relevant to—success by the other.

Completeness, Exclusivity, and Objectivity. The value bases are a classification, a way of ordering concepts. There are only six (or seven if we include the nature "value" basis, with value understood in the metaphorical sense). McShea (1990) argues that the classification is complete in that every value theory falls into at least one of the six categories. Marxism is culturalist in that values are considered to be nothing more than cultural reflections of the means of production in a society. Utilitarianism has a reductionist human nature basis, identifying the pursuit of pleasure and avoidance of pain as our only morally significant sentiments. Ideal-observer theory (Firth 1952) is a modern human nature–based theory in that the observer implicitly requires a generalized human affective profile. In other cases, the correspondence is less obvious. See McShea (1990) for further discussion.

McShea also argues that the bases are exclusive. For us that means that while a given standard of success may claim to be founded on more than one value basis, the cost of doing so is inconsistency and incoherence. For example, a notion of success claiming to be founded both on human nature and culture falls apart in every case in which the value judgments of a culture contradict those of human nature. The contradiction cannot be finessed. Nor can some other value standard—like God—be invoked to resolve it. One solution, to shift value bases as needed, may seem pragmatic and has the added advantage of drawing on academic virtues like cleverness and intellectual flexibility. But the cost is incoherence. To invoke multiple value bases or to shift value bases is to have no value basis at all. McShea (1990, 29) writes:

> The possible bases are ultimately incompatible with each other; an ethical theory that attempts to stand on two or more of them is in trouble. We know, especially from the history of religious value theory, that when Reason,

> God, Nature, and the unique individual conscience are appealed to as though they must concur, endless paradoxes follow. The easygoing assumption that all or several bases can contribute faithfully to a single method for the determination of correct value judgments can be maintained only on the further assumption that they so concur only for "men of good will." It always turns out that "good will" is the sole real basis, and a worthless one at that. One of the tragedies of human political life is the impossibility of institutionalizing means for identifying "men of good will."

Finally, it may be obvious, but is perhaps worth stating anyway, that there is no objective way to choose among the value bases. A value basis is a way of framing the discussion of value questions, and not the sort of thing for which there could be evidence. When we choose among them, we do so only on the basis of their usefulness or appropriateness in the pursuit of some purpose of ours. Thus, in arguing against some of the value bases—God, reason, nature (in the literal sense), the unique individual, and culture—I have assumed that readers of this volume share to some extent a common purpose: the investigation of aspects of evolutionary success that are knowable and stable, both over time and among individuals. Based on these desiderata, I have argued that only two qualify, that only two are potentially useful, and that therefore we have only two to choose from: nature (in the metaphorical sense) and human nature.

In the next section, I turn to nature, discussing the sorts of things that the evolutionary process has been said to "value" in the metaphorical sense.

5. What Nature "Values"

Adopting a nature "value" basis, what might evolutionary success consist in? What does nature support or promote? I know of no list of candidate variables for evolutionary success per se, but there is a modest literature in macroevolution on large-scale directionality that might offer some help here. Figure 2.2 shows a list of candidates for trends extracted from that literature, offered here as a possible starting point for thinking about evolutionary success. Some of these arose in the ongoing discussion of evolutionary progress in macroevolutionary studies, but most arose as part of the broader and value-neutral discussion of trends. For all variables on the list, an evolutionary rationale has been offered for why a trend is expected. In most,

Variables Argued to Show a Large-Scale Trend

- Body size (Bonner 1988; Payne et al. 2009)
- Fitness, in the sense of extinction resistance (Van Valen 1973)
- Autonomy (Rosslenbroich 2014, this volume)
- Ability to sense the environment (Ayala 1974)
- Information about the environment (Adami Ofria, and Collier 2000)
- Intelligence (Jerison 1973)
- Evolvability (Altenberg 1995; Kirschner and Gerhart 1998)
- Versatility (Vermeij 1973, this volume)
- Developmental burden, entrenchment (Reidl 1977; Arthur 2010, Wimsatt 1986, Gould 1989)
- Energy expenditure and basal metabolic rate (Bambach 1993)
- Energy intensiveness (Vermeij 1987, 2013, this volume)
- Energy rate density (Chaisson 2010)
- Entropy
  - Energetic interpretation (Wicken 1987; Weber et al. 1989; Salthe 1993)
  - Statistical-mechanical interpretation (Brooks and Wiley 1988; Brooks et al. 1989)
- Ecospace occupation (Knoll and Bambach 2000)
- Ecological dominance, ability to control the environment (Huxley 1942)
- Diversity (Sepkoski 1978; Sepkoski et al. 1981)
- Complexity
  - Horizontal: number of part types (Valentine Collins, and Meyer 1994; McShea and Brandon 2010)
  - Vertical: number of levels of organization, nestedness, hierarchy (Pettersson 1996; McShea 2001)

Figure 2.2 A sampling of variables that have been said to show a trend in evolution at the large scale, over much or all of the history of life. These are presented here as candidates for what the evolutionary process "values" (in a metaphorical sense). Citations are mainly founding documents rather than most-recent studies.

natural selection is thought to be the driving force. For some, the driver could be diffusion away from a minimum, the spread of life away from a lower-bound toward higher values (McShea 1994). Both mechanisms imply that increase is favored or promoted by the evolutionary process. For only three variables has a trend actually been documented quantitatively at the scale of all life over its entire history: body size, energy rate density, and complexity in the sense of hierarchy.

For many of these variables, a trend has been documented nonquantitatively (e.g., ecospace occupation and energy intensiveness) or at some lesser temporal or taxonomic scale (e.g., complexity in the sense of part types in animals over the 500-million-year-long history of the group).

Importantly, the list is incomplete. Other variables not on this list could be favored by the evolutionary process: dispersal ability and geographic range, population size, adaptability, and many others. (See the introduction to this volume and indeed, all of the papers in it.)

This volume introduces a new project, with intriguing possibilities for new discoveries but along a road that is full of potholes and deep ditches. In the hope of helping future explorers to avoid these hazards, here are some lessons from the study of large-scale trends.

1. *One Value Basis but No Single Standard of Success.* As mentioned earlier, natural selection is not the only driver of directional change in evolution. In other words, fitness is not the only thing that the evolutionary process favors. Most of the entries in Figure 2.2 have some direct connection to fitness (e.g., ability to sense the environment, energy intensiveness, ecological dominance), but others have only an indirect connection or none at all. Developmental entrenchment arises from the way development is structured, and the increase in hierarchy may be the result of diffusion away from a lower bound. As discussed earlier, the ZFEL has no connection at all, producing a tendency for diversity and complexity to increase even if selection is entirely absent. Given the multiplicity of things that the process could favor, and the inevitable conflicts among them, it seems likely that no single standard of success will be possible. Unless we somehow devise a kind of grand unified theory of evolution (McShea 2015) that somehow combines all traits that trend into, say, a single equation, there will be no way to know how to weight them. And there will no evolutionary bottom line, no summary statistic that estimates an overall degree of success by the nature "value" basis. At the end of our evaluations, we will likely be able to say of a given lineage something like this: it is very successful by one standard, less so by another, and hardly at all by a third.

2. *Timescale and Taxon Dependence.* Complexity in the sense of part types increases at the scale of all animals, where a trend has been documented in the maximum number of cell types (Valentine, Collins, and Meyer 1994). But at a smaller taxonomic and temporal scale, within vertebrates, and in a different kind of structure, skull bones, there is a decreasing trend (Sidor 2001). The same goes for whatever variables we discover are relevant to success. If some variable increases, say, on the timescale of human history, that does not mean it must also increase on the timescale of human evolution. If it increases in the human family, the hominids, that does not mean it must also increase in the evolution of *Homo sapiens*.

3. *Trends and Tendencies.* Nature "valuing" something is a *tendency*, a propensity. Whether or not that tendency produces an actual *trend* is a separate matter. A trend is net directional change in some variable, such as the mean or the maximum. A tendency refers to a propensity or force—the oomph, so to speak—that can underlie a trend. Thus, a graph showing a rise in the stock market over the past 90 years is clear evidence for a trend but not clear evidence for a tendency, for the existence of factors—such as increasing productivity or increasing population—driving it upward. Stocks could, in principle, increase by chance alone, even on long timescales. Conversely, the existence of a tendency does not guarantee the occurrence of a trend. My leaning against my house imparts a tendency for it to slide away from me (or fall down), but contrary forces resist, so it does not budge. There is a tendency but no trend. The distinction is crucial to coherent discussion of the causes of trends. For present purposes, the lesson is that not everything nature "values," not everything the process promotes and supports, will show an actual trend, and not everything that shows a trend is what the process promotes.

4. *Maxima and Variances.* Consider a population consisting of multiple groups, and suppose that the number of groups is increasing over time. Suppose further that our standard of success is energy or resource usage at the group level, and that the maximum is increasing. In other words, the level of energy usage of the most successful group in existence rises over time. The envelope of group success is expanding upward, so to speak. Faced with these data, the temptation might be to conclude that some force, some upward tendency, is at work, driving energy usage higher. But in fact that need not be the case. For one thing, the trend could be due to chance. But there is another possibility, one that does not involve any increasing tendency. When the number of groups is increasing, the expectation in the absence of a tendency is that the system will sample an ever larger portion of the distribution of possibilities. That is, as the number of groups increases, rarer group phenotypes with more extreme energy usage will be sampled. Thus, in expanding systems, a rising maximum is the null expectation and therefore by itself tells us nothing about the existence of underlying tendencies. This issue has arisen in the study of long-term trends in hierarchy, intelligence, and more (McShea 1994), where maxima get attention all out of proportion to their significance. In the study of trends, null models are a useful way to test for underlying tendencies, and in particular to determine whether an observed trend lies outside the distribution expected from a random-walk

model (Hunt 2006). (Other approaches, such as those based on maximum likelihood, are also useful in this context.)

Intuition can be a poor guide for variances, as well as for maxima. As discussed, the ZFEL says that the accumulation of accidents among lineages produces a tendency for them to become more different, that is, for diversity to increase. Diversity is a variance measure, and the ZFEL applies to every feature that varies. So the expectation in the absence of selection is increase in all variances, from variance in body size and intelligence to variance in energy usage. A null model has been devised that takes this into account, enabling us to detect and measure any forces at work, such as selection, as departures from the null expectation (McShea, Wang, and Brandon 2019).

5. *Non-Trend Patterns.* Evolutionary success is inevitably about trends, in particular increasing trends, but this focus should not prevent us from seeing other temporal patterns. Interesting discoveries about evolutionary success await those on the lookout for decreasing trends, truncated trends, stasis, and periodicity as well. In large-scale evolution, interesting non-trend patterns include the large-scale ecological stasis that locked up certain paleocommunities on timescales of millions of years (Morris et al. 1995), the failure of hierarchy to increase since the advent of highly individuated bryozoan colonies about 480 million years ago (McShea 2001), and the now confirmed 27-million-year periodicity of mass extinctions (Melott and Bambach 2010). For present purposes, the message is that in addition to looking for success, by whatever standard, we could usefully attend to failure, stability, and cycling. The forces governing success are doubtless complex, and investigating their varying expression under different circumstances can only help us to understand them.

6. *Success by the Nature Standard Could Be Miserable.* I made this point earlier, but it is so routinely ignored in discussions of evolutionary progress that it is worth repeating here in this discussion of success. Using a nature-based "value" standard, not all increasing trends will be cause for celebration. That is, not all will be successes by a human-nature value standard. From Figure 2.2, evolutionary trends in diversity and ability to sense the environment sound like something we might value. But trends in ability to modify the environment, complexity in the sense of part types, and ecological dominance are probably a mixed bag. Consider just the human manifestations of these three trends. Modifying the environment to accommodate human economic needs often seems to come at the expense of our aesthetic needs. Complex technology raises our standard of living in a number of ways, but

this complexity often has unintended consequences. And ecological dominance puts diversity and perhaps ecological stability at risk.

The point is that human success by a nature "value" standard could be a disaster by the standard of a human nature value. Not all trends in what nature "favors" will sit well with the human affective profile. Of course, as discussed, the two standards are incommensurate (as is every value standard with respect to all others). We should not expect them to agree, nor is there anywhere to turn to resolve the inevitable contradictions.

6. Conclusion

Stephen Jay Gould (1988, 319) once wrote, in a paper on evolutionary progress, that progress is a "noxious, culturally embedded, untestable, nonoperational, intractable idea." Presumably he would have said the same about evolutionary success. I think if we take this as a statement about intellectual history, about how the discussion of progress has gone (and by implication, how the study of success could easily go), Gould was right. But at least some of these problems are fixable, both for progress and for success. Science is exceptionally good at devising operational measures for vague concepts, making them tractable, and making hypotheses about them testable. If we can devise operational measures for happiness (as in psychology) or complexity (as in biology), surely we can operationalize success, at least by the standard of what the process "values." Operationalization might also obviate the charge of noxiousness. And as for cultural embeddedness, I have argued that this can be avoided by adopting either a human nature or nature (metaphorical sense) value basis. Thus, in an upbeat frame of mind, a possible comeback to the Gouldian critique is this: evolutionary success *can be* a scientifically wholesome, value-basis-specific, testable, operational, and highly tractable idea.

Of course, problems remain. In a downbeat mood, we must admit that the process might turn out to "value" many things, and it will be difficult to know in what proportion to weight them, with the result that any assessment of overall success by the nature standard would remain elusive. Also, the human-nature value basis makes agreement on values possible in principle, but does not make it easy or even realizable in fact. Finally, success by the human nature standard will likely remain forever in tension with success by the nature-based "value" standard. And those seeking a single answer to the question of human success will be forever disappointed.

Acknowledgments

I thank Hugh Desmond and Grant Ramsey for helpful comments on earlier versions of this chapter.

References

Adami, C., C. Ofria, and T. C. Collier. 2000. "Evolution of Biological Complexity." *Proceedings of the National Academy of Sciences* 97: 4463–4468.

Altenberg, L. 1995. "Genome Growth and the Evolution of the Genotype-Phenotype Map." *Lecture Notes in Computer Science* 899: 205–259.

Arthur, W. 2010. *Evolution: A Developmental Approach*. Chichester: John Wiley & Sons.

Ayala, F. J. 1974. "The Concept of Biological Progress." In *Studies in the Philosophy of Biology*, edited by F. J. Ayala and T. Dobzhansky, 339–355. London: Macmillan.

Ayer, A. J. 1950. *Language, Truth, and Logic*. London: Gollancz.

Bambach, R. K. 1993. "Seafood through Time: Changes in Biomass, Energetics, and Productivity in the Marine Ecosystem." *Paleobiology* 19: 372–397.

Barrett, L. F. 2017. *How Emotions Are Made: The Secret Life of the Brain*. Boston: Houghton Mifflin Harcourt.

Bonner, J. T. 1988. *The Evolution of Complexity by Means of Natural Selection*. Princeton, NJ: Princeton University Press.

Brooks, D. R., J. Collier, B. A. Maurer, J. D. H. Smith, and E. O. Wiley. 1989. "Entropy and Information in Evolving Biological Systems." *Biology and Philosophy* 4: 407–432.

Brooks, D. R., and E. O. Wiley. 1988. *Evolution as Entropy*. 2nd ed. Chicago: University of Chicago Press.

Brown, D. E. 2004. "Human Universals, Human Nature & Human Culture." *Daedalus* 133: 47–54.

Chaisson, E. J. 2010. "Energy Rate Density as a Complexity Metric and Evolutionary Driver." *Complexity* 16: 27–40.

Ekman, P. 2009. "Afterword." In C. Darwin, *The Expression of the Emotions in Man and Animals*, 363–392. Oxford: Oxford University Press.

Firth, R. 1952. "Ethical Absolutism and the Ideal Observer." *Philosophy and Phenomenological Research* 12: 317–345.

Gould, S. J. 1988. "On Replacing the Idea of Progress with an Operational Notion of Directionality." In *Evolutionary Progress*, edited by M. Nitecki, 319–338. Chicago: University of Chicago Press.

Gould, S. J. 1989. *Wonderful Life: The Burgess Shale and Nature of History*. New York: W. W. Norton.

Haidt, J. 2007. "The New Synthesis in Moral Psychology." *Science* 316: 998–1002.

Herskovits, M. J. 1972. *Cultural Relativism: Perspectives in Cultural Pluralism.* New York: Random House.

Hume, D. (1740) 1978. *A Treatise of Human Nature.* Edited by L. A. Selby-Bigge, 2nd ed. Oxford: Oxford University Press.

Hunt, G. 2006. "Fitting and Comparing Models of Phyletic Evolution: Random Walks and Beyond." *Paleobiology* 32: 578–601.

Huxley, J. S. 1936. "Natural Selection and Evolutionary Progress." *Nature* 138: 603–605.

Huxley, J. S. 1942. *Evolution: The Modern Synthesis.* New York: Harper.

Jerison, H. J. 1973. *Evolution of the Brain and Intelligence.* New York: Academic Press.

Kirschner, M., and J. Gerhart. 1998. "Evolvability." *Proceedings of the National Academy of Sciences* 95: 8420–8427.

Knoll, A. H., and R. K. Bambach. 2000. "Directionality in the History of Life: Diffusion from the Left Wall or Repeated Scaling of the Right?" *Paleobiology* 26: 1–14.

Lewens, T. 2018. "The Faces of Human Nature." In *Why We Disagree about Human Nature,* edited by E. Hannon and T. Lewens, 1–17. Oxford: Oxford University Press.

McShea, D. W. 1994. "Mechanisms of Large-Scale Evolutionary Trends." *Evolution* 48: 1747–1763.

McShea, D. W. 2001. "The Hierarchical Structure of Organisms: A Scale and Documentation of a Trend in the Maximum." *Paleobiology* 27: 405–423.

McShea, D. W. 2015. "Bernd Rosslenbroich: On the Origin of Autonomy: A New Look at the Major Transitions in Evolution." *Biology & Philosophy* 30: 439–446.

McShea, D. W., and R. N. Brandon. 2010. *Biology's First Law.* Chicago: University of Chicago Press.

McShea, D. W., and M. A. Changizi. 2003. "Three Puzzles in Hierarchical Evolution." *Integrative and Comparative Biology* 43: 74–81.

McShea, D. W., S. C. Wang, and R. N. Brandon. 2019. "A Quantitative Formulation of Biology's First Law." *Evolution* 73: 1101–1115.

McShea, R. 1990. *Morality and Human Nature.* Philadelphia: Temple University Press.

Mead, M. 1928. *Coming of Age in Samoa.* New York: Blue Ribbon Books.

Melott, A. L., and R. K. Bambach. 2010. "Nemesis Reconsidered." *Monthly Notices of the Royal Astronomical Society* 407: L99–L102.

Midgley, M. 1978. *Beast and Man: The Roots of Human Nature.* London: Routledge.

Morris, P. J., L. C. Ivany, K. M. Schopf, and C. E. Brett. 1995. "The Challenge of Paleoecological Stasis: Reassessing Sources of Evolutionary Stability." *Proceedings of the National Academy of Sciences* 92: 11269–11273.

Oyama, S. 2000. *The Ontogeny of Information: Developmental Systems and Evolution.* Durham, NC: Duke University Press.

Payne, J. L., A. G. Boyer, J. H. Brown, S. Finnegan, M. Kowalewski, R. A. Krause Jr., S. K. Lyons, et al. 2009. "Two-Phase Increase in the Maximum Size of Life over 3.5 Billion Years Reflects Biological Innovation and Environmental Opportunity." *Proceedings of the National Academy of Sciences* 106: 24–27.

Pinker, S. 2003. *The Blank Slate: The Modern Denial of Human Nature.* New York: Penguin.

Ramsey, G. 2018. "Trait Bin and Trait Cluster Accounts of Human Nature." In *Why We Disagree about Human Nature,* edited by E. Hannon and T. Lewens, 40–57. Oxford: Oxford University Press.

Riedl, R. 1977. "A Systems-Analytical Approach to Macro-evolutionary Phenomena." *Quarterly Review of Biology* 52: 351–370.

Roopnarine, P. D., G. Byars, and P. Fitzgerald. 1999. "Anagenetic Evolution, Stratophenetic Patterns, and Random Walk Models." *Paleobiology* 25: 41–57.

Rosslenbroich, B. 2014. *On the Origin of Autonomy*. New York: Springer.

Ruse, M. 1986. *Taking Darwin Seriously: A Naturalistic Approach to Philosophy*. Oxford: Blackwell.

Sahlins, M. 2008. *The Western Illusion of Human Nature*. Chicago: Prickly Paradigm Press.

Salthe, S. N. 1993. *Development and Evolution*. Cambridge, MA: Bradford/MIT Press.

Sepkoski, J. J., Jr. 1978. "A Kinetic Model of Phanerozoic Taxonomic Diversity: I. Analysis of Marine Orders." *Paleobiology* 4: 223–251.

Sepkoski, J. J., Jr., R. K. Bambach, D. M. Raup, and J. W. Valentine. 1981. "Phanerozoic Marine Diversity and the Fossil Record." *Nature* 293: 435–437.

Shaw, D. J. 1998. *Reason and Feeling in Hume's Action Theory and Moral Philosophy*. New York: Edwin Mellen Press.

Sidor, C. A. 2001. "Simplification as a Trend in Synapsid Cranial Evolution." *Evolution* 55: 1419–1442.

Singer, D. J. 2015. "Mind the Is-Ought Gap." *Journal of Philosophy* 112: 193–210.

Valentine, J. W., A. G. Collins, and C. P. Meyer. 1994. "Morphological Complexity Increase in Metazoans." *Paleobiology* 20: 131–142.

Van Valen, L. M. 1973. "A New Evolutionary Law." *Evolutionary Theory* 1: 1–30.

Vermeij, G. J. 1973. "Biological Versatility and Earth History." *Proceedings of the National Academy of Sciences* 70: 1936–1938.

Vermeij, G. J. 1987. *Evolution and Escalation*. Princeton, NJ: Princeton University Press.

Vermeij, G. J. 2013. "On Escalation." *Annual Review of Earth and Planetary Sciences* 41: 1–19.

Weber, B. H., D. J. Depew, C. Dyke, S. N. Salthe, E. D. Schneider, R. E. Ulanowicz, and J. S. Wicken. 1989. "Evolution in Thermodynamic Perspective: An Ecological Approach." *Biology and Philosophy* 4: 373–405.

Wicken, J. S. 1987. *Evolution, Thermodynamics, and Information*. New York: Oxford University Press.

Wimsatt, W. C. 1986. "Developmental Constraints, Generative Entrenchment, and the Innate-Acquired Distinction." In *Integrating Scientific Disciplines*, edited by W. Bechtel, 185–208. Dordrecht: Martinus-Nijhoff.

3

Human Success as a Complex of Autonomy, Adaptation, and Niche Construction

Bernd Rosslenbroich

1. Introduction

"The greatest division in our field between a more scientific anthropology and a humanistic one lies largely along theoretical lines." With this sentence, Wiessner (2016, 154) commences a contribution to a special issue of the journal *Current Anthropology* titled "Reintegrating Anthropology." According to many anthropologists, a central challenge for anthropology—as the study of biological and cultural variation in human societies over time—has been to unify the "biological" and "humanistic" approaches to human behavior (Canter 2012; Fuentes 2004, 2009, 2015, 2016; Fuentes and Wiessner 2016; Ingold 2007; Midgley 2010; Marks 2012; Schultz 2009; Sommer 2015; Velmans 2012).

Throughout the 20th century, biological approaches to the evolutionary origin of humans focused to a large extent on neo-Darwinian principles (mainly in the form of the modern synthesis, or the synthetic theory of evolution). These principles maintain that evolutionary changes are best explained via the action of evolutionary forces—mainly, natural selection—on genetic variations. In the context of human evolutionary change, this means that all—or at least most—properties of humans are to be explained as conferring a fitness advantage on individuals. This view tends to offer a radically selfish account of human nature, emphasizing competition, aggression, and reproductive success.

This biological approach has its limitations: it appears to be difficult to explain the wide range of human culture merely in terms of heritable fitness benefits. It has been questioned whether humans are dominated by the drive

Bernd Rosslenbroich, *Human Success as a Complex of Autonomy, Adaptation, and Niche Construction* In: *Human Success*. Edited by: Hugh Desmond and Grant Ramsey, Oxford University Press. © Oxford University Press 2023.
DOI: 10.1093/oso/9780190096168.003.0003

to reproduce or to compete for mates, resources, or power and predominance in business and society (Fuentes 2004, 2017; Marks 2012; Tattersall 2004). These criticisms often point to how most modern societies understand persons as moral and social beings created to live in communities linked by relationships of mutual caring, responsibility, and culture. However, since it was for a long time difficult to explain these properties as having arisen from natural selection, an apparent dichotomy arose between neo-Darwinian descriptions (in terms of survival and reproduction) and more humanistic descriptions of humans (in terms of caring and culture).

Evolutionary anthropologists have long attempted to reconcile this dichotomy. An influential approach has been that of Richerson and Boyd's "gene-culture coevolution" (Boyd and Richerson 1985; Richerson and Boyd 2005; see Demps and Richerson, chapter 8 in this volume). According to this conception of human evolution, genetic evolution can be adapted to the cultural environment, so that cultural changes can drive genetic evolution (and not just vice versa). This central importance of culture (also through technology and social learning mechanisms such as imitation) led to increasing social complexity in the genus *Homo* throughout the Pleistocene.

More recent approaches build on Boyd and Richerson's work, such as Hewlett's (2016) *Evolutionary Cultural Anthropology* and Laland's (2017) focus on niche construction. One distinct approach worth mentioning is that of Fuentes (2004, 2009, 2017), who has focused on the interplay between cooperation and competition. Competition, Fuentes argues, may also play a role, but he objects to how the neo-Darwinian approach assigns it primacy in driving evolutionary change. Instead, cooperation is a general pattern among primates, and human cooperation represents an intensified and complexified form of it. In his recent book, Fuentes (2017) emphasizes the importance of imagination and creative collaboration for human evolution.

In this chapter, I contribute an additional approach to reconciling biology with human culture. Like Fuentes, Richerson, Boyd, and Laland, I too draw on new developments in evolutionary biology (sometimes termed the "extended evolutionary synthesis"), where evolution is more than just survival and reproduction—or changes in allele frequencies. However, as developed in sections 3 through 7, my emphasis lies on a central but neglected feature that can be found throughout evolution: evolutionary change is not only to be characterized in terms of new adaptations to given environments but also in terms of changes in the degree of autonomy of individual organisms. This is especially evident during the major transitions in animal and human evolution (e.g., evolution of eukaryotes, multicellular organisms, or colonies). Using previous work (Rosslenbroich 2006, 2014), I will show how, during the major transitions, organisms increased in stability, robustness, self-regulation, homeostasis, and flexibility. In birds and mammals, this brought about the potential for more flexible and self-determined behaviors. I will especially attend to the implications for human evolution: humans have a special and far-reaching combination of features of autonomy, which serves as the basis for our physical, behavioral, mental, and cultural versatility and creativity.

On this basis, in sections 8 and 9, I argue that the feature of autonomy connects humans to nature and prehuman history, but at the same time—in its special combination and elaboration—allowed the evolution of human cultural characteristics. Against this backdrop, I propose that "human success" cannot be defined only in terms of survival and reproduction, but must include references to autonomy, independence, individualization, and perhaps to what is experienced as human freedom. Survival and reproduction may provide a baseline for success, but this biological basis allows humans to successfully react to environmental and social challenges in an extremely diverse and creative manner. Moreover, there is a strong component of agency in human evolution—as well as individual human life—to develop and extend autonomous abilities. This can include the incentive to develop individual independence, self-determination, cultural and social productivity, manipulative and technical capabilities, education, and much more. I propose that this trajectory in evolution generated, impelled, and enabled culture. To achieve individual goals and to contribute to these collective objectives is experienced as success. As this is the continuation of an evolutionary trajectory, there is no contradiction to the biological background; rather it is a reasonable continuation of evolution.

2. A Revolution in Evolution

It is important to differentiate among various components of evolutionary theory (Mayr 1982). Describing evolutionary patterns, trends, and processes is quite different from considerations about the factors that may drive the whole process. While the reconstruction of the evolutionary past of animals, plants, and humans becomes increasingly reliable and detailed, controversies about causal factors that might drive these changes have not ceased (Conway Morris 2000; Pigliucci and Müller 2010).

The synthetic theory of evolution in particular exhibited one-sidedness, with its focus on random mutation and selection and the view that the main outcome of evolution is divergence caused by different adaptations (for overviews, see Laland et al. 2014, 2015; Futuyma 2017; Lewens 2019). If this were true it would follow that all properties of organisms, including humans, are a result of former or actual adaptations, providing fitness benefits. And herein lies the epistemological flaw: the assumptions of the theory are first fixed, and then nature is scanned in order to find properties that confirm the theory. If the only factor taken into consideration is survival, then all characteristics of organisms must in fact contribute to it. The characteristics of humans must also be described from this perspective, which some recent theories are trying to do.

However, an alternative procedure for understanding evolutionary transitions would be, first, to look at the actual phenomena, patterns, and directions in the history of life and then to analyze the forces that might have generated them (Wake 1986). Of course, this is done in anthropology as well, but it often seems to lead to characteristics different from those of fitness and survival. Obviously this is the reason why there are anthropologists who wish to affirm evolution but are dissatisfied with current "neo-Darwinian" hegemony (Schultz 2009).

Beginning around the year 2000, the one-sidedness of the synthetic theory was increasingly recognized in evolutionary biology, leading to the emergence of the "extended evolutionary synthesis" (Bateson et al. 2017; Laland, Odling-Smee, and Turner 2014; Laland et al. 2014, 2015; Jablonka and Lamb 2005; Noble 2011, 2017; Pigliucci and Müller 2010; Shapiro 2011; West-Eberhard 2003).

Besides the repeatedly formulated doubts that the assumed random process would be able to create order within the evolutionary process, and the fact that the synthetic theory had little to say about the origin of macroevolutionary innovations, a number of empirical findings fueled the onset of these new considerations. One enigma arose with the growing knowledge of comparative molecular biology: it became increasingly difficult to explain the immense diversity of life despite its deep and pervasively similar molecular architecture (Gerhart and Kirschner 1997). In addition, modern molecular biology shows that many of the old assumptions about the genetic system, such as the system as a preformed blueprint of the organism or as the expected accumulation of mutations for evolutionary transitions, were incorrect (Parrington 2015; Neumann-Held and Rehmann-Sutter 2006). Perhaps it is most fundamental to explain the immense diversity of life despite its deep and pervasively similar molecular and genetic architecture. In this context, the disregard of the phenotype and the concentration mainly on the genotype, without knowing much about the relation between the two, is being questioned (Gerhart and Kirschner 1997; Shapiro 2011, 2017; Conway Morris 2000).

In addition to this work, increasingly discernible insights into the origin of evolutionary innovations have emerged. Although the result is still fragmentary, there are several surprises. Symbiosis, for example, can deliver a new state of a system within a single macroevolutionary step. Thus there seem to be systemic shifts in evolution and not just gradual processes (Margulis and Sagan 2002).

Other examples come from cell biology, comparative genetics, and developmental biology, showing that novelties can be generated by new combinations of conserved structures and functions. The genome, at least in some parts, is obviously not so much a result of random mutations as of the conservation of core functions together with new arrangements of building blocks (Carroll, Grenier, and Weatherbee 2005; Gerhart and Kirschner 1997; Kirschner and Gerhart 2005; Shapiro 2011), and these combine with epigenetic functions (Jablonka and Lamb 2005). These findings propose that evolution can occur through the epigenetic dimension of heredity even if temporarily nothing is happening on the genetic level. West-Eberhard (2003) suggested phenotypic plasticity as one of the key factors in evolution and that genes are followers rather than leaders in evolutionary change.

Another approach proposes niche construction as a significant evolutionary principle (Jones 2005; Laland and Sterelny 2006; Odling-Smee 2010; Odling-Smee, Laland, and Feldman 2003; Sterelny 2005). Niche construction is the building of niches by organisms and the mutual dynamic interaction between organisms and environments. Walsh (2015) makes a strong case for a role for the organism's activity as a whole and states that agency is a central factor in evolutionary changes. He holds "that organisms participate in evolution as agents" (xii). These new considerations on evolution will be of profound significance for anthropology (Fuentes 2004, 2016; Lewens 2017; Schultz 2009; see also Fuentes, chapter 10 in this volume).

3. The Theory of Autonomy in Evolution

If the methodology argued for above—"first . . . recognize and define patterns" (Wake 1986, 47)—is taken seriously, several different patterns can be found in the course of evolution (Rosslenbroich 2006). The major transitions—such as the generation of the first complex cells, the generation of multicellularity, the transition to land-living animals from a special group of fishes, the generation of reptiles and birds, and the long transition to the mammalian type of organization—are especially challenging to such an approach (Calcott and Sterelny 2011). This includes the transition to humans. As it becomes evident that the phenotype and the context of the system are more relevant for evolution than had been thought, patterns of phenotypic change are experiencing a renewed interest.

Our own biological studies from this point of view revealed that there is one especially prominent pattern. During the major transitions, the capacity for individual autonomy changed and was gradually enhanced and extended. Enhanced autonomy is understood as a property that increasingly allows a system to maintain its functions against internal and external perturbations and uncertainties, and at the same time offers widened possibilities for self-generated, flexible reactions and active behaviors. Physiological stability and versatility lead to enhanced capacities for the self-control of life functions and activities. Generally formulated, this means a widening of possibilities. Thus, there are organisms that are more subject to the direct physical, chemical, and biological conditions of their surroundings, and others that can act more on their own behalf because they are more active, flexible, and selective in their interaction with the environment.

These processes are described as changes in relative autonomy because numerous interconnections with the environment and dependencies upon it are retained. Also, it is not a linear trend, but rather an outcome of all the diverse processes which are operative during evolutionary changes.

Several biological elements can contribute in different degrees to changes of autonomy (Figure 3.1). They are not general rules or continuous trends. Rather, they function as a set of resources that can—singly or in combination with each other—change the capacity for autonomy. One such element is *spatial separation from the environment*, such as with cell membranes, cell walls, or integuments. To different degrees, they all serve to keep the environment outside the organism and to regulate and direct the exchange with it.

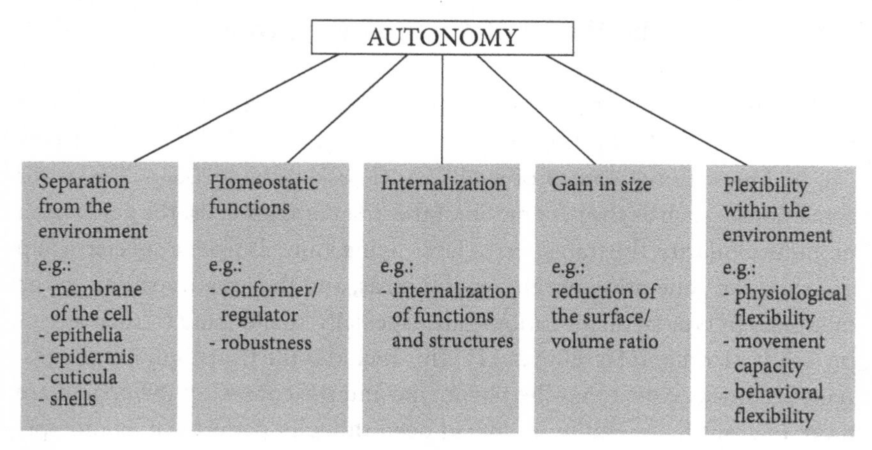

Figure 3.1 Set of resources that affect autonomy.

Homeostatic functions are means to establish and enhance internal functional stability. Another element is the displacement of morphological structures or functions from an external position into an internal position within the organism, here summarized as *internalization*. Multiple processes of internalization are involved in building up the inner anatomy of organisms, ontogenetically as well as phylogenetically. A *gain in size* during many transitions leads to a reduction of the surface-to-volume ratio. This means that in larger animals there is less direct contact with the immediate environment relative to the existing body mass. The smallest known cells, bacteria, have a large surface for environmental exchange. In larger bodies, this direct exchange capacity is reduced relative to the body mass.

These elements are prerequisites for establishing a certain amount of physiological *flexibility* within a given environment, that is, the capability of organisms to generate flexible functional responses to conditions and changes in their environment. Finally, this principle can be widened to include all forms of behavioral flexibility, emancipating organisms from mere short-term reactions to environmental factors.

These changes in autonomy can be described with morphological, physiological, and ethological criteria. We followed this pattern on different levels of evolution, from the evolution of the eukaryotic cells, through the origin of multicellularity, the generation of animal phyla, to the evolution of vertebrates, including birds, mammals, and humans. An overview is provided in Rosslenbroich (2014).

In a discussion of the concept of autonomy, McShea (2015, 440) argues that at the heart of the discourse

> is a notion of an organism as an entity, a whole, that arises from its parts, and that simultaneously governs the behavior of those parts. Parts and whole are both cause and consequence of each other. This back and forth between parts and whole sets the entity apart from its environment, to some extent, a separation that creates a kind of autonomy for the entity, a self that is distinct from its environment and that has the robustness to persist in the face of changes in that environment. In organisms, typically, this autonomy is achieved in part by boundaries, which insulate it, and further by the entity's ability to respond to environmental changes with internal changes that compensate. . . . Autonomy is the emergence of an entity, a local peak in "entity-ness," that accompanies a local increase in internally driven homeostasis.

This view is not intended to replace conventional evolutionary theories, nor is it claimed to be some sort of driving force. Instead, increasing autonomy is presented as a recurring pattern during macroevolutionary events. However, it is proposed that a more complete integration of the available knowledge of dependence on and independence from environmental factors is an important element for the further understanding of macroevolution. Presumably, this principle should be studied in its relation to other possible patterns in macroevolution, like complexity, size, and entropy (McShea 1998; Rosslenbroich 2006).

An understanding of what evolution has generated in the long run will be an essential piece of the jigsaw puzzle, which the new evolutionary biology has to put together. How can we understand evolution if we don't even know what it produced? Of course, we know that prokaryotes and single-celled eukaryotes were the earliest organisms living in this world and that later there were bees, mammals, and birds. But what are the general characteristics that changed? What is the qualitative difference between the nervous system of a polyp and that of an octopus, a lamprey, or a dolphin?

Diversity and changes of fitness are not the only issues when it comes to understanding the major transitions. Otherwise, all organisms, including ourselves, would be single-celled, probably with a wide divergence of colorful and muddled variants, as single cells, especially prokaryotes, have a tremendous capacity for survival. And then, what led to the appearance of human beings and their capacity for culture and civilization, for arts and humanities? Is it just an accident of evolutionary variations, or can we find out more about this event?

4. Some Examples from Animal Evolution

Let's consider some examples from animal evolution, where changes in the capacity for autonomy first appeared, and then turn to the case of primates (including humans).

A single-celled organism distinguishes itself from the surrounding fluid medium by actively creating its boundaries. Inside the boundary of a cell, many reactions and chemical transformations occur, and genetic and epigenetic instructions make it possible to generate a basic form of autonomy and self-determination.

During the generation of multicellularity, an epithelial boundary is formed, which is organized as integuments. The cells, together with the occluding junctions between them, seal the internal space from the environment so that the passage of substances can be controlled. The composition of fluids in the inner spaces can be regulated, and concentration gradients compared to the environment can be created. The extracellular matrix within the boundary effectively allows cells to create their own intercellular conditions, which regulate and protect them from the external milieu. However, there are large differences in the degrees of buffering of physical and chemical changes in this intercellular space. Some organisms have only basic regulative functions for the extracellular matrix; others develop organs such as nephridia, which regulate this composition. In summary, it can be stated that the general characteristics of multicellularity, as they evolved in early metazoans, essentially enhanced the possibilities of the physiological regulation of internal compartments and tissues. These (and some further) phenomena demonstrate that the transition from single cells to metazoans included different degrees of a relative emancipation from direct influences from the environment and a relative stabilization against its fluctuations.

The so-called Cambrian explosion involved the radiation of multicellular animal phyla, beginning around 540 million years ago. It was characterized by a huge variation of body forms, thus marking a phase of profound innovations leading to changes in biological activities. Looking at the fossil record, together with knowledge about recent phyla, it can be said that a set of resources, which can increase autonomy, were all involved in these innovations: increasing body size; reinforcement of the environmental separation through skins, cuticles, and shells; extension of homeostatic possibilities and the internalization of organs and functions (for example, respiratory organs, circulation systems, nephridial organs); and diversity of effective motor systems, so that movement capacities expanded and a much stronger mechanical impact on the environment became possible. The generation and extension of nervous systems enabled multicellular organisms to channel information from the environment by creating a dam between the external world and the interior milieu of the organism, and to react in a modulated and more or less self-determined way. The different body plans of the phyla are characterized by the varying combinations of this set of resources. The phyla started their further path through evolution with these different prerequisites, allowing for different possibilities of independence and flexibility.

A specialty of the basic organization of chordates and vertebrates was the construction of a central axis with affixed muscles, which makes the motoric system quite effective. The crucial innovation was the chorda-myomere system. The stiffening through the chorda—and later through the vertebral column of the fishes—offers a strong elastic insertion to the muscles. The system is made very effective by the antagonism of the muscles to the axis (which releases elastic forces), the antagonism of the muscles on both sides of the axis, and the concentrated transmission of forces on just one axis. The elasticity of the axis enables continuous propulsion, and the concentration of forces permits high acceleration as well as directed and fast movements with stamina. With the evolution of the vertebral column, further skeletal elements were generated, which stiffen and stabilize the axis; at the same time, the system of joints keeps it flexible.

Thus, the basis of vertebrates was a mechanical principle that from the outset had the potential to develop constructions for the high autonomy of movements. This potential was elaborated to different degrees in water. Then, and especially during the transition to land, it was the decisive feature in stiffening the whole body to counteract gravity, another prerequisite for further emancipation from environmental conditions.

With the formation of the vertebral column, a whole series of features developed, which allowed for the active lifestyle we usually associate with vertebrates. Pertinent to these features are specialized sensory organs and a brain, an extended metabolic capacity, an elaborated circulatory system, breathing organs with a large capacity, organs for digestion that deliver energy for high activity levels, a glomerular kidney, a complex endocrine and immunological system, as well as cartilage and bones functioning as strong building materials. Lifting the trunk above the ground, against the forces of gravity, generated new possibilities for the use of the lever principle of legs. Increases in the flexibility and efficiency of movement, especially in mammals and birds, were reached by reorganizations of the limbs and the vertebral column, essentially supported by endothermy, the autonomy of body temperature.

5. Nervous Systems and Behavior

The appearance and evolution of nervous systems are a particularly clear example of the emergence and modification of functions and organs, which support flexibility within the environment. The complexity and interconnectedness of central nervous systems are the basis of more malleable actions and reactions. And this relation leads to enhanced forms of flexible behavior.

Single cells can already perceive some factors in their environment that are relevant to them. In most cases, they react immediately by changing their movement or function. Contrary to this, metazoans have neurons with receptors on their surfaces. Here, incoming signals are transmitted into an excitation of the neuron, not into direct changes of a behavior. The excitation then influences other neurons, whereby a reaction is indirectly initiated. Here, an additional level of integration is present. First, the sensible neuron integrates incoming stimuli and may or may not activate action potentials. If an excitation occurs, the signals are processed in one or several postconnected neurons, and depending on the result, a reaction may or may not be initiated. With the generation of more complex nervous systems, the capacity for this indirect and self-determined reaction is expanded. An increasing *uncoupling* or *detachment of signals and reactions* is introduced, so that the signals are processed in neuronal intermediate steps before a reaction follows. The reaction is not directly initiated by the sensory input but is uncoupled from it and prepared by way of more or less complex processing. This modulation enables more malleable reactions of different degrees to environmental stimuli and thus is a feature of autonomy.

The principle of uncoupling is crucial for comparisons of the neuronal and behavioral abilities of animals. In general, there has been an evolution from diffuse nerve nets to concentrations of nervous tissues in ganglia and brains with an increase in the number of interneurons. The larger number of interneurons between input and output correlates with enhanced behavioral possibilities.

In the literature, there are conflicting views on the relation between brain organization in vertebrates and differences in their behavior. In general, however, there is an overall trend in relative brain size over the course of vertebrate evolution, although this increase is not linear but complex and diverse (Roth and Wullimann 2001; Striedter 2005). The differences in size, as well as features of component parts and the neuronal fine-wiring of different brains, led to different behavioral possibilities. For mammals Romer (1967, 1635) suggested, "If we were to attempt to define a mammal briefly, it could perhaps be done in two words—activity and intelligence."

According to Byrne (1995), behavioral flexibility can be regarded as an essential component of the intelligence of animals. Flexibility of behavior allows the individual to find its own solution for problems and tasks. Highly encephalized animals can even act independently of external necessities, as is the case during play in mammals and many birds (Burghardt 2005; Bekoff and Byers 1998). The essential point is that during play, extremely flexible combinations and variations of movements and behaviors, including new actions, can be generated and practiced.

Many mammals and some birds are able to generate internal representations, which can be dealt with more or less independently from the perceived, external world—that is, they are uncoupled from it. A cat that follows a mouse and sees it disappearing behind a curtain is able to foresee that the mouse will reappear on the other side of the curtain. Thus, the cat is able to draw conclusions about the mouse even if it gets no direct signals from the mouse. The cat must have some sort of an inner picture of the mouse, even when the cat cannot see it. The internal representation generates expectations regarding the mouse.

In Rosslenbroich (2014) the complex abilities of mammals and birds such as learning, play, imitation, tool use, insight, and self-awareness, which are well documented in ethology, are discussed from the viewpoint that flexibility is part of behavioral autonomy.

6. The Evolution of Humans

Humans have a special combination of autonomy features. This combination generates some essential degrees of flexibility and possibilities. First, there are those features we share with all mammals: skin, which simultaneously closes relatively tightly against the environment and is highly flexible and light (many tight-closing skins in animals become stiff and heavy); endothermy combined with a high aerobic capacity, enabling movements with endurance and to a large extent emancipating mammals from variations in environmental temperatures; effectively stabilized fluid management, including refined processes for homeostasis and highly efficient renal functions; a medium body size, which supports homeostatic functions but does not generate a larger burden for movements; and an extremely refined immune system.

In addition, there are features of autonomy we share with other primates. The extremities of nonhuman as well as human primates have an exceptionally wide range of movements in all directions. The shoulder and the pelvic girdle allow wide-ranging excursions. Also, there are extensive possibilities to rotate the distal extremities, the range of movements in the hand and foot as well as the independence of single fingers and toes from each other. There are special adaptations that can reduce this flexibility among nonhuman primates, but in general there is a trend throughout the evolution of primates to retain and to increase these varied possibilities. However, they lead not only to a broad spectrum of locomotory possibilities but also to new functions independent of locomotion (Wilson 1999; Kivell et al. 2016).

The thumb gains increasing independence, first being able to stretch out farther and then to rotate and become opposable to the other fingers. This feature supports exact grasping abilities not only during locomotion but also during the manipulation of objects. The hands increasingly take over further functions supported by an upright posture during sitting and squatting, which is typical for primates and emancipates the hands for different tasks.

All of these features of autonomy, which were generated during the evolution of primates, are elaborated to a special degree in humans. The hands are completely freed from locomotory functions because of our upright posture and have incredible flexibility and dexterity (think of a pianist or a surgeon).

This all arises from a combination of flexion of the fingers at the joints between the finger bones and the wrist bones and finger rotation, mainly of the thumb (Wilson 1999; Kivell et al. 2016; Trinkaus 1992, 2016; Nowell and Davidson 2010; Lovejoy et al. 2009). These patterns of mobility of the human shoulder and arm are of course shared with apes. However, the degree of mobility of the human hand is far greater than in all other primates (Napier 1993; Trinkaus 2016). Furthermore, this condition is supported by fine neural control and hand-to-eye coordination.

Remains of human evolution show that the upright posture is an exceedingly old innovation (Klein 2009; Niemitz 2010). Thus, the first steps toward the emancipation of the forehand evolved before most of the enlargement of the brain took place. However, the flexibility we now possess was not present to the same extent in early hominins. The hands of *Australopithecus afarensis* show an anatomical pattern with only slight modifications for manipulation. The first major changes in human manipulative prowess appeared with the emergence of the genus *Homo* (Klein 2009).

A precise grip was accompanied by further elaboration of the pyramidal tract, which in no other primate is as prominent as in humans, and this is the neurological basis for the precise movements of human hands (Striedter 2005; Yoshida and Isa 2018; Welniarz, Dusart, and Roze 2017).

Upright posture and walking were innovations that enabled hominins to explore wider areas and finally to leave Africa. Humans are able to walk with stamina and are easily able to manage distances of more than 30 kilometers per day. The combination of upright walking and hands, which can carry babies and other objects and which become increasingly able to perform techniques to provide humans with food and other necessities of life, enabled humans to undertake considerable migrations.

7. Evolution of the Human Brain

Humans are unique among primates in terms of brain size and brain organization (Sherwood et al. 2012; Striedter 2005; Roth and Wullimann 2001). A significant increase in brain mass occurred in the hominin lineage from its origination. Beginning principally with *Homo erectus*, brain expansion in hominins occurred at a more rapid pace. According to Singer (2006), our cultural competence results from the evolutionary development of certain cognitive functions. One of these is our ability to generate abstract, symbolic metarepresentations of cognitive contents by subjecting the results of first-order cognitive operations to further cognitive processing of a higher order.

This competence requires the ability to bind the distributed cognitive processes together and to represent the results of these binding operations anew at higher processing levels. Because these higher-order descriptions keep track of the brain's own cognitive operations, they can be considered the substratum for our ability to be aware of our own sensations and intentions as well as those of others. In turn, this awareness is probably the origin of our unique ability to use a symbolic communication system and to generate a theory of mind. In my terms, this is another level of the principle of detachment, which I traced throughout the evolution of the organization of nerves: the ability to generate cognitive contents on a metalevel and neuronal connections, which are not just in line with first-order representations.

Essential for representations is the cerebral cortex. The cerebellum is responsible for the fine-tuning of learned movements. Both the cortex and the cerebellum were responsible for most of the enlargement of the skull during the evolution of hominins and underwent major reorganizations (Singer 2006; Striedter 2005; Sherwood, Subiaul, and Zawidzki 2008).

One function of the cortex is to generate representations of objects and events from the environment (Sternberg 2009; Pitt 2020; Kriegel 2014; Kind 2014). Internal representations in the brain can be used instead of the object itself to, for example, prepare an action. These can be called "detached representations" and can be compared to an engineer's design for a bridge. The plan is detached from the real world, which delivers the necessary data. Then the plan is worked out, all elements are combined on a trial basis, and only if the combinations promise success will the plan be implemented. Similar to this planning on paper, the brain can simulate actions in its representations prior to realization.

Gärdenfors (2003) makes a distinction between "cued representations" and "detached representations." Cued representations stand for a perception from the present environmental situation. A detached representation, on the other hand, is not linked to such a present perception and can be used independently of external signals. He further subdivides detached representations into "dependent representations," which are dependent on an external referent, even if the object is not immediately present in the surroundings, and "independent representations," which have no relation with any external referent. Fantasies, imaginations, and abstract notions are representations, which can be handled independently. Detached representations are necessary for planning and for all other higher cognitive functions. Gärdenfors's hypothesis is that this also represents the sequence of the evolution of these abilities (cued representations → dependent representations → independent representations), and that the development of thinking can be described as the increasing detachment of representations.

In the same sense, sequences of actions can be planned regarding the future. This anticipatory planning is yet another example of how human thought can be detached from present situations. The generation of a concept of time might be closely connected to our ability to plan. Therefore, the human capacity to think has at least one origin in the principle of detachment that was generated throughout evolution.

The increase of secondary areas (association areas) of the brain is especially important in mammals. The prefrontal cortex, in particular, is understood as the area in which associations are generated, and that enables thinking and planning. These areas are not bound to monomodal functions, but rather are the basis for multiple connections. They provide the physiological basis for the plasticity of behavior. In primates, especially in man, the prefrontal cortex and the association areas are much larger than in other mammals and are involved in learning, creativity, and much more (Sherwood, Subiaul, and Zawidzki 2008; Striedter 2005). Striedter argues that the enlargement of the lateral prefrontal cortex probably helped to increase the ability of humans to suppress reflexive responses to stimuli and to enhance behavioral freedom.

In humans, detachments by means of the high grade of encephalization lead to far-reaching flexibility of behavior. Learning is a central component during child development. There is only a small amount of fixed behavior remaining on which we can rely. On the other hand, there is a learning ability that lasts a lifetime. This is also widened to cultural learning, which is thought to enable cumulative cultural evolution (Tomasello 2009).

Imitation (or imitation learning) is a type of social learning that is thought to play an especially important role in cultural inheritance (Heyes 2012; Richerson and Boyd 2005). It is in social learning that the observer acquires new abilities and skills on the basis of a highly elaborated physical and, especially, neuronal flexibility. The range and precision of our imitation of body movements far outstrip anything found elsewhere in the animal kingdom. In learning and using language and speech, and then growing far beyond mere imitation, we are not only able to communicate with others in a highly flexible and complex manner but also can communicate about past and present and about subjects completely detached from the external world.

As described before, high encephalization is related to play behavior (Byrne 1995; Burghardt 2005; Bekoff and Byers 1998). It allows action to be independent from external necessities and to form novel combinations of behavioral elements. Therefore, it is characteristic that play is such an important part of human culture. In children, play correlates with creativity, behavioral plasticity, physical fitness, and increased abilities to innovate (Hamayon 2016; Bateson and Martin 2013; Brown and Vaughan 2011). But adults also engage in play: from chess to boules, we cultivate our emancipation. In sports from baseball, skiing, or tennis to those of the Olympic Games, and in the acrobatic skills of circus performers, play combines with the cultivation of the autonomy of movements.

Flexibility and relative independence may be part of what is experienced by humans as the potential of freedom (Eibl-Eibesfeldt 1999; Hassenstein 1969; Heilinger 2007; Heldmaier and Neuweiler 2003; Neuweiler 2008). Biological restrictions still impose limitations, not only on our physiological needs but also on our behavior. However, at the same time, our biological organization enables the generation of a world of play, speech, and culture.

With all this, a feature emerged that is clearly unparalleled within the animal world: the capacity for self-control and willful, conscious behavior, with all the related qualities that make a responsible person.

8. Nature and Culture

At the beginning of this chapter, I made the point that it is possible to reconcile biology and the description of human cognitive and cultural abilities. Now it may be possible to make this proposal more explicit: autonomy theory shows that nature delivers—by evolution—the prerequisites for culture, since culture is built on increasingly flexible and self-determined capacities. These features of autonomy are prepared during prehuman evolution, but humans elaborate them in such an amount that new qualities are generated that are not present among animals.

This concept has the advantage that it neither reduces all human abilities to their biological underpinnings nor ignores our evolutionary past. Our evolutionary past and the biological prerequisites of our existence accompany us and also generate constraints. But the special combination of features of autonomy generates a world of abilities and dispositions that support the elaboration of culture, so that cultural impact and individual reasoning go beyond our biological roots in many respects.

The point is not that humans are dissociated from their biological roots. Rather, the biological roots are the basis we constantly use and act on. The relative autonomy of our physical and physiological organization forms the prerequisite for all those features that are specifically human. Our nature is not determining, but rather enabling.

Regarding human behavior, different levels may be discerned. There are ancient, conservative features retained through evolution. They include behaviors that are beneficial for survival, reproduction, and self-assertion. In addition, further levels with different degrees of versatility and creativity also evolved. With extensive variations, they are the basis of our cultural life. And then there are those levels of cognitive abilities that allow the discovery of new ideas in technology, science, the arts, as well as social life. Abilities at this level cannot be reduced to ancient behaviors. At the same time, however, they are not contradictory to biology.

Within such a concept, the myth of human nature without culture, which is criticized by Marks (2012), vanishes: nature and culture are mutually dependent, and the characteristics of this complex shift over time. There is no "beastly nature" to which culture is somehow added. There are, however, different states of autonomous abilities in the defined sense. This perspective supports Marks's view that "the idea that you can analytically separate human biology from culture and meaningfully study only human biological evolution, or that there is a 'natural history' of being human that is not a 'natural/cultural' history" (141) is fundamentally flawed.

The attempts to overcome the dichotomy between evolutionary factors and a more humanistic description of humans, which I cited above, may—in a preliminary sketch—acquire further foundation and evidence from the proposed theory. Hewlett's (2016) interaction between human behavior, culture, and biology gets its backbone from an understanding of the potential that nature provides. The abilities to which Laland (2017) points, such as learned and socially transmitted activities with accelerating cycles of feedback, are possible only with a highly malleable organism. Intelligence, language, teaching, and cooperation have their underpinning in autonomized human organization. Naturally, genes are one level (of the many other levels) of the organism to generate the abilities described above, so that autonomy theory creates a bridge with "gene-culture coevolution" (Richerson and Boyd 2005; Boyd and Richerson 1985; Demps and Richerson, chapter 8 in this volume). And the same holds true for Fuentes's (2004, 2009, 2017) considerations on neurological complexity, complex information sharing, social interactions, mutual reliance, and imagination.

Vermeij (chapter 5, this volume) argues that no other species has so completely freed itself from the influences of the physical and biological environment and that no other species has achieved such a degree of autonomous change.

9. Autonomy Theory and Human Success

Evolution is a complex of factors that are only rudimentarily understood. If a trajectory toward enhanced autonomy is part of this complex, then its features need to be regarded as a measure of human success. This is particularly necessary as human organization is characterized by a special and far-reaching combination of resources that affect autonomy. "Success" defined as "degree of autonomy" may include progress in resilience and flexibility toward environmental fluctuations and influences, enhanced abilities of self-determination, extensions in technological and cultural abilities, enhanced creativity in a large variety of fields, and social developments which affect individual ranges of personal development and degrees of freedom.

However, as argued before, autonomy is always relative and is related to the other factors involved. As a preliminary sketch, the relationship between autonomy, adaptation, and niche construction may be considered. In general, increasing autonomy changes the conditions for adaptation (in the sense of an adjustment to items in the environment), as there is an extended versatility toward environmental constraints as well as ability to deal actively with the environment (Walsh 2015). In order to develop agriculture during Neolithic times, for example, it was necessary to adapt to the current environmental conditions. At the same time, humans changed the environment quite profoundly so that niche construction was an important component during these events (Zeder 2016, 2017; Boivin et al. 2016). Growing food in a controlled manner, domesticating animals, and building houses were all steps in emancipating humans from the vagaries of the immediate offerings of the natural environment. The three factors mentioned are all involved in such an event, and all three need to be considered in order to understand human success.

The three factors are also present in any society, in the relationship of the individual person to their social surroundings, which are constructed by the members themselves. To some extent it is necessary to adapt to the conditions, habits, and laws of the family and the community. But at the same time, there are persons with individual ideas, interests, and abilities who generate changes. Part of human social life is to adjust the three factors to each other, and much of human history is a continuous struggle for different combinations of them. In the course of these struggles society and family conditions are permanently (re)constructed and changed. If these changes generate possibilities and opportunities for individual and social developments, this is regarded as success.

Autonomy theory delivers a concept that allows us to understand the evolutionary context and origin of such abilities in humans. They do not suddenly appear in humans as stand-alone adaptations, but rather are grounded in a structure of autonomy and thus have a long history in nature, during which different constellations of the complex of autonomy, adaptation, and environmental construction were generated. Various features contribute to changes in these constellations: the extreme increase in brain capacity along with the ability to generate detached representations, an increasing problem-solving ability, manual dexterity, a broad range of movement abilities, symbolic communication, cooperation, extensive learning capacities, and much more.

Reproductive success is a necessary condition for the development of autonomy, but increases in the latter can also affect the former. The flexible and intelligent action and reaction of early *Homo sapiens* to environmental challenges has often been considered a main factor in its ability to live in very diverse habitats and to spread all over the world (see Antón, chapter 6 in this volume, but also Fuentes 2004, 2017; Gibson 2005; Richerson and Boyd 2005; Zihlman and Bolter 2004). Gibson (2005, 23) writes, "Creativity, versatility and advanced learning capacities are primary hallmarks of the human mind." And Fuentes (2009, 190) sees a "specific role for flexibility and plasticity in behavioral response as a baseline as opposed to assumptions of optimality striving or single trait maximization."

Nonetheless, autonomy goes beyond mere reproductive success. Increased autonomous abilities may also be properties that humans strive for as ends in themselves, and hence the achievement of increased autonomy can be experienced as "success." These abilities could be the autonomous ability to manipulate objects, invent new techniques, and much more, thus constantly changing the complex described. Within this complex, there seem to be many feedback factors among behavior, culture, physical evolution, and environmental conditions. Enhanced creativity is also part of this trajectory. Fuentes (2017, 22) argues that creativity is at the root of many developments in humans: "You can't have that kind of variation if there isn't a spark that is specifically creative, that is, more than just a response to environmental pressures."

This aspect of autonomy, namely as a goal for human action, connects biological autonomy to political and ethical autonomy. Although this connection can only be tentatively suggested here, it does seem that political and ethical autonomy have long been valued. The striving for autonomy by ethnic groups or individuals has been a constant element in human endeavors, and its fulfillment has been experienced as success. Today the longing for increased personal autonomy is widespread, even though it often leads to tensions with established societal norms. This longing includes the efforts of humans to create individual lives and collective societies, whose future lies at least in part in their own hands. It also includes the realization of human rights, humanity, and respect for others, even across different cultures and value systems. Understanding and respecting humans as autonomous individuals is a core element at least in Western culture and societies (Pauen and Welzer 2015).

By means of autonomy, we detach many necessities from purely biological requirements and lift them into the cultural realm without, however, losing their connection. The need for food, for example, is brought into social events such as family celebrations or a candlelight dinner. Even sexual intercourse today is increasingly detached from reproduction and is part of a very intense encounter with the beloved. In this way, successful cooperation between groups or individuals, besides establishing a peaceful society in which people can live contentedly, may also be a sign of overall success. Not only do humans survive; at the same time they attempt to establish a life situation that supports their social and cultural needs, constantly adjusting the factors of the complex described.

To conclude, the concept of autonomy seems to bring together biological criteria with a humanistic framework in a way that can shed new light on human success. While I can touch upon only these possible characteristics of success that are grounded in autonomous abilities, future anthropological work might be able to identify many more aspects from the perspective of the framework of autonomy.

References

Bateson, P., N. Cartwright, J. Dupré, K. Laland, and D. Noble. 2017. "New Trends in Evolutionary Biology: Biological, Philosophical and Social Science Perspectives." *Interface Focus* 7: 20170051. http://dx.doi.org/10.1098/rsfs.2017.0051.

Bateson, P., and P. Martin. 2013. *Play, Playfulness, Creativity and Innovation*. Cambridge: Cambridge University Press.

Bekoff, M., and J. A. Byers, eds. 1998. *Animal Play: Evolutionary, Comparative and Ecological Perspectives*. Cambridge: Cambridge University Press.

Boivin, N. L., M. A. Zeder, D. Q. Fuller, A. Crowther, G. Larson, J. M. Erlandson, T. Denham, and M. D. Petraglia. 2016. "Ecological Consequences of Human Niche Construction: Examining Long-Term Anthropogenic Shaping of Global Species Distributions." *Proceedings of the National Academy of Sciences of the United States of America* 113 (23): 6388–6396. https://doi.org/10.1073/pnas.1525200113.

Boyd, R., and P. J. Richerson. 1985. *Culture and the Evolutionary Process*. Chicago: University of Chicago Press.

Boyd, R., and P. J. Richerson. 2005. *The Origin and Evolution of Cultures*. Oxford: Oxford University Press.

Brown, S., and C. Vaughan. 2011. *Play: How It Shapes the Brain, Opens the Imagination, and Invigorates the Soul*. London: Penguin.

Burghardt, G. 2005. *The Genesis of Animal Play: Testing the Limits*. Cambridge, MA: Bradford Books.

Byrne, R. W. 1995. *The Thinking Ape: Evolutionary Origins of Intelligence*. Oxford: Oxford University Press.

Calcott, B., and K. Sterelny. 2011. *The Major Transitions in Evolution Revisited*. Cambridge, MA: MIT Press.

Canter, D. 2012. "Challenging Neuroscience and Evolutionary Explanations of Social and Psychological Processes." *Contemporary Social Science* 7 (2): 95–115.

Carroll, S. B., J. K. Grenier, and S. D. Weatherbee. 2005. *From DNA to Diversity: Molecular Genetics and the Evolution of Animal Design*. Oxford: Blackwell.

Conway Morris, S. 2000. "Evolution: Bringing Molecules into the Fold. *Cell* 100: 1–11.

Eibl-Eibesfeldt, I. 1999. *Grundriß der vergleichenden Verhaltensforschung*. München: Piper.

Fuentes, A. 2004. "Cooperation or Conflict? It's Not All Sex and Violence: Integrated Anthropology and the Role of Cooperation and Social Complexity in Human Evolution." *American Anthropologist* 106 (4): 710–718.

Fuentes, A. 2009. *Evolution of Human Behavior*. Oxford: Oxford University Press.

Fuentes, A. 2015. "Integrative Anthropology and the Human Niche: Toward a Contemporary Approach to Human Evolution." *American Anthropologist* 117 (2): 1–14. doi:10.1111/aman.12248.

Fuentes, A. 2016. "The Extended Evolutionary Synthesis, Ethnography, and the Human Niche: Toward an Integrated Anthropology." *Current Anthropology* 57 (Suppl 13): S13–S26. doi:10.1086/685684.

Fuentes, A. 2017. *The Creative Spark: How Imagination Made Humans Exceptional*. New York: Dutton.

Fuentes, A., and P. Wiessner. 2016. "Reintegrating Anthropology: From Inside Out. An Introduction to Supplement 13." *Current Anthropology* 57 (Suppl 13): S3–S12.

Futuyma, D. J. 2017. "Evolutionary Biology Today and the Call for an Extended Synthesis." *Interface Focus* 7: 20160145. http://dx.doi.org/10.1098/rsfs.2016.0145.

Gärdenfors, P. 2003. *How Homo Became Sapiens: On the Evolution of Thinking*. Oxford: Oxford University Press.

Gerhart, J., and M. Kirschner. 1997. *Cells, Embryos, and Evolution: Toward a Cellular and Developmental Understanding of Phenotypic Variation and Evolutionary Adaptability*. Malden, MA: Blackwell.

Gibson, K. R. 2005. "Epigenesis, Brain Plasticity, and Behavioral Versatility: Alternatives to Standard Evolutionary Psychology Models." In *Complexities: Beyond Nature and Nurture*, edited by S. McKinnon and S. Silverman, 23–42. Chicago: University of Chicago Press.

Hamayon, R. 2016. *Why We Play: An Anthropological Study*. Chicago: Hau Books.

Hassenstein, B. 1969. "Aspekte der 'Freiheit' im Verhalten von Tieren." *Universitas* 24: 1325–1330.

Heilinger, J. C., ed. 2007. *Naturgeschichte der Freiheit*. Berlin: de Gruyter.

Heldmaier, G., and G. Neuweiler. 2003. *Vergleichende Tierphysiologie*. Vol. 1: *Neuro- und Sinnesphysiologie*. Berlin: Springer.

Hewlett, B. S. 2016. "Evolutionary Cultural Anthropology: Containing Ebola Outbreaks and Explaining Hunter-Gatherer Childhoods." *Current Anthropology* 57 (Suppl 13): S27–S37.

Heyes, C. 2012. "Grist and Mills: On the Cultural Origins of Cultural Learning." *Philosophical Transactions of the Royal Society B: Biological Sciences* 367: 2181–2191. doi:10.1098/rstb.2012.0120.

Ingold, T. 2007. "The Trouble with 'Evolutionary Biology'." *Anthropology Today* 23 (2): 13–17.

Jablonka, E., and M. J. Lamb. 2005. *Evolution in Four Dimensions: Genetic, Epigenetic, Behavioral, and Symbolic Variation in the History of Life*. Cambridge, MA: MIT Press.

Jones, D. 2005. "Niche Construction." *Nature* 438: 14–16.

Kind, A. 2014. "The Case against Representationalism about Moods." In *Current Controversies in Philosophy of Mind*, edited by U. Kriegel, 118–140. New York: Routledge.

Kirschner, M. W., and J. C. Gerhart. 2005. *The Plausibility of Life: Resolving Darwin's Dilemma*. New Haven, CT: Yale University Press.

Kivell, T. L., P. Lemelin, B. G. Richmond, and D. Schmitt, eds. 2016. *The Evolution of the Primate Hand: Anatomical, Developmental, Functional, and Paleontological Evidence*. New York: Springer.

Klein, R. G. 2009. *The Human Career: Human Biological and Cultural Origins*. 3rd ed. Chicago: University of Chicago Press.

Kriegel, U. 2014. "Two Notions of Mental Representation." In *Current Controversies in Philosophy of Mind*, edited by U. Kriegel, 161–179. New York: Routledge.

Laland, K. 2017. *Darwin's Unfinished Symphony: How Culture Made the Human Mind*. Princeton, NJ: Princeton University Press.

Laland, K., F. J. Odling-Smee, and S. Turner. 2014. "The Role of Internal and External Constructive Processes in Evolution." *Journal of Physiology* 592 (11): 2413–2422.

Laland, K. N., and K. Sterelny. 2006. "Seven Reasons (Not) to Neglect Niche Construction." *Evolution* 60: 1751–1762.

Laland, K., T. Uller, M. Feldman, K. Sterelny, G. B. Müller, A. Moczek, E. Jablonka, et al. 2014. "Does Evolutionary Theory Need a Rethink?" *Nature* 514: 161–164.

Laland, K. N., T. Uller, M. W. Feldman, K. Sterelny, G. B. Müller, A. Moczek, E. Jablonka, and F. J. Odling-Smee. 2015. "The Extended Evolutionary Synthesis: Its Structure, Assumptions and Predictions." *Proceedings of the Royal Society B* 282: 20151019.

Lewens, T. 2017. "Human Nature, Human Culture: The Case of Cultural Evolution." *Interface Focus* 7: 20170018. http://dx.doi.org/10.1098/rsfs.2017.0018.

Lewens, T. 2019. "The Extended Evolutionary Synthesis: What Is the Debate About, and What Might Success for the Extenders Look Like?" *Biological Journal of the Linnean Society* 127 (4): 707–721. https://doi.org/10.1093/biolinnean/blz064.

Lovejoy, C. O., S. W. Simpson, T. D. White, B. Asfaw, and G. Suwa. 2009. "Careful Climbing in the Miocene: The Forelimbs of Ardipithecus ramidus and Humans Are Primitive." *Science* 326: 70. doi:10.1126/science.1175827.

Margulis, L., and D. Sagan. 2002. *Acquiring Genomes: A Theory of the Origins of Species*. New York: Basic Books.

Marks, J. 2012. "The Biological Myth of Human Evolution." *Contemporary Social Science* 7 (2): 139–157.
Mayr, E. 1982. *The Growth of Biological Thought: Diversity, Evolution, and Inheritance.* Cambridge, MA: Harvard University Press.
McShea, D. W. 1998. "Possible Largest-Scale Trends in Organismal Evolution: Eight 'Live Hypotheses.'" *Annual Review of Ecology and Systematics* 29: 293–318.
McShea, D. W. 2015. "Bernd Rosslenbroich: On the Origin of Autonomy: A New Look at the Major Transitions in Evolution." *Biology and Philosophy* 30: 439–446. doi:10.1007/s10539-015-9474-2.
Midgley, M. 2010. *The Solitary Self: Darwin and the Selfish Gene.* Durham, UK: Acumen.
Napier, J. 1993. *Hands.* Princeton, NJ: Princeton University Press.
Neumann-Held, E. M., and C. Rehmann-Sutter, eds. 2006. *Genes in Development: Re-Reading the Molecular Paradigm.* Durham, NC: Duke University Press.
Neuweiler, G. 2008. *Und wir sind es doch—die Krone der Evolution.* Berlin: Wagenbach.
Niemitz, C. 2010. "The Evolution of the Upright Posture and Gait—a Review and a New Synthesis." *Naturwissenschaften* 97 (3): 241–263.
Noble, D. 2011. "Neo-Darwinism, the Modern Synthesis and Selfish Genes: Are They of Use in Physiology?" *Journal of Physiology* 589 (5): 1007–1015.
Noble, D. 2017. "Evolution Viewed from Physics, Physiology and Medicine." *Interface Focus* 7: 20160159. http://dx.doi.org/10.1098/rsfs.2016.0159.
Nowell, A., and I. Davidson, eds. 2010. *Stone Tools and the Evolution of Human Cognition.* Boulder: University Press of Colorado.
Odling-Smee, F. J. 2010. "Niche Inheritance." In *Evolution: The Extended Synthesis*, edited by M. Pigliucci and G. B. Müller, 175–207. Cambridge, MA: MIT Press.
Odling-Smee, F. J., K. N. Laland, and M. W. Feldman. 2003. *Niche Construction: The Neglected Process in Evolution.* Princeton, NJ: Princeton University Press.
Parrington, J. 2015. *The Deeper Genome: Why There Is More to the Human Genome Than Meets the Eye.* Oxford: Oxford University Press.
Pauen, M., and H. Welzer. 2015. *Autonomie: Eine Verteidigung.* Frankfurt: S. Fischer.
Pigliucci, M., and G. Müller. 2010. *Evolution—The Extended Synthesis.* Cambridge, MA: MIT Press.
Pitt, D. 2020. "Mental Representation." In *The Stanford Encyclopedia of Philosophy*, edited by E. N. Zalta. https://plato.stanford.edu/archives/spr2020/entries/mental-representation.
Richerson, P. J., and R. Boyd. 1998. "The Evolution of Human Ultrasociality." In *Indoctrinability, Ideology and Warfare*, edited by I. Eibl-Eibesfeld and F. K. Salter, 71–96. New York: Berghahn Books.
Richerson, P. J., and R. Boyd. 2005. *Not by Genes Alone: How Culture Transformed Human Evolution.* Chicago: University of Chicago Press.
Romer, A. S. 1967. "Major Steps in Vertebrate Evolution." *Science* 158: 1629–1638.
Rosslenbroich, B. 2006. "The Notion of Progress in Evolutionary Biology—the Unresolved Problem and an Empirical Suggestion." *Biology and Philosophy* 21: 41–70.
Rosslenbroich, B. 2014. *On the Origin of Autonomy: A New Look at the Major Transitions in Evolution.* Cham, Switzerland: Springer.
Roth, G., and M. F. Wullimann. 2001. *Brain Evolution and Cognition.* New York: Wiley-VCH.
Schultz, E. 2009. "Resolving the Anti-antievolutionism Dilemma: A Brief for Relational Evolutionary Thinking in Anthropology." *American Anthropologist* 111 (2): 224–237.

Shapiro, J. A. 2011. *Evolution: A View from the 21st Century*. Upper Saddle River, NJ: FT Press Science.

Shapiro, J. A. 2017. "Biological Action in Read-Write Genome Evolution." *Interface Focus* 7: 20160115. http://dx.doi.org/10.1098/rsfs.2016.0115.

Sherwood, C. C., A. L. Bauernfeind, S. Bianchi, M. A. Raghanti, and P. R. Hof. 2012. "Human Brain Evolution Writ Large and Small." *Progress in Brain Research* 195: 237–254.

Sherwood, C., F. Subiaul, and T. W. Zawidzki. 2008. "A Natural History of the Human Mind: Tracing Evolutionary Changes in Brain and Cognition." *Journal of Anatomy* 212: 426–454. doi:10.1111/j.1469-7580.2008.00868.x.

Singer, W. 2006. "The Evolution of Culture from a Neurobiological Perspective." In *Evolution and Culture,* edited by S. C. Levinson and P. Jaisson, 181–201. Cambridge: Cambridge University Press.

Sommer, M. 2015. *Evolutionäre Anthropologie: Eine Einführung*. Hamburg: Junius Verlag.

Sterelny, K. 2005. "Made by Each Other: Organisms and Their Environment." *Biology and Philosophy* 20: 21–36.

Sternberg, R. J. 2009. *Cognitive Psychology*. Belmont, CA: Wadsworth, Cengage Learning.

Striedter, G. 2005. *Principles of Brain Evolution*. Sunderland, MA: Sinauer.

Tattersall, I. 2004. "Emergent Behaviors and Human Sociality." In *The Origins and Nature of Sociality*, edited by R. W. Sussman and A. R. Chapman, 237–248. New York: de Gruyter.

Tomasello, M. 2009. *Why We Cooperate*. Cambridge, MA: MIT Press.

Trinkaus, E. 1992. "Evolution of Human Manipulation." In *The Cambridge Encyclopedia of Human Evolution,* edited by S. Jones, R. Martin, and D. Pilbeam, 346–349. Cambridge: Cambridge University Press.

Trinkaus, E. 2016. "The Evolution of the Hand in Pleistocene Homo." In *The Evolution of the Primate Hand: Anatomical, Developmental, Functional, and Paleontological Evidence*, edited by T. L. Kivell, P. Lemelin, B. G. Richmond, and D. Schmitt, 545–571. New York: Springer.

Velmans, M. 2012. "The Evolution of Consciousness." *Contemporary Social Science* 7 (2): 117–138.

Wake, D. B. 1986. "Directions in the History of Life." In *Patterns and Processes in the History of Life*, edited by D. M. Raup and D. Jablonsky, 47–67. Berlin: Springer.

Walsh, D. 2015. *Organisms, Agency, and Evolution*. Cambridge: Cambridge University Press.

Welniarz, Q., I. Dusart, and E. Roze. 2017. "The Corticospinal Tract: Evolution, Development, and Human Disorders." *Developmental Neurobiology* 77 (7): 810–829.

West-Eberhard, M. J. 2003. *Developmental Plasticity and Evolution*. Oxford: Oxford University Press.

Wiessner, P. 2016. "The Rift between Science and Humanism: What's Data Got to Do with It?" *Current Anthropology* 57 (Suppl 13): S154–S166.

Wilson, F. R. 1999. *The Hand: How Its Use Shapes the Brain, Language, and Human Culture*. New York: Vintage Books.

Yoshida, Y., and T. Isa. 2018. "Neural and Genetic Basis of Dexterous Hand Movements." *Current Opinion in Neurobiology* 52: 25–32. doi:10.1016/j.conb.2018.04.005.

Zeder, M. 2016. "Domestication as a Model System for Niche Construction Theory." *Evolutionary Ecology* 30: 325–348. doi:10.1007/s10682-015-9801-8.

Zeder, M. 2017. "Domestication as a Model System for the Extended Evolutionary Synthesis. *Interface Focus* 7: 20160133. http://dx.doi.org/10.1098/rsfs.2016.0133.

Zihlman, A. L., and D. R. Bolter. 2004. "Mammalian and Primate Roots of Human Sociality." In *The Origins and Nature of Sociality*, edited by R. W. Sussman and A. R. Chapman, 23–52. New York: de Gruyter.

4

Human Success

A Contextual and Pluralistic View

Marion Hourdequin

1. Introduction

In his 1990 book, biologist Edward O. Wilson argues that the concepts of success and dominance are frequently invoked in biology, that they are often used interchangeably, and that they have real significance for the biological sciences. Thus, although the terms are "risky" and laden with cultural connotations, "it is important for more than lexicographic reasons to try to render more precise the usage of success and dominance" (19). Wilson's book focuses on the success and dominance of social insects, but both his admonition and his recommendation apply equally with respect to *human* success. The concept of human success is complex and multifaceted, with relevance not only to the biological realm but to more general understandings of how humans conceive of themselves qua human beings and what we, as a species, are striving or should strive for. To begin to grasp the significance of the concept of human success within and beyond biology, it will be helpful to begin by considering the concept of success more generally.

Success is a teleological concept: it is defined in relation to some goal, end, or outcome. If the goal of business is to profit, a successful business is one that is profitable. If the goal of teaching is to facilitate student learning, a successful teacher is a teacher whose students learn well. If the goal of soccer is to win, a successful soccer team is one that wins frequently, or at least more often than not. Where goals are plural or contested, the same will be true of conceptions of success. Is the only or primary goal of business to make a profit? Some say yes; others argue that business should aim to provide products or services that "[make] people's lives better" (Denning 2011). Relative to the first goal—profit—the comparative success of various businesses can be assessed straightforwardly, by looking at their bottom lines. Relative to the second,

Marion Hourdequin, *Human Success* In: *Human Success*. Edited by: Hugh Desmond and Grant Ramsey, Oxford University Press. © Oxford University Press 2023. DOI: 10.1093/oso/9780190096168.003.0004 the measurement of success is more complicated, and scores on the success-as-profit measure won't necessarily correlate with success in making lives better. A business that successfully fabricates "needs" and induces people to buy their products may be highly profitable, but some such businesses may actually make people's lives worse.

Philosophers, surprisingly, have written relatively little about success per se. This is perhaps for the reason suggested above: success makes sense only in a context, in relation to some goal. Thus, one can find philosophical accounts of the success of science (Wright 2014) or philosophical success (Van Inwagen 2006; Kelly and McGrath 2017) and discussions of success in humanitarian intervention (Dobos 2016). However, more general discussions of human success are less common (although see Solomon 2006 for one example).[1] Because success can be hard to pin down, it is tempting to turn to science for answers, and in the case of *human* success, one might think biology can offer some insight, whether for individual human beings or for our species as a whole.

This chapter examines two evolutionary approaches to human success at the species level: approaches based on *persistence* and approaches focused on *proliferation*. Although such conceptions may be helpful in understanding certain ecological and evolutionary patterns, I argue that both are inadequate as general accounts of human success. These evolutionary approaches fail to square with widespread intuitions regarding human success, and they lack the conceptual resources with which to consider the *ethical* dimensions of human success. Therefore, in conceptualizing human success it would be a mistake to restrict ourselves to an evolutionary frame. More broadly, I argue that trying to settle on a single conception of human success would be a mistake.

Using case studies from the work of conservationist Aldo Leopold and contemporary responses to climate change, I show that there are important perspectives, questions, and disagreements about human success that evolutionary conceptions do not explicitly address. Therefore, accounts of human success that are intended for a particular purpose—such as a scientific explanation of humans' ability to occupy diverse environments across the globe—should make that purpose clear and leave space for conceptions of human success that address different or broader goals. Additionally, just as there are different reasonable conceptions of what constitutes a successful life for individual human beings, there are different reasonable conceptions of what would constitute success for *Homo sapiens* as a species. Although the Anthropocene has begun to generate conversations about the minimal conditions for human success, it is not clear that humans should or need to settle on a single conception of human success.

2. Biological Success: Evolutionary Views

From a biological perspective, an organism's evolutionary success can be defined in relation to organismic fitness and measured as a function of an individual organism's realized fitness, that is, their actual reproductive outcomes.[2] In this case, biological fitness is the telos or goal that grounds assessments of success. Extending this model, one might think that if organismic success is determined in relation to the achievement of some organism-level telos, species-level success must require a species-level goal or telos. Species-level success might be measured for one species in relation to others (relative fitness), or in connection to a metric that doesn't require interspecies comparisons (absolute fitness). On this view, *human* success must be defined relative to some species-level goal or end, such as the persistence or proliferation of the human species over time, which might be assessed in either relative (e.g., persistence as compared to other species) or absolute (e.g., total persistence time) terms (cf. Wilson 1990). Within this evolutionary framework, however, there are deep complications in understanding species-level fitness. For example, how does one determine when a species begins and ends? Although the "death" of a species may be clear in cases of extinction,[3] the "birth" of a species is generally quite fuzzy. And if a species radically transforms over time, should the transformed species count as a continuation of the old species or as a successor to it? This latter issue is relevant in the human case, as some visions of human enhancement through technological means could produce radically different kinds of beings, which some call *posthumans*, suggesting at least in name that these modified creatures would constitute a new, successor form of life (see, e.g., Bostrom 2008).

There are innumerable questions to explore within an evolutionary conception of human success. However, this is far from the only possible conception of human success, and I want to argue that it would be a mistake to restrict ourselves to the evolutionary frame. This is because evolutionary success is not the only kind of species-level success that does or should matter to human beings. Although evolutionary success is not entirely *independent* of individual human success or the success of human beings as a species, human success—broadly conceived—cannot be reduced to evolutionary success.

To see this, consider human success as species persistence: few people would likely count the *mere* persistence of *Homo sapiens* as the epitome of success for our species, just as few would take the mere survival of an individual to fully encompass success at the individual level. Philosophical accounts affirm this as well: moral and political philosophies such as those of Plato, Aristotle, Confucius, Laozi, Gandhi, Arendt (1958), Rawls (1971), and Nussbaum (2006) focus not on mere survival but on the quality of human lives and societies. Thus, persistence through time—which might be a reasonable measure of *evolutionary* success—is insufficient to fully capture the wider range of conceptions of human success.

A second way of understanding evolutionary success at the species level might focus on the proliferation of one species relative to another (measured by species population size or density, for example) or differential expansion in the appropriation of space, energy, or resources (see Rosenberg and Bouchard 2015; Van Valen 1989). But this approach, too, fails to encompass important conceptions of human success, such as those that envision human success as involving limits on human proliferation relative to other species, not only to ensure our own survival but to ensure the survival and flourishing of *other* species (see, e.g., Naess 2005). Given that many contemporary ecological problems—from pollution and climate change to mass extinction—are the result of differential human expansion that disrupts ecological systems and displaces nonhuman species, increasing human dominance (through proliferation or differential expansion relative to other species) appears neither necessary nor sufficient for human success from a broader point of view.

The prospect of measuring human success through differential expansion raises another concern in restricting conceptualizations of success to a species-level evolutionary frame. This is because evolutionary conceptions of success may lack the conceptual resources through which to consider the *ethical* dimensions of human success.[4] The differential expansion view brings this concern into focus because the appropriation of resources, including the exclusion of other groups' or species' access to those resources, forms the core of this conception of success, yet the ethical notion of *equity* focuses squarely on problematizing the unbridled appropriation of resources at others' expense. From an ethical perspective, therefore, the differential expansion approach appears problematic. Applying the differential expansion approach at the intraspecies level for human beings, for example, would suggest that the displacement and disruption of some human groups by others would count as "success" on the part of the dominant and displacing group, a view that is in tension with most philosophical conceptions of justice (other than a Thrasymachean one) and therefore with an ethical account of success.

In response to these concerns,.an objector might argue that persistence- and proliferation-based accounts are intended as *biological* notions of success and should be understood and evaluated as such, from a scientific point of view. On this view, ethical concerns need not be considered when defining and assessing human success from a biological or evolutionary perspective. A *different* conception of human success may be appropriate in the ethical realm, or perhaps it might be possible to develop an all-things-considered conception of human success that takes ethics into account, but an evolutionary account of human success need not address ethical concerns.

This is an important point. Different success concepts are appropriate for different contexts, and an evolutionary conception of human success might have a different purpose or role than an ethical conception. If success is defined in relation to some goal, outcome, or end, then a biological conception of human success might be relativized to the evolutionary outcome of proliferation or persistence, whereas an ethical conception of success might be relativized to certain ethical values or goals. From an ethical perspective, some views might overlap with biological conceptions of success. For example, some might think that ensuring the persistence of *Homo sapiens* over time is an important—or even the most important—ethical goal for human beings, and others might think that maximizing the total number of human lives is an ethically valuable goal. Still others might step away from persistence and proliferation, arguing that coexistence with other species over time is a better goal, even if it requires constraints on human proliferation or increases the likelihood that our own species will wink out of existence. What is critical with respect to ethical conceptions of human success, however, is that they cannot be justified directly by appeal to evolutionary success, and there are risks in conflating evolutionary and ethical conceptions of success, for reasons noted earlier. Those developing and applying evolutionary conceptions of human success should thus take care to clarify the basis, role, and limits of the conceptions they employ.

Even so, evolutionary accounts of human success may be hard to keep distinct from conceptions of human success in other domains, given the ongoing interchange between evolutionary biology and society, and the blurring of descriptive and normative claims at the interface between them. Beginning in the late 19th century, evolutionary notions of success diffused widely into the social realm,[5] where they were used to justify various social policies. In many cases, the consequences were morally problematic. For example, Social Darwinists prescribed reduced assistance to the poor based on the idea that social welfare policies interfere with the natural process of evolution, undermining the capacity of human beings and human communities to flourish in the long run (e.g., Sumner 1914; for a careful recent discussion of Social Darwinism, see Paul 2003). In the 20th century, similar ideas helped to justify social practices and policies, such as "fitter families" contests, antimiscegenation laws, forced sterilization of those deemed "unfit," racial science, and more (Selden 2005; Roberts 2011). In her recent book, *Beyond Biofatalism*, Gillian Barker (2015) argues that ideas from the contemporary field of evolutionary psychology similarly bleed into the social realm. She suggests that certain concepts in evolutionary psychology have both descriptive and normative elements, "allowing some thinkers to present normative positions as if they are descriptive or derive directly from descriptive facts and are authenticated by evolutionary science" (9). The concept of human success, too, entangles the descriptive and the prescriptive and thus may be susceptible to the same sleights of hand. Given the interplay of science with other realms of human thought and action, it may be helpful to consider human success from a broader point of view and to be aware of the risks and limitations of developing conceptions of success in an evolutionary vacuum.

3. Whither Human Success?

If the arguments offered above are compelling, it may be helpful to look not only to evolutionary biology as a basis for conceptualizing human success but to broader notions of what it might mean for human beings, as a species, to be successful. In approaching this task, we might consider what conceptions of human success are *for*. Why seek an understanding of human success, and what aspects or dimensions of "success" do we, as investigators, hope to comprehend? What role(s) might particular conceptions of human success play in scientific inquiry, in guiding human action, or in other domains?

From a scientific perspective, a conception of human success might help to focus both scientific questions and scientific explanations. Beginning with the observation that humans have been successful—in some senses, at least—one might try to make precise the specific way(s) in which humans have been successful in order to explain that success. For example, one might seek to understand how human beings came to proliferate as widely as they have and to appropriate as many resources as they have (which surely has had some role in human proliferation). Or one might be interested in humans' capacity to make and remake their own environments in diverse and creative ways, to cooperate with strangers, or to intentionally manipulate the lives and futures of other living things. Each of these aspects of the human species might be considered a distinctive form of "success," subject to evolutionary, ecological, sociocultural, or other forms of explanation. One might wonder, however, whether and how labeling various human capacities as forms of "success" adds explanatory value. What is gained from asking *What has made human beings so successful as a species, as demonstrated by their proliferation across the globe?* as opposed to *What has enabled the human species to proliferate across the globe?* Although I don't have an answer to this question, it seems that invoking success draws in a forward-looking and normative component in ways that narrower questions about specific capacities do not, and in doing so, success seems to introduce more to *contest*. Explaining more specific phenomena or capacities may be more straightforward, less explicitly value-laden, and therefore less controversial. For example, one might agree that humans have proliferated across the globe and share the view that explaining this proliferation is scientifically worthwhile, while disagreeing that the global proliferation of *Homo sapiens* need be understood in terms of success.

These considerations raise questions of when and how notions of success should be invoked in scientific explanation. This is particularly important because it is not clear that the kinds of "human successes" that might provoke scientific inquiry necessarily match the kinds of human successes that individual human beings or broader human communities might seek, from an ethical or all-things-considered perspective. As noted above, the prodigious proliferation of human beings, or the human species' appropriation of a large share of the Earth's natural resources, may not necessarily count as success, at least not in an unqualified way. Thus, the context of and purpose for accounts of human success are important to clarify, especially if one seeks a conception of human success not simply to identify and explain particular patterns in the human species over time, but to guide human behavior or set goals for the future.

The significance of this point comes into focus if one looks at contemporary debates surrounding the idea of the Anthropocene. In recent years, Paul

Crutzen, Christian Schwägerl, Will Steffen, and many others have argued that the Earth has entered a new geological epoch—*the Anthropocene*—marked by pervasive human influence on Earth systems (Steffen, Crutzen, and McNeill 2007; Crutzen and Schwägerl 2011). Yet whether this epoch should be accepted, celebrated, or lamented is a matter of fractious and ongoing debate (Haraway 2016; Cuomo 2017; Santana 2019). If human beings have modified the planet so extensively that it is now appropriate to name the next geologic epoch after ourselves, should this count as a form of human success and a reason for celebration? Or should human intervention of this extent and magnitude be viewed as a form of human failure and a matter for lamentation? Or, to consider a third possibility, is the existence of pervasive human influence on the planet simply something to accept, with the question of success most productively focused on the future, on what would count as success for *Homo sapiens*, as a species, *now*?

4. Contested Conceptions of Success in the Anthropocene

Although Aldo Leopold (1887–1948) died more than a half-century before the term "Anthropocene" was coined, his work engaged questions that remain central to thinking through conceptions of human success today. Leopold did not focus specifically on the concept of human success, but he did examine the related idea of *progress*, a central organizing notion of his time. Leopold was born in the late 19th century, following decades of aggressive westward expansion by the United States, guided by the doctrine of manifest destiny, which linked European settlement of North America to ideas of progress, civilization, and a natural, God-given right. Leopold challenged prevailing notions of progress, however, and in doing so, implicitly challenged underlying ideas of human success. His critique helps illuminate the relationship between progress and success, and it brings out ethical dimensions of these ideas, which remain at the center of debates surrounding the Anthropocene and responses to contemporary challenges such as global climate change.

There are two passages in *A Sand County Almanac* that bring into focus Leopold's (1949) critique of prevailing notions of human progress. In "Arizona and New Mexico," Leopold writes explicitly—and cynically—about the conceptions of progress that cleared the way for recreational and economic development in the U.S. West or that advance narrow human interests at the expense of other living things. In the short section titled "Escudilla," about a mountain in eastern Arizona, Leopold laments the "various emissaries" of progress, one of whom was the government trapper sent to find "any destructive animals in need of slaying" (135). The trapper kills the lone old bear that haunted the mountain, leaving Leopold wondering "who wrote the rules for progress" and whether "making Escudilla safe for cows" is enough to justify taking the last grizzly living there (135). Throughout the section titled "Arizona and New Mexico," Leopold questions the narrowness and short-sightedness of "human progress," arguing for a broader, more ecologically and evolutionarily informed view. He does not hesitate to turn the critical lens on himself, and in the famous passage "Thinking Like a Mountain," describes his enthusiastic and unthinking slaughter of a mother wolf: "In those days, we had never heard of passing up a chance to kill a wolf. In a second we were pumping lead into the pack, but with more excitement than accuracy. . . . When our rifles were empty, the old wolf was down, and a pup was dragging a leg into impassable slide rocks" (130). Leopold attributes the killing to his own narrow and thoughtless view: "I was young then, and full of trigger-itch; I thought that because fewer wolves meant more deer, that no wolves would mean hunters' paradise" (130). This form of thinking, Leopold argues, is not restricted to his youthful encounter with the wolf. He sees it in the predator extermination campaigns throughout the western United States; he sees it in notions of progress that focus on human interests without considering broader ecological consequences; and he sees it in economic modes of valuation that cannot account for the noneconomic value of animals, plants, and ecological systems as a whole. In the end, Leopold suggests that a new ethic is needed, one in which human beings shift their role from that of "conqueror" to that of "plain member and citizen" of the biotic community. Human success, for Leopold, lies not in human expansion at the expense of other species but in developing a land ethic in which humans consider themselves one part of a broader ecological community and seek to develop relations with that community that enable mutual flourishing. Leopold has thus been interpreted as an important articulator of a non-anthropocentric ethic (Callicott 1989) or, one might say, a non-anthropocentric conception of human success, in which humility and receptivity, rather than domination and control, play a central role (cf. Norlock 2011).[6] Interestingly, Leopold (1949, 202) describes the development of ethics as "a process of ecological evolution," involving the development of new modes of cooperation.

Contemporary discussions of humanity's role in the Anthropocene reflect a distinct split between those who favor a Leopoldian-style view and those who embrace a more active, directive, and managerial approach (Hourdequin 2017). Of course, not all are happy to accept demarcation of the Anthropocene as a distinctive age of human influence, and among those who do, there remain deep philosophical differences in how human success might be constructively conceived. Whereas the normative implications of accounts of human success in scientific contexts might be limited or unclear, in debates over the Anthropocene they are often quite explicit. Some conceptualize a "good Anthropocene" as one in which humans embrace their role as "planetary managers," charged with looking after our own species and others alike (see, e.g., Marris 2011a, 2011b; Ellis 2011). This view tends to support an interventionist approach to environmental management, which might include genetic modification to help species adapt to climate change or novel predators (see, e.g., Rohwer 2018), or intentional, global-scale climate intervention through solar geoengineering (Keith 2013). Others find this vision deeply hubristic, infused with unwarranted confidence in humans' capacity for successful and benevolent planetary management (Hamilton 2013; Hulme 2014). Those in this latter camp generally favor *reducing* human intervention and influence and approaching humans' planetary role with greater humility. Still others question the appropriateness of treating the Anthropocene and the human response to it monolithically. These critics ask who, exactly, is the *Anthropos* in the Anthropocene (see, e.g., Cuomo 2017)? And who are the "planetary managers" invoked by the epoch's boosters? This line of criticism argues that human influence on the planet is the result of uneven appropriation and use of resources: not all human beings have contributed or currently contribute equally to this epoch of human impact. Thus, this view challenges grouping all human beings together with respect to both the designation of the Anthropocene and responses to it.

Debates about whether to designate a new Anthropocene epoch, how to conceptualize it, and what—if anything—humans should do differently in response to its designation, demonstrate deeply divergent views about how to conceptualize and evaluate human success. What's more, these debates themselves are occurring within a modest and distinct subset of human beings worldwide. This is where the *Whose Anthropocene?* question comes in, and correspondingly, *Whose success?* Who gets to determine, define, and shape human responses to the Anthropocene, with whose success in mind, and how defined? For example, non-anthropocentrists might question whether a "good Anthropocene" is possible if this epoch is one that leaves human beings at the center of things, in terms of influence, control, and concern. Others might ask whether discussions of the Anthropocene even have all human beings in mind, and if so, whether and how the diverse cultures, values, perspectives, and ways of life among human beings planet-wide are being fully taken into account in conceptualizing human success in the Anthropocene. Is it even possible to conceptualize equitable management at the planetary scale, and if so, what might that look like? Divergent conceptions of success in the Anthropocene (and beyond) have real consequences, and one way to see this is to look at the case of global climate change, with particular attention to climate engineering.

5. Climate Change, Geoengineering, and Human Success

> It's an engineering problem, and it has engineering solutions. (Rex Tillerson, speaking on climate change, quoted in Council on Foreign Relations 2012)

> Geoengineering—specifically stratospheric aerosol injection—is not only risky, but supports powerful economic interests, protects an inherently ecologically harmful social formation, relegates the fundamental social-structural changes needed to address climate change, and is rooted in a vision of a nature as a set of passive resources that can be fully controlled in line with the demands of capital. (Gunderson, Petersen, and Stuart 2018, 1)

If the hallmark of the Anthropocene is pervasive, planetary-scale human influence, then anthropogenic climate change is at the heart of the Anthropocene. Since the late 1800s, when human beings began burning significant quantities of fossil fuels, carbon dioxide concentrations in the atmosphere have risen from under 300 parts per million (ppm) to over 400 ppm, exceeding levels experienced at any time in the past 400,000 years (NASA 2022). In that same time period, global mean surface temperature has risen approximately 1.01°C (NASA 2022), and the effects of changing climate are being felt across the world, as precipitation patterns change, the frequency and intensity of storms increase, oceans warm, permafrost thaws, and coastal erosion threatens shorelines and their inhabitants.

In the early stages of international discussion surrounding climate change, the conversation focused on mitigation: drawing down greenhouse gas emissions and, correspondingly, concentrations of CO_2, methane, and other greenhouse gases in the atmosphere. As emissions continued to rise, attention to climate adaptation began to grow. It is now clear that no matter how quickly emissions reductions occur, a substantial amount of climate change is "locked in" for the coming decades, and human societies and ecological systems will need to adapt to new climate parameters. In the past decade, however, geoengineering has emerged as a third potential response to climate change. Proposed geoengineering (aka "climate engineering" or "climate intervention") strategies typically fall into two broad groups: (1) those that aim to remove carbon dioxide from the atmosphere and sequester it so it cannot contribute to climate warming (known as carbon dioxide removal strategies, or CDR), and (2) those that aim to compensate for elevated greenhouse gas concentrations by reflecting sunlight back into space (known as solar radiation management, or SRM) (Shepherd et al. 2009). Although solar radiation management could be undertaken using a distributed approach, such as painting rooftops white, one of the leading proposals is global in scale. By injecting sulfate aerosols into the stratosphere, it may be possible to mimic the global cooling effects of a large-scale volcanic eruption, like that of Mt. Pinatubo in 1991 (Shepherd et al. 2009; National Research Council 2015).

Global-scale sunlight reflectance strategies for geoengineering tend to be controversial because they do not reduce greenhouse gases in the atmosphere and may have unintended consequences of various kinds—though one could argue that large-scale volcanic eruptions like Pinatubo offer a "proof of concept" (the Earth cooled by about 0.5°C for a year following this eruption, which spewed 20 million tons of sulfate into the upper atmosphere; USGS 1997), the quantity of sulfates needed to sustain a significant cooling effect over decades would be much greater than that of a single volcanic eruption. SRM also raises the worry that masking the effects of increasing atmospheric CO_2 will reduce incentives for mitigation, undermining the long-term prospects for stabilizing and reducing greenhouse gas concentrations—even though there is widespread consensus among climate scientists and policy experts that SRM alone offers a wholly inadequate response to climate change in the long run (Shepherd et al. 2009; National Research Council 2015).

Discussions of SRM geoengineering feature multiple metaphors and frames, and these frames arguably reflect different underlying conceptions of human success not only with respect to climate change specifically but in relation to humanity more generally. Proponents of SRM research and development have framed solar geoengineering as an "insurance policy," "plan

B," and an approach that has the potential to ameliorate climate-related distributive injustice for the "global poor." (See Scott 2012 for a discussion of framing, Fragnière and Gardiner 2016 for critique of the "plan B" framing, and Horton and Keith 2016 for discussion of distributive justice and climate engineering.) Those who are warier describe the small group of scientists working on solar climate engineering as the "geo-clique," characterize solar geoengineering as a "techno-fix," and raise questions about hubris, moral corruption, and compound injustice (see, e.g., Gardiner 2010; Preston 2012; Hamilton 2013). Framing plays a complex role in ongoing discussions of geoengineering, among both experts and broader publics, because different frames highlight different aspects of geoengineering as a climate response. Bellamy and colleagues (2012, 605) argue that using a climate emergency frame may increase the public acceptability of geoengineering "through implicit implication of necessity": in an emergency, extreme measures may be required. They also suggest that appraisals of geoengineering often limit the range of choices being assessed, offering comparisons between multiple forms of geoengineering but not between geoengineering and other climate responses (608). More broadly, narrow framings of geoengineering may represent and promote a narrow range of visions for the future (605).[7]

Underlying disagreements about geoengineering are divergent perspectives on the role of technology in climate change and human futures more generally. Arguably, these divergent perspectives reflect divergent conceptions of human success. Those who see geoengineering as a viable response to climate change typically see climate change in a relatively technologically oriented or technocratic frame. John Barry (2016) points out that even the distinction between CDR strategies and SRM strategies embeds a technologically centered point of view. These approaches are frequently contrasted by noting that CDR addresses the root cause of climate change, whereas SRM does not. However, as Barry notes, drawing the contrast in this way—describing greenhouse gases as the "root cause" of climate change—seems to suggest greenhouse gases "somehow magically [appear], presenting a clear and present danger to the earth's climate system" (116), without addressing the underlying social, economic, and structural reasons that greenhouse gases have accumulated and continue to accumulate in the atmosphere. The vision of human success underlying this technologically and scientifically centered view locates the problem in the physical environment and the solution in technologies that modify the physical environment rather than locating climate problems in social, economic, political, and cultural realms. For those in the latter camp, a successful response to climate change would not merely bring global temperatures down; it would involve fundamental reorientations of values and economies, and even a wholesale reenvisioning of global capitalism (Klein 2014).

Returning to geoengineering, we can understand why those who see climate change as primarily a problem of global temperature tend to view SRM as an optimization problem (Ban-Weiss and Caldeira 2010) and—if successfully optimized—as the best way to address climate-related distributive injustice (Horton and Keith 2016). On this view, human success in developing an appropriate response to climate change is primarily a matter of intervening "optimally" to tweak the right climate parameters, dialing back the climate to something that more closely resembles a pre-climate-change state. Ideally, such intervention could benefit people globally, particularly the most vulnerable, by bringing temperature and precipitation parameters closer to their pre-climate-change state (Horton and Keith 2016; Irvine et al. 2019). However, research in climate-vulnerable communities has shown that many people are concerned not only or primarily about the climatic effects of SRM but also about the social and political impacts of a technology over which they may have limited control. In interviews with people in the South Pacific, sub-Saharan Africa, and the North American Arctic, Carr and Yung (2018, 127) found that people were concerned that geoengineering research might overlook local needs, worsen global inequalities, and/or "make vulnerable populations even more dependent upon the decisions and actions of more powerful actors in distant places." Thus, claims that geoengineering technologies "could enable surprisingly uniform benefits"[8]—insofar as they center on physical parameters alone—offer what others may see as a very limited view of human success, or of a successful response to climate change more specifically. Nevertheless, modeling studies focused on physical parameters tend to receive significantly greater attention than social and ethical concerns in relation to solar geoengineering.

What all of this suggests is that evaluations of the adequacy of geoengineering as a response to climate change depend on broader conceptions of the "problem." If the climate problem is understood simply as rising mean global surface temperature, then geoengineering looks like a more promising solution than if the problem is unsustainable and inequitable human relationships with one another and with ecological systems across the globe. In addition, although the geoengineering debate is not about human success per se, it further illustrates how conceptions of success are best understood as relative to particular goals, frames, and visions. The discourses of geoengineering also show how particular conceptions of success may box out others, making it difficult to inject different perspectives in ways that shape how problems and solutions are framed.

6. Conclusion: The Value of Plural and Contested Conceptions of Human Success

The concept of human success more generally depends on what is understood to be the goal or aim for human beings as a species. It thus presupposes that there *is* some common goal. However, this in itself is contestable not only because there are surely many different conceptions of what a species-wide goal might be but also because not everyone may agree that it makes sense even to try to articulate a conception of success for *Homo sapiens as a species*. Of course, given the degree of global interconnectedness—ecological, social, cultural, economic, and political—in the 21st century, some might argue that a common conception of human success, at least a thin one, is needed to enable coordinated actions that support the persistence of human life on Earth. This generates pressure toward planetary conceptions of human success, or at least the preconditions for them, which are often taken to require the maintenance of conditions that enable our species to persist. The idea of a "safe operating space for humanity," for example, establishes "planetary boundaries" that human beings must not "transgress" to avoid shifting the Earth out of a "desirable Holocene state" (Rockström et al. 2009). These boundaries concern nine domains, including climate change, biodiversity loss, ocean acidification, chemical pollution, and use of freshwater.

Although the concept of a safe operating space for humanity considers biodiversity and ecosystem health, it is worth noting that these boundaries are defined primarily in relation to *human* sustainability. For example, biodiversity is described as instrumentally valuable, with the "planetary boundary" on biodiversity pegged to the level of biodiversity needed to sustain ecosystem resilience, presumably in support of stable human communities. In the end, Rockström and colleagues (2009, 475) conclude that "the evidence so far suggests that, as long as the thresholds are not crossed, humanity has the freedom to pursue long-term social and economic development."

A global conception like that of Rockström and colleagues (2009) might be attractive on the grounds that human success has a minimum floor. On this view, human persistence is a precondition for human success. Thus, even if there are plural conceptions of human success, and even if it would be inappropriate to seek a thick conception of human success applicable to our species as a whole, the "safe operating space" concept helps clarify the necessary (if not sufficient) conditions for human success. However, although at first this argument may seem both compelling and unobjectionable, any "thin" universal conception of human success will face a twofold challenge. One is to avoid slippage from thin to thick notions of human success and to avoid the impingement of purportedly thin concepts—like the "safe operating space" idea—on more particular, thicker ones. The other is to avoid assuming what may at first seem obvious, that indefinite persistence is a necessary condition for human success. A human life can be successful even if it comes to an end (as for all of human history, human lives inevitably have), and it seems plausible that human beings as a species could be "successful" in key respects, despite the finitude of our species. Perhaps *premature* extinction would undermine human success, or at least premature extinction that is anthropogenic—for example, a nuclear war that brought about human extinction in 2100—but clarifying what counts as premature extinction is a task in its own right. My point is not to argue for any particular conception of human success but to suggest that even the most basic and seemingly obvious conditions for human success (such as continued persistence) can be called into question, and even apparently innocent conceptions of human success have normative assumptions built into them that deserve critical examination.

There may be both scientific and social value in seeking to conceptualize and understand human success. From a scientific perspective, one might be interested to understand how humans have proliferated so widely, and how *Homo sapiens* have been able to survive and thrive in such diverse conditions across the planet. Because we live in a deeply interdependent and globalized world, it may be valuable to consider and discuss what might constitute "success" for human beings, as a species, going forward, and what would be required to work toward it. Perhaps the United Nations sustainable development goals could be seen as one effort to articulate a common vision of human "success" or human "progress." The United Nations describes these as "a shared blueprint for peace and prosperity for people and the planet, now and into the future" (United Nations Department of Economic and Social Affairs 2015). Of course, the sustainable development goals themselves are contested, and there are multiple reasonable conceptions of what such goals should entail.

Global-scale economies and global-scale environmental and social challenges create pressure in favor of convergent goals and convergent definitions of success. However, they also create power dynamics in relation to success, where institutions, governments, and social, political, and economic systems impose certain conceptions of success and criteria for success on individuals, social groups, and even humanity as a whole. Thus, if the concept of human success is to be an inclusive one, it should remain contested, subject to multiple interpretations and to discussions of whether and how various conceptions of human success are useful (or not). Being explicit and remaining circumspect about the role and function of success concepts in various domains—scientific and otherwise—can contribute to this pluralistic approach. If success is a teleological concept, defined in relation to some goal or end, then contestation about human success will also, necessarily, involve contestation about the goal or end for *Homo sapiens*. This contestation will call into question whether there is or should be a common goal or end against which human success might be judged, whether and how the measure(s) of human success might change over time, and whether success for human beings might encompass not only our own persistence and flourishing but that of the other diverse forms of life with which we share the planet.

Acknowledgments

I would like to thank Hugh Desmond, Grant Ramsey, two anonymous reviewers, and audience members at the 2019 Human Success: Evolutionary Origins and Ethical Implications Conference at KU Leuven for helpful comments and suggestions on earlier versions of this chapter.

Notes

1. Browsing the popular literature, one finds plenty of self-help books outlining the path to success, but these texts generally presuppose some conception of success. For example, there's *The Law of Success: The Master Wealth-Builder's Complete and Original Lesson Plan for Achieving Your Dreams* by Napoleon Hill ([1928] 2008) and *Million Dollar Habits: 27 Powerful Habits to Wire Your Mind for Success, Become Truly Happy, and Achieve Financial Freedom* by Stellan Moreira (2017). These books clearly focus on financial success and link financial success to happiness. There are also books about

success in relationships, success in leadership, spiritual success, and so on. Mostly, these books focus on *individual* success.

2. This might be operationalized, for example, by measuring the surviving offspring a particular individual produces over its lifetime. Some philosophers of biology argue that trait fitness rather than organismic fitness is the key fitness concept (Sober 2013); however, Pence and Ramsey (2015, 1082) suggest that the fitness of organisms is conceptually important in evolutionary biology, as well as *metrologically* important, that is, important as a "quantitative measure in evolutionary studies," and the concept of organismic fitness has played an important role in evolutionary theory since its inception.
3. Even extinction is complicated, however. The concept of functional extinction, for example, does not depend on the loss of the very last surviving individual of a species; functional extinction occurs when a species' population falls below a level at which it can continue to play a particular ecological role (Sellman, Säterberg, and Ebenman 2016). See also Thom van Dooren's (2014) *Flight Ways*, in which he argues that extinction is best understood as an extended process, with a "dull edge."
4. Cultural evolutionary theory and multilevel selection theories do offer some interesting and promising resources for reconciling complex forms of altruism with evolution and can perhaps help to explain the evolution of human moral capacities. For further discussion of cultural evolution, gene-culture coevolution, and group selection, see Richerson and Boyd (2008).
5. A number of authors have argued convincingly that the diffusion is not unidirectional, from science to society. Instead, there are feedback loops. In the case of Darwinian evolutionary biology, Gregory Radick (2003) and others have written on the influence of social theory on the development of Darwin's theory of evolution by natural selection. See also Hull (2005) and Larson (2006).
6. Not all interpreters of Leopold agree that he should be read as a non-anthropocentrist. Bryan Norton (1988, 2005), for example, has advanced a pragmatic interpretation of Leopold's work.
7. Kyle Whyte (2018) echoes this point, arguing that current discourses provide little space to discuss the conditions that led to consideration of geoengineering in the first place, which include persistent structural inequities and historical and ongoing systems of domination among human beings—particularly as exemplified in relations between modern nation-states and Indigenous peoples.
8. David Keith, quoted in Harvard John A. Paulson School of Engineering and Applied Sciences (2019).

References

Arendt, H. 1958. *The Human Condition*. Chicago: University of Chicago Press.

Ban-Weiss, G. A., and K. Caldeira. 2010. "Geoengineering as an Optimization Problem." *Environmental Research Letters* 5 (3): 034009. https://doi.org/10.1088/1748-9326/5/3/034009.

Barker, G. 2015. *Beyond Biofatalism: Human Nature for an Evolving World*. New York: Columbia University Press.

Barry, J. 2016. "Bio-fueling the Hummer? Transdisciplinary Thoughts on Techno-optimism and Innovation in the Transition from Unsustainability." In *Transdisciplinary Perspectives on Transitions to Sustainability*, edited by E. Byrne, G. Mullally, and C. Sage, 106–123. New York: Routledge.

Bellamy, R., J. Chilvers, N. E. Vaughan, and T. M. Lenton. 2012. "Appraising Geoengineering." Tyndall Centre for Climate Change Research, Working Paper 153. https://tyndall.ac.uk/wp-content/uploads/2021/09/TWP-153.pdf.

Bostrom, N. 2008. "Why I Want to Be Posthuman When I Grow Up." In *Medical Enhancement and Posthumanity*, edited by B. Gordijn and R. Chadwick, 107–137. N.p.: Springer.

Callicott, J. B. 1989. *In Defense of the Land Ethic: Essays in Environmental Philosophy*. Albany, NY: SUNY Press.

Carr, W. A., and L. Yung. 2018. "Perceptions of Climate Engineering in the South Pacific, Sub-Saharan Africa, and North American Arctic." *Climatic Change* 147 (1–2): 119–132.

Council on Foreign Relations. 2012. "CEO Speaker Series: A Conversation with Rex W. Tillerson." June 27. https://www.cfr.org/event/ceo-speaker-series-conversation-rex-w-tillerson.

Crutzen, P. J., and C. Schwägerl. 2011. "Living in the Anthropocene: Toward a New Global Ethos." *Yale Environment 360*, January 24. http://e360.yale.edu/feature/living_in_the_anthropocene_toward_a_new_global_ethos/2363.

Cuomo, C. J. 2017. "Against the Idea of an Anthropocene Epoch: Ethical, Political and Scientific Concerns." *Biogeosystem Technique* (4): 4–8.

Denning, S. 2011. "Is the Goal of a Corporation to Make Money?" *Forbes*, September 26. https://www.forbes.com/sites/stevedenning/2011/09/26/is-the-goal-of-a-corporation-to-make-money/#3f371ae654ed.

Dobos, N. 2016. "Idealism, Realism, and Success in Armed Humanitarian Intervention." *Philosophia* 44 (2): 497–507.

Ellis, E. 2011. "The Planet of No Return." *Breakthrough Journal* 2 (Fall): 37–44.

Fragnière, A., and S. M. Gardiner. 2016. "Why Geoengineering Is Not 'Plan B.'" In *Climate Justice and Geoengineering*, edited by C. J. Preston, 15–32. Lanham, MD: Rowman & Littlefield.

Gardiner, S. M. 2010. "Is 'Arming the Future' with Geoengineering Really the Lesser Evil? Some Doubts about the Ethics of Intentionally Manipulating the Climate System." In *Climate Ethics: Essential Readings*, edited by S. M. Gardiner, S. Caney, D. Jamieson, and H. Shue, 284–312. New York: Oxford University Press.

Gunderson, R., B. Petersen, and D. Stuart. 2018. "A Critical Examination of Geoengineering: Economic and Technological Rationality in Social Context." *Sustainability* 10 (1): 269.

Hamilton, C. 2013. *Earthmasters: The Dawn of the Age of Climate Engineering*. New Haven, CT: Yale University Press.

Haraway, D J. 2016. *Staying with the Trouble: Making Kin in the Chthulucene*. Durham, NC: Duke University Press.

Harvard John A. Paulson School of Engineering and Applied Sciences. 2019. "Finding the Right 'Dose' for Solar Geoengineering." Phys.org, March 11. https://phys.org/news/2019-03-dose-solar-geoengineering.html.

Hill, N. (1928) 2008. *The Law of Success: The Master Wealth-Builder's Complete and Original Lesson Plan for Achieving Your Dreams*. New York: Penguin.

Horton, J., and D. Keith. 2016. "Solar Geoengineering and Obligations to the Global Poor." In *Climate Justice and Geoengineering: Ethics and Policy in the Atmospheric Anthropocene*, edited by C. Preston, 79–92. Lanham, MD: Rowman & Littlefield International.

Hourdequin, M. 2017. "The Ethics of Ecosystem Management." In *The Oxford Handbook of Environmental Ethics*, edited by S. M. Gardiner and A. Thompson, 449–462. New York: Oxford University Press.

Hull, D. L. 2005. "Deconstructing Darwin: Evolutionary Theory in Context." *Journal of the History of Biology* 38 (1): 137–152.

Hulme, M. 2014. *Can Science Fix Climate Change? A Case against Climate Engineering*. Malden, MA: Polity.

Irvine, P., K. Emanuel, J. He, L. W. Horowitz, G. Vecchi, and D. Keith. 2019. "Halving Warming with Idealized Solar Geoengineering Moderates Key Climate Hazards." *Nature Climate Change* 9 (4): 295–299.

Keith, D. 2013. *A Case for Climate Engineering*. Cambridge, MA: MIT Press.

Kelly, T., and S. McGrath. 2017. "Are There Any Successful Philosophical Arguments?" In *Being, Freedom and Method: Themes from van Inwagen*, edited by J. A. Keller, 324–339. New York: Oxford University Press.

Klein, N. 2014. *This Changes Everything: Capitalism vs. the Climate*. New York: Simon & Schuster.

Larson, B. M. H. 2006. "The Social Resonance of Competitive and Progressive Evolutionary Metaphors." *BioScience* 56 (12): 997–1004.

Leopold, A. 1949. *A Sand County Almanac and Sketches Here and There*. New York: Oxford University Press.

Marris, E. 2011a. *Rambunctious Garden: Saving Nature in a Post-Wild World*. New York: Bloomsbury.

Marris, E. 2011b. "We Are Planet Managers." *New York Times*, May 23. https://www.nytimes.com/roomfordebate/2011/05/19/the-age-of-anthropocene-should-we-worry/we-are-planet-managers.

Moreira, S. 2017. *Million Dollar Habits: 27 Powerful Habits to Wire Your Mind for Success, Become Truly Happy, and Achieve Financial Freedom*. Independently published.

Naess, A. 2005. "The Basics of Deep Ecology." *Trumpeter* 21 (1): 61–71.

NASA. 2022. "Global Climate Change: Vital Signs of the Planet." https://climate.nasa.gov. Accessed November 3, 2022.

National Research Council. 2015. *Climate Intervention: Reflecting Sunlight to Cool Earth*. Washington, DC: National Academies Press. https://doi.org/10.17226/18988.

Norlock, K. J. 2011. "Building Receptivity: Leopold's Land Ethic and Critical Feminist Interpretation." *Journal for the Study of Religion, Nature and Culture* 5 (4): 493–512.

Norton, B. G. 1988. "The Constancy of Leopold's Land Ethic." *Conservation Biology* 2 (1): 93–102.

Norton, B. G. 2005. *Sustainability: A Philosophy of Adaptive Ecosystem Management*. Chicago: University of Chicago Press.

Nussbaum, M. C. 2006. *Frontiers of Justice: Disability, Nationality, Species Membership*. Cambridge, MA: Harvard University Press.

Paul, D. B. 2003. "Social Darwinism and Eugenics." In *The Cambridge Companion to Darwin*, edited by J. Hodge and G. Radick, 214–239. New York: Cambridge University Press.

Pence, C. H., and G. Ramsey. 2015. "Is Organismic Fitness at the Base of Evolutionary Theory?" *Philosophy of Science* 82: 1081–1091.

Preston, C. J. 2012. "Solar Radiation Management and Vulnerable Populations: The Moral Deficit and Its Prospects." In *Engineering the Climate: The Ethics of Solar Radiation Management*, edited by C. J. Preston, 77–94. Lanham, MD: Lexington Books.

Radick, G. 2003. "Is the Theory of Natural Selection Independent of Its History?" In *The Cambridge Companion to Darwin*, edited by J. Hodge and G. Radick, 143–167. New York: Cambridge University Press.

Rawls, J. 1971. *A Theory of Justice*. Cambridge, MA: Harvard University Press.

Richerson, P. J., and R. Boyd. 2008. *Not by Genes Alone: How Culture Transformed Human Evolution*. Chicago: University of Chicago Press.

Roberts, D. 2011. *Fatal Invention: How Science, Politics, and Big Business Re-create Race in the Twenty-First Century*. New York: New Press.

Rockström, J., W. Steffen, K. Noone, Å. Persson, F. S. Chapin III, E. F. Lambin, T. M. Lenton, et al. 2009. "A Safe Operating Space for Humanity." *Nature* 461 (7263): 472–475.

Rohwer, Y. 2018. "A Duty to Cognitively Enhance Animals." *Environmental Values* 27 (2): 137–158.

Rosenberg, A., and F. Bouchard. 2015. "Fitness." In *The Stanford Encyclopedia of Philosophy* (Fall), edited by E. N. Zalta. https://plato.stanford.edu/archives/fall2015/entries/fitness/.

Royal Society. 2009. *Geoengineering the Climate: Science, Governance and Uncertainty*. London: Royal Society.

Santana, C. 2019. "Waiting for the Anthropocene." *British Journal for the Philosophy of Science* 70 (4): 1073–1096. https://doi.org/10.1093/bjps/axy022.

Scott, D. 2012. "Insurance Policy or Technological Fix? The Ethical Implications of Framing Solar Radiation Management." In *Engineering the Climate: The Ethics of Solar Radiation Management*, edited by C. J. Preston, 151–168. Lanham, MD: Lexington Books.

Selden, Steven. 2005. "Transforming Better Babies into Fitter Families: Archival Resources and the History of the American Eugenics Movement, 1908–1930." *Proceedings of the American Philosophical Society* 149 (2): 199–225.

Sellman, S., T. Säterberg, and B. Ebenman. 2016. "Pattern of Functional Extinctions in Ecological Networks with a Variety of Interaction Types." *Theoretical Ecology* 9 (1): 83–94.

Sober, E. 2013. "Trait Fitness Is Not a Propensity, but Fitness Variation Is." *Studies in the History and Philosophy of the Biological and Biomedical Sciences* 44: 336–341.

Solomon, R. C. 2006. "On Success." *Philosopher's Magazine*, 3rd quarter, 20–26. https://doi.org/10.5840/tpm20063540.

Steffen, W., P. J. Crutzen, and J. R. McNeill. 2007. "The Anthropocene: Are Humans Now Overwhelming the Great Forces of Nature?" *AMBIO: A Journal of the Human Environment* 36 (8): 614–622.

Sumner, W. G. 1914. *The Challenge of Facts and Other Essays*. Edited by A. G. Keller. New Haven, CT: Yale University Press.

United Nations Department of Economic and Social Affairs. 2015. "Sustainable Development: The Seventeen Goals." https://sdgs.un.org/goals. Accessed November 3, 2022.

USGS (U.S. Geological Survey). 1997 "The Cataclysmic 1991 Eruption of Mt. Pinatubo, Philippines." https://pubs.usgs.gov/fs/1997/fs113-97/. Accessed November 3, 2022.

Van Dooren, T. 2014. *Flight Ways: Life and Loss at the Edge of Extinction*. New York: Columbia University Press.

Van Inwagen, P. 2006. *The Problem of Evil: The Gifford Lectures Delivered in the University of St. Andrews in 2003*. Oxford: Oxford University Press.

Van Valen, L. 1989. "Three Paradigms of Evolution." *Evolutionary Theory* 9: 1–17.

Whyte, K. P. 2018. "Indigeneity in Geoengineering Discourses: Some Considerations." *Ethics, Policy & Environment* 21 (3): 289–307. doi:10.1080/21550085.2018.1562529.

Wilson, E. O. 1990. *Success and Dominance in Ecosystems: The Case of the Social Insects*. Vol. 2. Oldendorf: Ecology Institute.

Wright, J. 2014. *Explaining Science's Success: Understanding How Scientific Knowledge Works*. New York: Routledge.

PART II
EXPLAINING HUMAN SUCCESS

5

The Origin and Evolution of Human Uniqueness

Geerat J. Vermeij

1. Introduction

If there is one thing science has taught us about ourselves, it is that we humans are animals. Anatomically and phylogenetically, we are primates, an evolutionary branch of mammals with a long history dating back more than 50 million years to the Early Cenozoic era. We are historical documents, having inherited many of our characteristics from our ancestors and modifying them as our lineage became adapted to our environment and to ourselves.

This fact has made it fashionable in some circles to claim that there is nothing special about us, just as there is nothing unique about Earth as a planet, the sun as a star, or the Milky Way as a galaxy. According to this view, we are not necessarily superior to, say, the Eurasian Neanderthals or Denisovans whom modern humans have largely replaced (Roberts and Thorpe 2014; Villa and Roebroeks 2014; Kolodny and Feldmann 2017). Instead, according to these arguments, chance and random variation figure prominently in our success.

A variation on this theme is that our brains and bodies were shaped and stabilized by our Pleistocene surroundings, well before agriculture and civilization became established, with the implication that this premodern legacy bequeathed aspects of human nature that limit our ability to cope with the technology we ourselves created (Nesse and Williams 1994; Dennett 2006). The common thread connecting these arguments is that adaptation, whether it be genetic or cultural, was ineffective in human history, either denying us uniqueness or contributing little to our success.

The urge to put us in our ordinary place in the cosmos and to minimize the power of adaptation, however, takes revisionism too far and overlooks

Geerat J. Vermeij, *The Origin and Evolution of Human Uniqueness* In: *Human Success*. Edited by: Hugh Desmond and Grant Ramsey, Oxford University Press. © Oxford University Press 2023. DOI: 10.1093/oso/9780190096168.003.0005 the many ways in which the modern human species is unique in the history of life. Just as a pile of stones and timbers can be fashioned into a great cathedral, so our animal legacy and the principles governing its evolution have served as the foundation for unparalleled cognitive, social, and economic achievements that have made us the most powerful species ever to roam the Earth. With our genetically inherited and culturally acquired characteristics, we continue to adapt to rapidly changing circumstances, most of them of our own making. Limitations on adaptation still apply, and our rates of social evolution may be insufficient to avoid the problems created by our monopolistic position, but we are forging a cultural environment that makes possible both much faster adaptation and less reliably adaptive cultural change than is the case in the primarily gene-based evolution in other animals and plants. The universe was not created for us, and the Earth and the sun are not at the center of the universe, but we are undeniably and very consequentially unique in our part of the cosmos. We are the products of evolution, but our species has transcended all life-forms in the history of our planet.

As living things that survive and reproduce by acquiring and defending resources that are converted into biological work, we and all other organisms are members of ecosystems, or economies, emergent structures in which energy and materials are produced (or made available), consumed, and redistributed (Vermeij 2004, 2009, 2018; Vermeij and Leigh 2011). Whether simple, as in a primordial microbial mat, or complex, as in the modern human world, those economies are built on interactions—competition, predation, parasitism, mutualism, and cooperation—among organisms that are adapted to their circumstances. Adaptation—a good fit between organism and environment—evolves by natural selection at the genetic level and by direct effects that living things have on their circumstances. Organisms affect each other as well as the enabling factors that limit or enhance access to and availability of resources and therefore the extent to which adaptation can be achieved (Vermeij 2013). The principles governing economies and adaptation are invariant even if the outcomes of change are enormously diverse both in space and over time. An economic perspective on evolution, centered on the roles of competition and cooperation, is essential if we are to understand the evolutionary trajectory of our own species and to place our unique attributes in a broader biological context.

How, when, and where did the transformation from a run-of-the-mill primate to a species with profoundly consequential unique traits take place? These questions have received close scrutiny from scholars for hundreds of years, with increasingly compelling scenarios about the origins of toolmaking, language, social institutions, agriculture, industrialization, and the scientific revolution (Harari 2015). Even so, there is room for a more explicitly evolutionary account of the circumstances that favored our lineage, and not other lineages elsewhere or at another time in Earth history, to take the leap—or more accurately the many leaps—to economic hegemony. Are the particulars just accidents of history, or are there aspects that are predictable in the context of economic principles and ecological history? This chapter is my attempt to come to grips with these issues.

I come to this inquiry as an outsider on two counts. First, and most obviously, I am neither an anthropologist nor an archaeologist, nor could I be mistaken for an economist or social scientist. Instead, my background is paleobiology and evolutionary biology. Starting with a lifelong love affair with molluscs and plants, I have sought to uncover and understand large-scale patterns in the history of life, with an emphasis on interactions among living things for locally scarce resources and the adaptations resulting from those interactions. The enemies and allies of organisms have collectively shaped much of this history by serving as agents of natural selection in environments that organisms respond to, control, and modify. My perspective on humanity is therefore to place the evolution of our species in the larger panorama of the tumultuous yet knowable history of life as a whole and the principles that govern that history. This outsider's approach allows me to be dispassionate about the many controversies that swirl around every aspect of the social and archaeological sciences.

Second, I am uncomfortable in large social settings and have long had an aversion to mass movements. Perhaps counterintuitively, this personal preference for solitude or a small circle of trusted and loved acquaintances has given me an appreciation for the central role that social conventions and institutions play in all aspects of human behavior and history. I would like to think that my molluscan social skills, together with a keen interest in and careful observation of humans and their interactions, keep my mind open to new ideas and skeptical of ideology.

2. What Makes Humans Unique?

Nearly all human traits have precedents in our ancestors. This should come as no surprise; after all, evolution is descent with modification, and a vast body of evidence demonstrates that we descended from anthropoid apes as the consequence of adaptive evolution. Furthermore, lineages unrelated to our own have also approached some of the traits that we think of as human: large brain, flexibility in brain function, social organization, a sense of morality, sophisticated means of communication, the capacity to modify the environment to benefit individuals and groups, and tool use (de Waal 1996). One or more of these characteristics typifies some parrots, crows, beaked whales, elephants, and New World monkeys. What makes humans unique is that these traits, as well as some capacities with no antecedents or convergent expression in other lineages, have developed much further in our species than in any other.

Studies of primate genomes have made it clear that our uniqueness does not reside in our genes. Yes, some gene sequences are loosely associated with distinctly human traits, but these genetic differences are not qualitatively any more remarkable than those that distinguish other species from each other. In any case, there is a very long chain—or, more accurately, an intricate network—connecting a protein-coding or protein-regulating gene with an organism's form, physiology, and behavior. Yet it is these phenotypic expressions, emerging from the ways genes interact with each other and with conditions inside and outside the body, that are subject to the selective processes underlying evolutionary adaptive change. For these reasons, and despite the increasingly gene-centric ethos of much of human biology—and of evolution generally—I shall have little to say about genes in this essay.

In Table 5.1 I list the anatomical, cognitive, social, and economic traits that are either unique to humans or that have been embellished beyond levels observed in other animals. Although these characteristics are listed separately, they should not be thought of as independent of each other. Instead, the combination of positive feedbacks (mutual reinforcement among the conditions and traits that enable and compel change), common underlying connections and correlations among the genes that affect heritable characteristics, and functional trade-offs ensure that evolution affects many aspects of form and behavior simultaneously, a phenomenon that the evolutionary morphologist Keith Thomson (1969) called correlated progression.

For example, Lieberman (2007) cogently argues that the sequential actions in producing speech are controlled by the basal ganglion in the brain, which also controls other activities such as walking and long-distance bipedal running, a capacity unique to the genus *Homo*. Motor control, motor learning, and cognition are all intricately linked by the basal ganglion, cerebellum, and

Table 5.1 List of distinctive human traits with references

Hidden estrus: Wrangham et al. 1999; Prum 2017
Large globular brain and chin: Lieberman 1998; Seymour, Bosiocic, and Snelling 2016; Street et al. 2017; Gomez-Robles et al. 2017; Neubauer, Hublin, and Gunz 2018
Small postcanine teeth: Wrangham et al. 1999; Gomez-Robles et al. 2017
Large robust big toe: Rolian, Lieberman, and Hallgrímsson 2010
Opposable thumb and lateral fingers for precise manipulation: Niewoehner et al. 2003; Rolian, Lieberman, and Hallgrímsson 2010
Shoulder modifications for accurate long-distance throwing: Roach et al. 2013
Mostly naked skin: Jablonski 2004
Descent of tongue into larynx, descent of larynx, enabling speech: Lieberman 2007
Far-reaching sociality and cooperation: Bingham 2000; Gowdy and Krall, 2014; Gowdy and Krall 2016; Raghanti et al. 2018
Individuality: this chapter
Use of external technology and energy sources: Arthur 2009
Clothing: Henshilwood et al. 2001; Jablonski 2004
Cultivation of crops and animal husbandry: Diamond 1997; Purugganan and Fuller 2009
Spread of dangerous ideas and technology: Vermeij and Leigh 2011
Global economic monopoly: Vermeij and Leigh 2011; Vermeij 2018
Sex as entertainment: this chapter
Construction of compound tools: Ambrose 2001; Wadley, Hodgskiss, and Grant 2009; Hoffecker and Hoffecker 2017; Backwell et al. 2018; Shipton et al. 2018
Music: Conard et al. 2009; Mithen 2006; Wurz 2009
Combinatorial language: Deacon 1997; Mithen 2006; Lieberman 2007; Arbib, Liebal, and Pika 2008; Scott-Phillips 2015; Cardini 2017
Enhanced cognition through cooperation: dos Santos and West 2018

prefrontal cortex (Weaver 2005), all under the control of the gene FOXP2. The implication is that increased selection favoring endurance running beginning with *Homo erectus* (Bramble and Lieberman 2004) primed the integrated system to evolve the capacity to produce and understand speech. The uniquely human ability to use the arms in throwing objects accurately at high speeds and over long distances (Roach et al. 2013) is still another manifestation of correlated evolution associated with greater activity and cognition. Likewise, selection for a more robust big toe and short lateral toes, enabling long-distance running, affects the serially homologous thumb and lateral fingers of the hand, facilitating precision grasping and the fashioning and use of stone tools (Rolian, Lieberman, and Hallgrímsson 2010). These trends likely began in *Australopithecus* by at least 3.5 million years ago (Ma), but achieved their modern level of expression in *Homo*. Uniting all these factors is increased blood supply to an enlarging, increasingly globular, energy-hungry brain, as indicated by the size of the foramen for the carotid artery (Pontzer et al. 2016; Seymour, Bosiocic, and Snelling 2016; Neubauer, Hublin, and Gunz 2018). In short, features of anatomy, cognitive function, and sociality evolved together during at least the early phases of our lineage's history.

The concerted evolution of physical and cognitive traits in our lineages invites caution about pinpointing a single evolutionary breakthrough as the trigger for all that was to follow. Cause and effect are inextricably intertwined whenever living things interact, and interdependencies—among body parts, among genes, among species in an ecosystem, and among members of human society—make it impossible and imprudent to assume that a particular circumstance, whether it be environmental, anatomical, or behavioral, sets our species on a path toward uniqueness. As I discuss further below, competition within our species and between humans and other animals must have been a necessary prerequisite for the advantages that the evolution of greater cognitive abilities and sociality brought to our lineage. Anatomical changes were likely important during the early phases of our evolution, but positive feedbacks involving their evolution and forces favorable to increased sociality prevent a further dissection of circumstances leading to the human transformation.

Some genes-oriented scholars contend that humans continue to change physically thanks to natural selection. They point to increasing stature, increasing frequencies of genes with effects on the expression of sickle-cell anemia, bipolar disorder, and anxiety-related traits, and a rise in the incidence of incongruencies between the width of the birth canal and fetal head size as evidence that gene-based adaptive evolution continues to take place in the modern human species (Sato and Kawata 2018; Mitteroecker 2019). These same scholars, however, recognize that medical intervention can counteract or eliminate many of these conditions. This unique circumstance leads me to suggest that the observed genetic changes are due not to natural selection but to its absence. A change in gene frequency does not automatically imply the action of natural selection. Instead, it may simply reflect the lifting of constraint, allowing gene frequencies to vary in nonadaptive ways. In other words, technology has increasingly replaced phenotypic (and ultimately genetic) selection in the evolution of our species.

The transition from physical to a more explicitly social and cultural evolution (Foley 2016) was well under way by the time early members of our genus—*Homo habilis* and *H. ergaster* in Africa, *H. erectus* in Asia—evolved during the Early Pleistocene between 2.5 Ma and 1.8 Ma. This shift entailed increases in the use of external sources of energy, social innovations, and vast expansions in communication and transportation, propelling our lineage to global ecological dominance and ultimately leading to the development of a worldwide economy whose fundamental properties mirror those of whole ecosystems but whose pace and reach far exceed those seen in nature. These aspects of human uniqueness—external energy, social organization, individualism, and language—are treated sequentially below, but they are inextricably intertwined.

2.1 Energy Sources

The deliberate use of external sources of energy by humans began with fire. Only hominins, including Neanderthals, have tamed fire to functional ends (Sorensen, Claud, and Soressi 2018; Wrangham 2017). No other animals set controlled fires to keep warm, drive predators and insects away, prepare food, forge metals, and power engines. If Richard Wrangham (2017; Wrangham et al. 1999) is right, as I believe he is, cooking became part of human behavior with the evolution of *Homo ergaster*. Although cutting food into small pieces with tools reduces the energy required for chewing (Zink and Lieberman 2016), the energy savings are modest compared with the energy saved and gained by cooking tubers, leaves, and meat. Cooking also kills parasites and detoxifies plant compounds. According to Wrangham, the physical consequences of the advent of cooking include reductions in the size of jaws and teeth, decreased size differences between males and females, and larger brains (see also Organ et al. 2011). Greater emphasis on biparental care, communal eating and sharing food, and greater cooperation among females are among the social consequences of habitual cooking Wrangham proposes. Even hidden estrus may be a consequence of the use of fire for cooking.

Fire may be the first means by which humans exploited resources of energy and power outside themselves, but it was by no means the last. With animal domestication came the exploitation of animals to work in agriculture, wheeled transport, and large-scale war. We harness wind and water for power and tap fossil (and later nuclear) fuels to power the modern economy. This expanding repertoire of external energy sources has enabled our rise to economic dominance and propelled our species toward unparalleled sociality.

2.2 Social Organization

Compared with other animals, including other social primates, humans have achieved a level of ultrasociality—division of labor, cooperation among genetically unrelated individuals, and concerted action by coherent groups coupled with intense competition among individuals—that is unrivaled and still evolving (Dugatkin 1999; Nowak 2006; Stearns 2007; Gowdy and Krall 2014, 2016). Social insects have evolved even more highly integrated societies, but they have done so by producing genetically closely related infertile offspring from one queen and one or several males. The infertile workers function autonomously but are more like cells or tissues of a single organism than like genetically distinct actors that are capable of reproducing (Crespi 2014; Gowdy and Krall 2016). In primates, the extent of sociality correlates with large brain size, a long reproductive life span, and strong reliance on cultural transmission of learned information and social norms (Kaplan and Robson 2002; Street et al. 2017).

Much of human nature is shaped by the evolution of our ultrasociality. We all belong to groups—tribes, religions, clubs, gangs, corporations, educational institutions, nations, political parties, service organizations, and more—and it is these organized groups that are the most effective agents in human societies for accomplishing tasks and establishing norms that could not be achieved by any one individual. It is therefore essential that diverse individuals feel a strong sense of belonging, a compulsion to fit in. In doing so, we often make sacrifices for the sake of the group and adhere to beliefs that we might not accept as individuals but which are essential to group cohesion. In the extreme case of religion and some forms of state control, these beliefs often contradict scientifically verifiable facts, but group identity and group allegiance become more important arbiters of truth than all else. In our species, therefore, it is the social environment that has become the dominant agency to which we respond culturally; it is the social environment that encourages some behaviors over others, determines which norms are accepted and which are not, establishes rules for how we interact, and creates trust among strangers (Seabright 2004). Individuals are prepared to die for the sake of a nation or a religious conviction because the group has created incentives—promises of a good afterlife, high social status for victors and martyrs, and severe punishment by human or supernatural authority—all socially designed for the cohesion of the group (Hauert et al. 2007; Henrich et al. 2010). Our evolutionary environment has become us.

Although we are exceptional in that the human evolutionary environment is primarily cultural and social, we are far from alone in having evolved under a selective regime in which members of one's own species have become the dominant arbiters of reproductive success. Animals and plants in which eggs are fertilized inside the body or under the direct control of an individual must exercise mate choice, where individuals recognize and choose among potential mates, often at a distance. Mates have therefore become a new resource for which intense competition among individuals can ensue. Animals can engage directly in mate choice, but flowering plants often rely on animal vectors for pollination and thus effectively to choose mates. This reproduction-related imperative has led to the origin and elaboration of signals—displays of color, sound, odor, exaggerated body parts, and distinctive behavior (Prum 2017)—as well as bizarrely complex genitalia, nicely summarized in Menno Schilthuizen's (2014) recent book, *Nature's Nether Regions*. Survival-related adaptations remain important in these species, but they have been augmented and modified by sexual selection.

In most internally fertilizing animals, sexual attraction and mating behavior remained closely tied to survival-related traits. Even when courting displays and mating rituals are costly and extreme, they signify good prospects for survival and successful reproduction. Sex appears to be pleasurable in many animals, even when there is often sexual conflict between partners, as in many insects, vertebrates, and hermaphroditic snails. All of this pertains to humans as well, but in humans sex has taken on an additional role of pure entertainment that has become largely divorced from the act of procreation. Symptoms of this new role of sex include prostitution and pornography, both of which—like the entertainment industry as a whole—have flourished as societies became increasingly affluent.

Because sexual selection cannot proceed without signaling and communication among individuals of the same species, I suggest that mate choice provided the foundation for two important and seemingly contradictory trajectories in some animal lineages, including our own: the evolution of social organization by cooperation between genetically distinct individuals, and the emergence of individuality, where vast differences among members of the species have large effects on society as a whole.

That internal fertilization and mate choice are necessary conditions for the evolution of animal societies is indicated by the absence of social organization in species in which fertilization is external and gametes are spawned into the surrounding medium (see Vermeij and Grosberg 2018). Internal fertilization has the great advantage of protecting the fertilized egg from predation, a benefit enhanced further by the evolution of development of the embryo in well-protected seeds or eggs and ultimately by parental care. With parental care came the possibility of cooperation and eventually of social life, outcomes that enable social animals to compete more effectively than solitary ones (Wilson and Hölldobler 2005). These benefits are reflected in the fact that all ecosystems in the sunlit biosphere are dominated competitively by social animal species, especially on land (Vermeij 2017). Importantly, social organization further shields individuals from predation (Wilson and Hölldobler 2005), opening the evolutionary door for the emergence of individuality. Studies of orangutan behavior in Borneo show that this extra cushion of safety from predators allows individuals of this social species, each with different traits and abilities, to explore, paving the way for the emergence of individual creativity (van Schaik et al. 2016).

2.3 The Role of Individuals

As I noted earlier, modern human societies are unique in the animal kingdom in that they promote, recognize, and celebrate the role of individual contributions. Artists, writers, and composers as well as many scientists and inventors work alone or in very small groups. Like actors, athletes, politicians, and religious leaders, these creative members of society can become famous. Although some of their ambition for cultural and scientific achievement arises from intense curiosity and emotion, it is also fueled by powerful social incentives such as financial rewards, a high reputation, and prestigious prizes. The acceptance and adoption of art, literature, music, science, and technology depend on social attitudes, and all creative pursuits and individual initiatives play out in a social context. Societies differ in how they tolerate or encourage creativity and novelty, but the societies that have been economically most successful have combined robust social institutions and social stability with a climate favorable to individual expression.

It is this combination that makes the human social experiment so different from the societies of other species. In other animals, individuals are genetically and behaviorally more or less autonomous agents, but they interact more as members of a class than as individuals with unique characteristics. The distinction between self and nonself has a very long history in the animal kingdom (Grosberg and Quinn 1986; Grosberg and Hart 2000), but only a few great apes have achieved a level of individuality that enables them to recognize themselves in a mirror (de Waal 1996). Among humans, the public recognition of individual contributions and accomplishments likely attained its modern expression only after the beginning of agriculture and civilization. Even as societies become increasingly connected and integrated, competition among self-interested individuals for social status is as vigorous as ever, with the result that individuality continues to be widely admired and encouraged (Rosen 2005).

2.4 Language

No human capacity exemplifies our dual social and individual nature better than language. As a means of communication, it is an inherently social phenomenon, which unites groups and distinguishes among parts of society. At the same time, language permits individuals to distinguish themselves, to be creative in their expression, and to influence the actions of others.

The evolution of language and other combinatorial systems of communication and composition is to my mind the most far-reaching departure in the human species from norms established in other animals. Crickets, katydids, cicadas, frogs, and many fishes produce and receive a very limited number of highly stereotyped sounds that function principally in attracting and differentiating among potential mates, and in some cases to startle predators or threaten competitors (Senter 2008). Birds—especially songbirds—and beaked whales create and receive a substantially greater diversity of individual sounds (phonemes), sound sequences (syllables), and combinations of these sequences, and some birds can even produce two notes simultaneously, but the complexity and variability of these sounds does little more than communicate attractiveness to potential mates or threats to competitors. A few birds display rudimentary syntax—communication of different meanings using combinations of phonemes (Collier et al. 2014; Russell and Townsend 2017), as do some monkeys. Human language, however, far surpasses animal communication in its endless combinatorial possibilities, using linear sequences of phonemes and context-dependent sequences that carry added meaning (Scott-Phillips and Blythe 2013; Scott-Phillips 2015; Collier et al. 2014). The closest approach to human language is the combinatorial, context-dependent sequence of calls by Africa's Campbell's monkey (*Cercopithecus campbelli campbelli*). Individual calls have distinct meanings with respect to particular threats, but sequences take on additional meanings (Ouattara, Lemasson, and Zuberbuhler 2009). In the human species, language is intimately bound up with individual recognition, intentional communication to influence listeners, evaluation and transmission of knowledge, and learning from others.

As a system of vocal (and later of written) symbols, language offers a striking analogy to an organism's genome (Vermeij 2010). In the genome, any of four kinds of nucleotide are strung together to form DNA molecules, different sectors of which specify genes that translate to proteins, orchestrate development, regulate physiology, and control the expression of other genes. Individual nucleotides by themselves convey no information, but three of them, when connected in a line, correspond to an amino acid, and a string of these nucleotide codons specifies proteins, which are composed of a linear sequence of amino acids. Meaning and function therefore emerge in the genome only when nucleotides are organized into genes and amino acids are organized into proteins. Genes that do not specify proteins can regulate other genes by the position and context of these genes, implying a sort of genetic syntax. Similarly, although phonemes in isolation can communicate some information about the person uttering them, increased content emerges when they form words, sentences, and stories. Just as an effectively infinite number of combinations and sequences can be generated with just four types of nucleotides and 20 common amino acids, an infinite number of meanings and messages can be generated with a relatively small number of sounds, written symbols, and syntactical rules. Humans have likewise perfected other mechanisms for generating diversity combinatorially, including music (especially polyphonic music) and the construction of machines and other devices composed of multiple parts that are assembled rather than created through differentiation of a single progenitor structure (Arthur 2009). In all these combinatorial systems, emergent meaning becomes context-dependent, where the temporal and spatial position of symbols or parts adds or modifies meaning. Crucially, all these systems easily accommodate changes in the nature, number, ordering, and functions of parts, creating a degree of versatility that promotes adaptability.

The analogy between languages and genomes even extends to their origins, which in both systems involve sharing. To be useful, a language must be spoken (and written) and understood by more than one individual and must lead to an action or an altered emotional state. Languages as complex as those spoken by humans are inherently social. The early phases of the history of the genetic code also reside in a community, in this case of microbial organisms. Genes were widely shared and exchanged in what amounts to a community-wide gene pool (Woese 2000; Doolittle and Bapteste 2007). Once stabilized by natural selection, the genetic code remained essentially unchanged, even as gene pools began to evolve in lineages in which genes were passed down through the generations. The phonemes and written symbols of language have been far more diverse, but languages also incorporate an important heritable component, even though that heritability is cultural rather than genetic.

2.5 The Good and the Bad of Cooperation

Almost everyone who has written about human social behavior in recent years has emphasized our cooperative and altruistic nature and how this proclivity has been responsible for human advancement. Pinker (2011) is one of many authors who has rightly pointed to an overall decrease in violence between individuals and perhaps among nations, with the implications that human societies have over time become less combative and more cooperative. Although this trend is well documented by data from many sources, it seems to me equally undeniable that our social organization, with its rapid forms of communication and many competing interest groups, increasingly facilitates the contagious spread of false, misleading, and dangerous information, as well as of destructive technology. This problem—of propaganda, rumor, advertisement, hysteria, mob behavior, manipulation by demagogues, and uncritical acceptance of ideology—represents the dark side of the compulsion to conform, to fit in, to identify with a group. The underlying ideas of these social ills are cultural, spreading more rapidly and with less control than beneficial genes (Vermeij and Leigh 2011). The closest analogy in nature is the spread of invasive species, which is promoted by disruption of the diffuse controls that limit population sizes of species in ecosystems. The very forces that have made the human species so successful have also unleashed forces that tear at the social fabric. This situation could well represent an instance where our sociality has evolved beyond the point where cultural adaptations, including laws, regulations, and accepted social norms, can limit the damage.

The absence of effective control is perhaps most evident, and most difficult to rectify, at the global scale of the human imprint on Earth. Thanks to our technology and far-reaching sociality, the human species has achieved an ecological and evolutionary monopoly in the modern biosphere. Very few species in the wild can effectively adapt to us except as parasites (Perry 2014), pathogens (Wolfe, Panosian Dunavan, and Diamond 2007), or species living in our houses, farm fields, and cities. The species we hunt, fish, cut down, or otherwise exploit either cannot adapt to us at all (Fenberg and Roy 2008; Vermeij 2012) or have become smaller with shorter life spans; many of these animals and plants have become extinct simply because they lack the morphological, behavioral, or chemical defenses to compensate for our extraordinary individual and collective power, which exceeds that of any other species, living or extinct (see also Ripple and Van Valkenburgh 2010; Darimont et al. 2015). Most of the world's habitats have been substantially modified through our activities, and humans have for several decades been exploiting potentially renewable resources unsustainably (Vermeij and Leigh 2011; Vermeij 2018). The only hope for our species and for all the rest of life on Earth is that we find ways to limit our collective power and to reconcile in adaptive ways the conflict between the common good of the biosphere and the more parochial short-term interests of individuals, groups, and the totality of our species.

3. Where and When?

When did the unique social and individual attributes of modern humans arise? What were the circumstances that favored these sweeping transformations? Were these circumstances as unique as the attributes that set humans apart from all other animals?

There is little doubt that the lineage leading to the genus *Homo* is African and that *Homo* itself also originated in Africa. It is therefore in Africa where specialization for long-distance running, accurate throwing, early changes in brain size and function, and early steps toward ultrasocial behavior evolved. Although the modern human genome contains elements of Eurasian Neanderthals and Denisovans (Holliday, Gautney, and Friedl 2014), leading to the possibility that some of our cognitive and highly social characteristics could have arisen or been elaborated outside Africa, the trend toward an increasingly globular brain in humans is not the result of genetic introgressions from Neanderthal sources (Gunz et al. 2019). The archaeological evidence indicates that symbolic thought and abstract representation appeared earlier in Africa than elsewhere. This evidence includes shell beads in both southern and northern Africa (Henshilwood et al. 2002; Vanhaeren et al. 2006; Bouzougar et al. 2007), exceptionally polished stone tools indicating care to produce pleasing objects (Henshilwood et al. 2001), and engravings and ochre drawings on stones and ostrich egg shells (Bingham 2000; Henshilwood et al. 2003), all before 100 Ka. Figurative art (depiction of animals) is known outside Africa first in Borneo after 50 Ka (Aubert et al. 2018) and perhaps slightly later in Europe (Conard 2003). The capacity to sing and to move rhythmically as a form of early music likely arose in a social context among early members of *Homo* in Europe as well as Africa (Mithen 2006; Wurz 2009), even if evidence for musical instruments (flutes) does not appear until 40 Ka in southern Germany (Conard, Malina, and Münzel 2009). Stone awls in Blombos Cave, South Africa, with an age of at least 63 Ka, could indicate the earliest known adoption of clothing (Henshilwood et al. 2001). The production of compound tools and weapons—an advance over implements in one piece—was also perfected in Africa (Hoffecker and Hoffecker 2017; Backwell et al. 2018; Shipton et al. 2018). Linguistic evidence strongly indicates an African origin and subsequent expansion of phonemes (Atkinson 2011). The modern human species and its cognitive and technological characteristics thus evolved in Africa (Stringer 2012; Scerri et al. 2018).

Conditions favorable to the evolution of social behavior and to bipedal locomotion—prerequisites for the later elaboration of uniquely human traits—are found primarily on large land masses. Insects, birds, and mammals that have evolved complex social systems have invariably done so on continents. The reason is straightforward: cohesive social groups hold a competitive edge over animals that act alone. In smaller settings, this advantage might be smaller because the intensity of competition is lower and limits on population size and metabolic rates are more stringent (Leigh, Vermeij, and Wikelski 2009). The same continental bias applies to the evolution of bipedal locomotion in vertebrates—dinosaurs (and their flying descendants, the birds), some lizards, and various groups of marsupials, rodents, and primates. The adaptive predispositions that seem to have been necessary for the emergence of human uniqueness themselves arose on large continents, with Africa perhaps being most suitable.

What, if anything, is there about Africa that makes that continent so favorable for the evolution of a species like ours? I suggest that intense competition among large warm-blooded predators established an evolutionary environment in which social organization, which confers substantial competitive advantages, became both favorable and feasible for primates with a broad diet that includes meat. Data on frequencies of tooth breakage in large carnivores show that competition among predators that hunt aggressive prey is fierce in Africa (Van Valkenburgh and Hertel 1993) and was probably even more intense there during the Pleistocene, when Africa supported many predatory and herbivorous mammals larger than those today (Faith 2014). During the Plio-Pleistocene of East Africa, at least four predatory mammals—a lion ancestor, giant hyena, large bear, and saber-toothed cat—hunted large herbivores and presumably hominins (Faith, Rowan, and Du 2019). The threat from predators and potential competitors was therefore exceptionally high, requiring extraordinary countermeasures by a primate that, without weapons, could offer little resistance. Similarly vigorous competition likely prevailed in Asia and in Pleistocene North America (Van Valkenburgh and Hertel 1993). Moreover, there were social primates in Asia that could have been well placed to achieve humanlike sociality. Africa, however, might have been more favorable than Asia because facultatively bipedal social primates with hands free to take on novel grasping functions had existed in Africa for five to six million years, whereas there is no evidence for such primates in Asia. North America was home to quadrupedal social carnivores (canids) but not to primates or other candidate animals capable of grasping with limbs or other body parts. The more island-like continents of Australia, South America, and Madagascar would likewise have been unsuitable, in part because they lacked the sophistication and competitive vigor of native predators (Leigh, Vermeij, and Wikelski 2009). Hopping kangaroos evolved in Australia during the Pliocene, but they were grazers (Couzens and Prideaux 2018); large lemurs are known from the Pleistocene of Madagascar, but they were fruit-eaters and herbivores and were not bipedal. The only other viable candidates to serve as potential ancestors to humanlike animals would have been large Mesozoic bipedal theropod dinosaurs, but their forelimbs were small, and the global mass extinction at the end of the Cretaceous period put an end to that possibility.

Africa is not only large but also geographically varied. Genetic evidence has demonstrated that populations of large mammals such as giraffes, zebras, and elephants are subdivided to the point of forming geographically distinct species. Such subdivision seems extremely likely in the case of early *Homo sapiens*, fossils of which have been found from Morocco to South Africa (Scerri et al. 2018). In relatively mobile species such as these mammals, intermittent contact was likely, allowing for cultural exchange and hybridization in the case of our own lineage. It would be interesting to know if some of the great cultural transitions, such as that between the Middle and Later Stone Age in Africa, coincided with or followed times of contact among populations. Whatever the answer, the transitions toward modernity were probably not simultaneous continent-wide, but instead began in a restricted part of the continent. The highest genetic diversity among living human populations is found in hunter-gatherer groups in southern Africa, a finding that led Brenna Henn and her colleagues (2011) to suggest that the modern human lineage originated in that part of the continent. It is, however, also possible that these living populations are remnants of a geographically much more widespread group that was displaced in other parts of Africa by invaders from elsewhere. If that scenario is correct, then the conditions prevailing today in southern Africa might indicate how the remnant populations were able to persist rather than the environment where transitions to modernity began.

These reservations aside, I believe southern Africa remains a plausible candidate for the region where anatomically and culturally modern humans originated. Following Curtis Marean (2011, 2016), I suggest that a key transition toward modernity occurred with the evolution of territoriality, likely associated with the defense of family units or well-circumscribed foraging zones. Species with a well-developed social organization and with territorial behavior enforced by groups or individuals evolve only where food resources are predictably available, accessible, and abundant in well-defined areas (Geist 1978; Wilson and Hölldobler 2005; Emlen 2008). Coastal South Africa is an ideal place for the transition to human territoriality and sophisticated group dynamics because it supports a highly productive community of marine shore animals, including mussels, limpets, large topshells, and lobsters, which humans are known to have exploited as early as 125 Ka (Klein et al. 2004; Marean et al. 2007; Marean 2014; Klein 2016). Shores in more tropical parts of Africa are much less productive, with the result that potentially edible animals are smaller and less abundant. The restricted area of the seashore and the low mobility of the food animals living there made sectors of the temperate South African coast suitable for year-round human settlement. Territorial life would have enhanced within-group cooperation and between-group competition, the latter leading to the development of increasingly lethal weapons and ultimately to warfare among groups (Bowles 2009).

4. Cultural Evolution

Whereas the early crucial steps toward modern human cognition, language, social organization, and technology took place in Africa and were carried to parts of the Middle East and South Asia before 75 Ka, later events spanned the globe. Anatomical and cognitive modernity also originated in Africa and appears to have been instrumental in the dispersal of *Homo sapiens* not only to Eurasia but across the sea to Australia, Oceania, and the Americas (Hoffecker and Hoffecker 2017). Although the particulars varied from place to place according to local and regional circumstances (d'Errico and Stringer 2011), many major innovations arose independently in several parts of the world, including agriculture, urban life, writing, government in entities from small states to empires, money, industrial production, and science. Human societies varied in the timing and extent of these cultural developments. Abundant and accessible food sources provided opportunities for social innovation because some of the resources can support occupations not directly related to food production (Algaze 2001). Diamond (1997), for example, suggests that regions where domesticable plants and animals were readily available, as in the Middle East and China, were particularly favorable for the development of urban centers and for economic specialization outside herding and farming. Productive coastal ecosystems, such as those in Peru and the Persian Gulf, could likewise support civilizations despite the proximity of unproductive deserts (Algaze 2001). Others have pointed out that social norms, such as friendliness toward new ways of doing things, promoted innovation (Mokyr 1990; Snooks 1998). Long-distance trade, facilitated by marine transport and a landscape without significant barriers such as mountains, likewise stimulated, and was promoted by, a social ethos of innovation and creativity (Critchfield 1981). These circumstances were likely more common, or more easily achieved, than were those prompting early transitions to sociality and technology in Africa. The reason is that cultural evolution became increasingly independent of external conditions and increasingly contingent on relationships within and between human societies. In other words, cultural change took on a life of its own, often tracing predictable paths and building on established social foundations. Change was propelled by the enormous benefits of group membership and group action and of institutionalized cooperation. Social and technical innovation, in fact, accelerated as social institutions such as governments, schools, religions, trading relationships, legal systems, and corporations developed and stabilized (Vermeij and Leigh 2011). Advances in transportation and communication over the past few centuries have ensured that innovation and cultural change have become effectively global. In short, the human species has collectively made its global environment more productive, more conducive to, and more responsive to cultural change, and vastly more powerful.

Cooperation and conflict among individuals and among groups have become the nearly exclusive mechanisms of cultural evolution in our species. No other species has so completely freed itself from the influences of the physical and biological environment. No other species has achieved the degree of autonomous change, nearly all of which is driven by conditions operating within and among human societies.

The consequences of human evolution being driven almost completely by interactions among humans are profound. Having transcended ecological and evolutionary controls by other species, the human species has become a global monopoly, the first species to do so in Earth's history (Vermeij and Leigh 2011; Vermeij 2018). Economists have long appreciated the destructive nature of monopolies within human society, including reduced competition, prices that do not reflect market forces, and practices that are often at odds with the greater common good. In addition, and less well known, is the problem that monopolies cannot easily correct errors because there are no effective bodies overseeing and regulating their activities (Vermeij 2009; Vermeij and Leigh 2011). The rapid overexploitation and environmental destruction that the human species is bringing about worldwide on land, in the air, and in the sea are the most obvious—and predictable—outcomes of our global monopoly and our almost complete immunity to the selective forces that have kept all other species, living and fossil, in check.

5. Conclusions

Human uniqueness largely resides in our social nature, the combination of widespread cooperation in groups of distinctive individuals that communicate in a combinatorial language. The path toward our economic hegemony on Earth began modestly in Africa with anatomical shifts associated with bipedalism and brain evolution, coupled with a trend toward group sociality compared with our primate ancestors. The adoption of territoriality and year-round settlement, perhaps along the highly productive shores of temperate southern Africa, propelled humans toward cognitive social and technological modernity. Further cultural evolution and technological innovations led to the global spread of our species and made it increasingly independent of local and regional controls imposed by physical and biological challenges such as famines, volcanic eruptions, and disease, with the result that *Homo sapiens* became the dominant agent of change on our planet.

This historical account of human uniqueness reveals the roles that both chance and predictability play in the evolution of unprecedented historical phenomena. The most obvious imprint of chance of historical contingency is the coincidence of at least three more or less independent circumstances that set the stage for the hominin divergence from other primates: intense competition for locally scarce and dangerous prey, an ancestor with social tendencies, and an animal in which body parts (in our case the forelimbs) have become free to take on new functions for manipulating objects. By themselves, these circumstances are quite common and widespread, but the combination is extremely rare. In retrospect, it might have been predictable that Africa would be the continent where the appropriate combination of environment and evolutionary predisposition was most likely to develop. Positive feedback and intense selection in a social bipedal species then ensued, propelling our species along a predictable evolutionary path of cultural change. The most striking aspect of this predictability is that human economic history follows a sequence that parallels the history of the biosphere as a whole and obeys the same evolutionary and economic principles (Vermeij and Leigh 2011). The pace of change is enormously faster, the time interval over which the human economic experiment has played out is more than 1,000 times shorter, and the magnitude of changes humans have brought about is very much greater, but selective processes arising from competition and cooperation introduce fundamental, knowable, and predictable historical sequences in the affairs of nature and the affairs of our species. Individuals, species, and ecosystems come and go according to contingent circumstances, but the directions and consequences of change, powered by pervasive feedbacks, reveal an underlying emergent order (Vermeij 2019). Particulars matter, but it is the documentation and elucidation of the predictable aspects of history that pose some of the greatest challenges of evolutionary biology and the human-centered sciences.

References

Algaze, G. 2001. "Initial Social Complexity in Southwestern Asia: The Mesopotamian Advantage." *Current Anthropology* 42: 199–233.

Ambrose, S. H. 2001. "Paleolithic Technology and Human Evolution." *Science* 291: 1748–1753.

Arbib, M. A., K. Liebal, and S. Pika. 2008. "Primate Vocalization, Gesture, and Evolution of Human Language." *Current Anthropology* 49: 1053–1076.

Arthur, W. B. 2009. *The Nature of Technology: What It Is and How It Evolves.* New York: Free Press.

Atkinson, G. D. 2011. "Phonemic Diversity Supports a Serial Founder Effect Model of Language Expansion from Africa." *Science* 332: 346–349.

Aubert, M., P. Sesiawan, A. A. Oktaviana, A. Brumm, P. H. Sulistyarso, E. W. Saptomo, B. Istiawan. 2018. "Palaeolithic Cave Art in Borneo." *Nature* 564: 254–257.

Backwell, L., J. Bradfield, K. J. Carlson, T. Jamashvili, L. Wadley, and F. d'Errico. 2018. "The Antiquity of Bow-and-Arrow Technology: Evidence from Middle Stone Age Layers at Sibudu Cave." *Antiquity* 92: 289–301.

Bingham, P. M. 2000. "Human Evolution and Human History: A Complete Theory." *Evolutionary Anthropology* 9: 248–257.

Bouzougar, A., N. Barton, M. Vanhaeren, F. d'Errico, S. Collcutt, E. Rhodes, J.-L. Schwenninger. 2007. "82,000-Year-Old Shell Beads from North Africa and Implications for the Origins of Modern Human Behavior." *Proceedings of the National Academy of Sciences of the United States of America* 104: 9964–9969.

Bowles, S. 2009. "Did Warfare among Ancestral Hunter-Gatherers Affect the Evolution of Human Social Behaviors?" *Science* 324: 1293–1298.

Bramble, D. M., and D. E. Lieberman. 2004. "Endurance Running and the Evolution of *Homo*." *Nature* 432: 345–352.

Cardini, F.-E. 2017. "Arguing for a Conscious Emergence of Language." *Lingua* 194: 67–86.

Collier, K., B. Bickel, C. P. van Schaik, M. B. Manser, and S. W. Townsend. 2014. "Language Evolution: Syntax before Phonology?" *Proceedings of the Royal Society B* 281: 2014.0263.

Conard, N. J. 2003. "Palaeolithic Ivory Sculptures from Southwestern Germany and the Origins of Figurative Art." *Nature* 426: 830–832.

Conard, N. J., M. Malina, and S. A. Münzel. 2009. "New Flutes Document the Earliest Musical Tradition in Southwestern Germany." *Nature* 460: 737–740.

Couzens, A. M. C., and G. J. Prideaux. 2018. "Rapid Pliocene Radiation of Modern Kangaroos." *Science* 362: 72–75.

Crespi, B. 2014. "The Insectan Apes." *Human Nature* 25: 6–27.

Critchfield, R. 1981. *Villages.* New York: Doubleday.

Darimont, C. T., C. H. Fox, H. M. Bryan, and T. E. Reimchen. 2015. "The Unique Ecology of Human Predators." *Science* 349: 858–860.

Deacon, T. W. 1997. *The Symbolic Species: The Co-evolution of Language and the Brain.* New York: W. W. Norton.

Dennett, D. C. 2006. *Breaking the Spell: Religion as a Natural Phenomenon.* New York: Penguin.

d'Errico, F., and C. B. Stringer. 2011. "Evolution, Revolution or Saltation Scenario for the Emergence of Modern Culture." *Philosophical Transactions of the Royal Society B* 36S: 1060–1069.

de Waal, F. B. M. 1996. *Good Natured: The Origins of Right and Wrong in Humans and Other Animals*. Cambridge, MA: Harvard University Press.

Diamond, J. 1997. *Guns, Germs, and Steel: The Fate of Human Societies*. New York: W. W. Norton.

Doolittle, W. F., and E. Bapteste. 2007. "Pattern Pluralism and the Tree of Life Hypothesis." *Proceedings of the National Academy of Sciences of the United States of America* 104: 2043–2049.

dos Santos, M., and S. A. West. 2018. "The Coevolution of Cooperation and Cognition in Humans." *Proceedings of the Royal Society B* 285: 2018.0723.

Dugatkin, L. 1999. *Cheating Monkeys and Citizen Bees: The Nature of Cooperation in Animals and Humans*. New York: Free Press.

Emlen, D. J. 2008. "The Evolution of Animal Weapons." *Annual Review of Ecology, Evolution and Systematics* 39: 387–413.

Faith, J. T. 2014. "Late Pleistocene and Holocene Extinctions on Continental Africa." *Earth-Science Reviews* 128: 105–121.

Faith J. T., J. Rowan, and A. Du. 2019. "Early Hominins Evolved within Non-analog Ecosystems." *Proceedings of the National Academy of Sciences of the United States of America* 116: 21478–21483.

Fenberg, P. B., and K. Roy. 2008. "Ecological and Evolutionary Consequences of Size-Selective Harvesting: How Much Do We Know?" *Molecular Ecology* 17: 209–220.

Foley, R. A. 2016. "Mosaic Evolution and the Pattern of Transitions in the Hominin Lineage." *Philosophical Transactions of the Royal Society B* 371: 2015.0244.

Geist, V. 1978. "On Weapons, Combat, and Ecology." In *Advances in the Study of Communication and Effect*, vol. 4: *Aggression, Dominance and Individual Spacing*, edited by L. Krames, P. Pliner, and T. Calloway, 1–30. New York: Plenum.

Gomez-Robles, A., J. B. Smaers, R. L. Holloway, P. D. Polly, and B. A. Wood. 2017. "Brain Enlargement and Dental Reduction Were Not Linked in Human Evolution." *Proceedings of the National Academy of Sciences of the United States of America* 114: 468–473.

Gowdy, J., and L. Krall. 2014. "Agriculture as a Major Evolutionary Transition to Human Ultrasociality." *Journal of Bioeconomics* 16: 179–202.

Gowdy, J., and L. Krall. 2016. "The Economic Origins of Ultrasociality." *Behavioral and Brain Sciences* 39: e92.

Grosberg, R. K., and M. W. Hart. 2000. "Mate Selection and the Evolution of Highly Polymorphic Self/Nonself Recognition Genes." *Science* 289: 2111–2114.

Grosberg, R. K., and J. F. Quinn. 1986. "The Genetic Control and Consequences of Kin Recognition by the Larvae of a Colonial Marine Invertebrate." *Nature* 322: 456–459.

Gunz, P., A. K. Tilot, K. Wittfeld, A. Teumer, C. Shapland, J. G. M. van Erp, M. Dannemann. 2019. "Neanderthal Introgression Sheds Light on Modern Human Endocranial Globularity." *Current Biology* 29: 120–127.

Harari, Y. N. 2015. *Sapiens: A Brief History of Humankind*. New York: Harper/Harper Collins.

Hauert, C., A. Traulsen, M. A. Nowak, and K. Sigmund. 2007. "Via Freedom to Coercion: The Emergence of Costly Punishment." *Science* 316: 1905–1907.

Henn, B. M., C. R. Gignoux, M. Jobin, J. M. Granka, J. M. Macpherson, J. M. Kidd, L. Rodríguez-Botigué . 2011. "Hunter-Gatherer Genomic Diversity Suggests a Southern African Origin for Modern Humans." *Proceedings of the National Academy of Sciences of the United States of America* 108: 5154–5162.

Henrich, J., J. Ensminger, R. McElreath, A. Barr, C. Barnett, A. Bolyanatz, J. Camilo Cardenas . 2010. "Markets, Religion, Community Size, and the Evolution of Fairness and Punishment." *Science* 327: 1480–1484.

Henshilwood, C. S., F. d'Errico, C. W. Marean, R. G. Milo, and R. Yates. 2001. "An Early Bone Tool Industry from the Middle Stone Age at Blombos Cave, South Africa: Implications for the Origins of Modern Human Behaviour, Symbolism and Language." *Journal of Human Evolution* 41: 633–678.

Henshilwood, C. S., F. d'Errico, M. Vanhaeren, K. Van Niekerk, and Z. Jacobs. 2014. "Middle Stone Age Shell Beads from South Africa." *Science* 304: 404.

Henshilwood, C. S., F. d'Errico, R. Yates, Z. Jacobs, C. Tribolo, G. A. T. Duller, N. Mercier. 2002. "Emergence of Modern Human Behavior: Middle Stone Age Engravings from South Africa." *Science* 295: 1278–1280.

Henshilwood, C. S., and C. W. Marean. 2003. "The Origin of Modern Human Behavior: Critique of the Models and Their Test Implications." *Current Anthropology* 44 (5): 627–651.

Hoffecker, J. F., and I. T. Hoffecker. 2017. "Technological Complexity and the Global Dispersal of Modern Humans." *Evolutionary Anthropology* 16: 284–299.

Holliday, T. W., J. R. Gautney, and L. Friedl. 2014. "Right for the Wrong Reasons: Reflections on Modern Human Origin in the Post-Neanderthal Genome Era." *Current Anthropology* 55: 696–724.

Jablonski, N. 2004. "The Evolution of Human Skin and Skin Color." *Annual Reviews of Anthropology* 33: 585–623.

Kaplan, H. S., and A. J. Robson. 2002. "The Emergence of Humans: The Coevolution of Intelligence and Longevity with Intergenerational Transfers." *Proceedings of the National Academy of Sciences of the United States of America* 99: 10221–10226.

Klein, R. G. 2016. "Shellfishing and Human Evolution." *Journal of Anthropological Archaeology* 44: 198–205.

Klein, R. G., G. Avery, K. Cruz-Uribe, D. Halkett, J. E. Parkington, T. Steele, T. P. Volman, and R. Yates. 2004. "The Ysterfontein 1 Middle Stone Age Site, South Africa, and Early Human Exploitation of Coastal Resources." *Proceedings of the National Academy of Sciences of the United States of America* 101: 5708–5715.

Kolodny, O., and W. Feldmann. 2017. "A Parsimonious Neutral Model Suggests Neanderthal Replacement Was Determined by Migration and Random Species Drift." *Nature Communications* 8: 1040.

Leigh, E. G., Jr., G. J. Vermeij, and M. Wikelski. 2009. "What Do Human Economies, Large Islands and Forest Fragments Reveal about the Factors Limiting Ecosystem Evolution?" *Journal of Evolutionary Biology* 22: 1–12.

Lieberman, D. E. 1998. "Sphenoid Shortening and the Evolution of Modern Human Cranial Shape." *Nature* 393: 158–162.

Lieberman, P. 2007. "The Evolution of Human Speech: Its Anatomical and Neural Bases." *Current Anthropology* 48: 39–66.

Marean, C. W. 2011. "Coastal South Africa and the Coevolution of the Modern Human Lineage and the Coastal Adaptation." In *Trekking the Shore: Changing Coastlines and the Antiquity of Coastal Settlement*, edited by N. Bicho, J. A. Haws, and L. G. Davis, 421–440. New York: Springer.

Marean, C. W. 2014. "The Origins and Significance of Coastal Resource Use in Africa and Western Eurasia." *Journal of Human Evolution* 77: 17–40.

Marean, C. W. 2016. "The Transition to Foraging for Dense and Predictable Resources and Its Impact on the Evolution of Modern Humans." *Philosophical Transactions of the Royal Society B* 371: 2015.0239.

Marean, C. W., M. Bar-Matthews, J. Bernatchez, E. Fischer, P. Goldberg, A. I. R. Herries, Z. Jacobs. 2007. "Early Human Use of Marine Resources and Pigment in South Africa during the Middle Pleistocene." *Nature* 449: 905–908.

Mithen, S. 2006. *The Singing Neanderthals: Language, Mind and Body*. Cambridge, MA: Harvard University Press.

Mitteroecker, P. 2019. "How Human Bodies Are Evolving in Modern Societies." *Nature Ecology and Evolution* 3: 124–126.

Mokyr, J. 1990. *The Lever of Riches: Technological Creativity and Economic Progress*. New York: Oxford University Press.

Nesse, R. M., and G. C. Williams. 1994. *Why We Get Sick: The New Science of Darwinian Medicine*. New York: Random House.

Neubauer, S., J.-J. Hublin, and P. Gunz. 2018. "The Evolution of Modern Human Brain Shape." *Science Advances* 4: eaao5961. doi:10.1126/sciadv.aao5961.

Niewoehner, W. A., A. Vergstrom, D. Eichele, M. Zuroff, and J. T. Clark. 2003. "Manual Dexterity in Neanderthals." *Nature* 422: 395.

Nowak, M. A. 2006. "Five Rules for the Evolution of Cooperation." *Science* 314: 1560–1563.

Organ, C., C. L. Nunn, Z. Machanda, and R. W. Wrangham. 2011. "Phylogenetic Rate Shifts in Feeding Time during the Evolution of *Homo*." *Proceedings of the National Academy of Sciences of the United States of America* 108: 14555–14559.

Ouattara, K., A. Lemasson, and K. Zuberbuhler. 2009. "Campbell's Monkeys Concentrate Vocalizations into Context-Specific Call Sequences." *Proceedings of the National Academy of Sciences of the United States of America* 106: 22026–22031.

Perry, G. H. 2014. "Parasites and Human Evolution." *Evolutionary Anthropology* 23: 218–228.

Pinker, S. 2011. *The Better Angels of Our Nature: Why Violence Has Decreased*. New York: Penguin.

Pontzer, H., M. H. Brown, D. A. Raichlen, H. Dunsworth, B. Hare, K. Walker, A. Luke. 2016. "Metabolic Acceleration and the Evolution of Human Brain Size and Life History." *Nature* 533: 390–392.

Prum, R. O. 2017. *The Evolution of Beauty: How Darwin's Forgotten Theory of Mate Choice Shapes the Animal World—and Us*. New York: Doubleday.

Purugganan, M. D., and D. Q. Fuller. 2009. "The Nature of Selection during Plant Domestication." *Nature* 457: 843–848.

Raghanti, P. A., M. K. Edler, A. R. Stephenson, E. L. Munger, B. Jacobs, P. R. Hof, C. C. Sherwood, R. L. Holloway, and C. O. Lovejoy. 2018. "A Neurochemical Hypothesis for the Origin of Hominins." *Proceedings of the National Academy of Sciences of the United States of America* 115: E1108–E1116.

Ripple, W. J., and B. Van Valkenburgh. 2010. "Linking Top-Down Forces to the Pleistocene Megafaunal Extinction." *BioScience* 60: 516–526.

Roach, N. B., M. Venkadesan, M. J. Rainbow, and D. E. Lieberman. 2013. "Elastic Energy Storage in the Shoulder and the Evolution of High-Speed Throwing in *Homo*." *Nature* 498: 483–486.

Roberts, A. M., and S. K. S. Thorpe. 2014. "Challenges to Human Uniqueness: Bipedalism, Birth and Brains." *Journal of Zoology* 292: 281–289.

Rolian, C., D. Lieberman, and B. Hallgrímsson. 2010. "The Coevolution of Human Hands and Feet." *Evolution* 64: 1558–1568.

Rosen, S. P. 2005. *War and Human Nature*. Princeton, NJ: Princeton University Press.

Russell, A. F., and S. W. Townsend. 2017. "Communication: Animal Steps on the Way to Syntax?" *Current Biology* 27: R753–R755.

Sato, D. X., and M. Kawata. 2018. "Positive and Balancing Selection on SLC18A1 Gene Associated with Psychiatric Disorders and Human-Unique Personality Traits." *Evolution Letters* 2: 499–510.

Scerri, E. M., M. G. Thomas, A. Manica, P. Gunz, J. T. Stock, C. Stringer, M. Grove . 2018. "Did Our Species Evolve in Subdivided Populations across Africa, and Why Does It Matter?" *Trends in Ecology and Evolution* 33: 582–594.

Schilthuizen, M. 2014. *Nature's Nether Regions: What the Sex Lives of Bugs, Birds, and Beasts Tell Us about Evolution, Biodiversity, and Ourselves*. New York: Viking.

Scott-Phillips, S. C. 2015. "Nonhuman Primate Communication, Pragmatics, and the Origins of Language." *Current Anthropology* 56: 56–66.

Scott-Phillips, S. C., and R. A. Blythe. 2013. "Why Is Combinatorial Communication Rare in the Natural World, and Why Is Language an Exception to This Trend?" *Journal of the Royal Society Interface* 10: 20130520.

Seabright, P. 2004. *The Company of Strangers: A Natural History of Economic Life*. Princeton, NJ: Princeton University Press.

Senter, P. 2008. "Voices of the Past: A Review of Paleozoic and Mesozoic Animal Sounds." *Historical Biology* 20: 255–287.

Seymour, R. S., V. Bosiocic, and E. P. Snelling. 2016. "Fossil Skulls Reveal That Blood Flow Rate in the Brain Increased Faster Than Brain Volume during Human Evolution." *Royal Society Open Science* 3: 160305.

Shipton, C., P. Roberts, W. Archer, S. J. Armitage, C. Bita, J. Blinkhorn, C. Courtney-Mustaphi. 2018. "78,000-Year-Old Record of Middle and Later Stone Age Innovation in an East African Tropical Forest." *Nature Communications* 9: 1832.

Snooks, G. D. 1998. *The Laws of History*. London: Routledge.

Sorensen, A. C., E. Claud, and M. Soressi. 2018. "M. Neandertal Fire-Making Technology Inferred from Microwar Analysis." *Scientific Reports* 5: 10965.

Stearns, S. C. 2007. "Are We Stalled Part Way through a Major Evolutionary Transition from Individual to Group?" *Evolution* 61: 2275–2280.

Street, S. E., A. F. Navarrete, S. M. Reader, and K. N. Laland. 2017. "Coevolution of Cultural Intelligence, Extended Life History, Sociality, and Brain Size in Primates." *Proceedings of the National Academy of Sciences of the United States of America* 114: 7908–7914.

Stringer, C. B. 2012. "What Makes a Modern Human?" *Nature* 485: 33–35.

Thomson, K. S. 1969. "The Biology of the Lobe-Finned Fishes." *Biological Reviews* 44: 91–154.

Vanhaeren, M., F. d'Errico, C. Stringer, S. L. James, J. A. Todd, and H. K. Mienis. 2006. "Middle Paleolithic Shell Beads in Israel and Algeria." *Science* 312: 1785–1788.

van Schaik, C. P., K. Burkart, L. Damerius, S. I. F. Forss, K. Koops, M. A. Noordwijk, and C. Schuppli. 2016. "The Reluctant Innovator: Orangutans and the Phylogeny of Creativity." *Philosophical Transactions of the Royal Society B* 371: 2015.0183.

Van Valkenburgh, B., and F. Hertel. 1993. "Tough Times at La Brea: Tooth Breakage in Large Carnivores of the Late Pleistocene." *Science* 261: 456–459.

Vermeij, G. J. 2004, *Nature and Economic History*. Princeton, NJ: Princeton University Press.

Vermeij, G. J. 2009. "Comparative Economics: Evolution and the Modern Economy." *Journal of Bioeconomics* 11: 105–134.

Vermeij, G. J. 2010. *The Evolutionary World: How Adaptation Explains Everything from Seashells to Civilization*. New York: Thomas Dunne.

Vermeij, G. J. 2012. "The Limits of Adaptation: Humans and the Predator-Prey Arms Race." *Evolution* 66: 2007–2014.

Vermeij, G. J. 2013. "On Escalation." *Annual Review of Earth and Planetary Sciences* 41: 1–19.

Vermeij, G. J. 2017. "How the Land Became the Locus of Major Evolutionary Innovations." *Current Biology* 27: 3178–3182.

Vermeij, G. J. 2018. "Building a Healthy Economy: Learning from Nature." In *Economia: Methods for Reclaiming Economy*, edited by O. Mink and W. Oosterhuis, 53–76, 111–119. Eindhoven: Baltan Laboratories.

Vermeij, G. J. 2019. "Power, Competition, and the Nature of History." *Paleobiology* 45: 517–530.

Vermeij, G. J., and R. K. Grosberg. 2018. "Rarity and Persistence." *Ecology Letters* 21: 3–8.

Vermeij, G. J., and E. G. Leigh Jr. 2011. "Natural and Human Economies Compared." *Ecosphere* 2: 39.

Villa, P., and W. Roebroeks. 2014. "Neandertal Demise: An Archaeological Analysis of the Modern Human Superiority Complex." *PLoS One* 9: e96424.

Wadley, L., S. Hodgskiss, and M. Grant. 2009. "Implications for Complex Cognition from the Hafting of Tools with Compound Adhesives in the Middle Stone Age, South Africa." *Proceedings of the National Academy of Sciences of the United States of America* 106: 9590–9594.

Weaver, A. H. 2005. "Reciprocal Evolution of the Cerebellum and Neocortex in Fossil Humans." *Proceedings of the National Academy of Sciences of the United States of America* 102: 3576–3580.

Wilson, E. O., and B. Hölldobler. 2005. "Eusociality: Origin and Consequences." *Proceedings of the National Academy of Sciences of the United States of America* 102: 13367–13371.

Woese, C. 2000. "Interpreting the Universal Phylogenetic Tree." *Proceedings of the National Academy of Sciences of the United States of America* 97: 8392–8396.

Wolfe, N. D., C. Panosian Dunavan, and J. Diamond. 2007. "Origins of Major Human Infectious Diseases." *Nature* 447: 269–273.

Wrangham, R. 2017. "Control of Fire in the Paleolithic: Evaluating the Cooking Hypothesis." *Current Anthropology* 58: S303–S313.

Wrangham, R. W., J. Jones Holland, G. Laden, D. Pilbeam, and N. Conklin-Brittain. 1999. "The Raw and the Stolen: Cooking and the Economy of Human Origins." *Current Anthropology* 40: 566–594.

Wurz, S. 2009. "Interpreting the Fossil Evidence for the Evolutionary Origins of Music." *Southern African Humanities Journal* 21: 395–417.

Zink, K. D., and D. E. Lieberman. 2016. "Impact of Meat and Lower Palaeolithic Food Processing Techniques on Chewing in Humans." *Nature* 531: 500–503.

6

Wanderlust

A View from Deep Time of Dispersal, Persistence, and Human Success

Susan C. Antón

1. Introduction

Humans are highly successful, widely dispersed, biocultural organisms. Most everything in this statement is easily defined—but what, exactly, do we mean by *highly successful*? Often what we recognize as highly successful for humans is based not on flourishing in a particular adaptive role or even persisting over a long period of evolutionary time, but rather on moving beyond the expectations of a primate of our size. Yet, I argue, this definition of success is a low bar for humans and even most species of genus *Homo*. Such an expectation fails to acknowledge the set of adaptive advantages in place before *H. sapiens* came on the scene or to take into consideration any broader definition of success for a species with our extensive niche construction (see Fuentes, chapter 10, this volume). A present-day analogy is when we compare individual success but ignore the very different family backgrounds of the individuals compared. Should we be more impressed by a child who has had every resource and educational advantage and who grows up to be an average white-collar professional than we are by the child raised in poverty and a crime-ridden neighborhood who stays out of jail, raises a family, and remains non-criminally employed? As in these examples, I suggest that by ignoring the original context/background of *H. sapiens*, we use the wrong measure of success. We should expect more of our species. A perspective from the fossil record suggests that the time to recognize the success of moving beyond the expectations for a primate of our size is long past. Early *Homo* took the first several of these steps, and *H. erectus* took further steps by exhibiting great ecological tolerance, dispersing broadly, and persisting

Susan C. Antón, *Wanderlust* In: *Human Success*. Edited by: Hugh Desmond and Grant Ramsey, Oxford University Press.
 DOI: 10.1093/oso/9780190096168.003.0006

across their range for the better part of two million years. To gauge it accurately, the present-day success of our species should be measured against how far we have come from this base and whether we are clever enough to persist for another two million years.

Certainly, recent humans defy many ecological laws and can in these ways be considered successful. We break Damuth's (1981) law by living at much higher population densities than should be feasible for a mammal of our size. We are an exception to the generation-time law (Ginzburg and Colyvan 2004) by "stacking offspring" and creating much shorter generation times than should be present in a large-brained primate of our size. We grow larger brains than even a primate of our size should. And, likely related to this, we take larger prey than predicted for a single terrestrial carnivore of our size (see figure 1 of Williams 2019). Indeed, we fill nonprimate ecological niches and occupy a much greater range of climate zones than the typical primate. Our dispersal around the world makes us the only truly global primate and a dominant force in nearly every habitat on the planet. The precise details of when, how, and why these abilities arise are interesting, complicated to illuminate, and contested. But each is linked to our ability to use cultural, technological, or behavioral means to extend our phenotype and abilities and to create or cocreate our niche. Many of these exceptions first occurred in the distant past, even before the advent of our species. It is there that we should look for explanations of the origins of these abilities. Only with this background knowledge can we accurately gauge our species' success.

Here I focus on the origin of one particular facet of our success: our ability to disperse broadly. While *H. sapiens* is the only truly global primate, globalization began when hominins first left the African continent. I make the case below that this ability to disperse extensively beyond the previously existing ecological niche, and to persist over deep time across this range, entails important qualities that undergird the success of *H. sapiens*. Frustratingly, historical explanations for the cause and importance of this dispersal often devolve to single prime-mover explanations—usually an environmental factor or technology—from which all other things develop almost as a matter of course.

2. Beyond the Savanna

Since the beginning of paleoanthropology, the expansion of African grasslands has dominated our thinking about nearly everything related to hominin evolution and human origins (e.g., Dart 1925; Robinson 1954). The so-called savanna hypothesis emerged as a global but nebulous explanation of what started us down a particular path (to bipedality) and eventually led to all manner of other changes (including meat-eating, toolmaking, and migration; see, for example, Robinson 1966). The role of the savanna adaptation took many forms, leading Dart to the "killer ape" hypothesis, and others to more moderate dietary hypotheses. The key in all instances was a stark directional transition between thick forest (arboreal) life and open savanna (terrestrial) life. Robinson (1966), for example, argued that the reduction of forested areas from the Miocene to the mid-Pliocene resulted in fragmented environments that placed forest-dwelling ancestral apes under resource pressure. He argued that the first expansion of open areas corresponded with the evolution of bipedalism (in the form of *Australopithecus*), a form of locomotion that facilitated movement between forest patches. Nonetheless, the increasingly open environments were seen as a challenge to vegetarian bipeds. To meet this challenge, *Australopithecus* supplemented their diet with small protein packets (meat and insects). *Homo* arose, almost inevitably, as a lineage that took fuller advantage of the growing number of animal resources available in a savanna environment. And both these hominins accessed these resources first by using and then by making stone tools. Access to meat and marrow sustained an enlarged brain in *Homo*, which in turn was critical for survival, resulting in an evolutionary feedback loop. In its original form, the savanna hypothesis privileged global environmental indicators of a directional trend in cooling and drying, taking little note (either theoretical or empirical) of the local environments in which particular hominins lived (e.g., Dart 1925; Robinson 1966). That is, the savanna provided not so much a specific selective pressure in which individuals survived (or not) as much as it formed a kind of directional climatic backdrop that, once established, explained vast swaths of human evolution via feedback mechanisms essentially *intrinsic* to the organism (see also Potts 2013).

Although the savanna hypothesis is less a unified hypothesis than a loose collection of a few different ideas linking grasslands to sequential shifts in human evolution (see Bobe and Behrensmeyer 2004), there is certainly geological evidence of a worldwide trend of extensive cooling and drying of the Mio-Pliocene environment, especially over the past six million years (e.g., Uno et al. 2016). This global pattern does imply that at least some African environments included less tree cover and more open grassland. However, the definition of "savanna," which comprises anything from 5% to

80% tree cover (Cerling, Wynn, et al. 2011), is so broad as to be unhelpful in considering selective pressures on specific individuals. The presence of warm-season grasses and sedges (those that use the Hatch-Slack cycle or C_4 pathway to fix carbon, often referred to as C_4 grasses) is picked up as a signal from about 10 million years ago onward (Uno et al. 2016). From this point in the Miocene onward, grasslands steadily increase in prevalence, and C_4 resources increase as components of mammalian diets. Yet all indicators seem to suggest that it is the period from four million years to one million years ago in eastern Africa that sees the increasing dominance of these grasslands (Bobe and Behrensmeyer 2004; Uno et al. 2016). Furthermore, the landscape was likely not dominated by such open areas until after 2.5 million years ago. Additionally, C_4 grasslands became dominant at different times in different areas across the continent (see Cerling 1992; Cerling, Wynn, et al. 2011; Uno et al. 2016). Thus, the environmental shift is not a stark one from forest to plains but a process of creating myriad different environmental opportunities on the landscape. Furthermore, the association of the earliest hominins was not with open areas but, as in the case of *Ardipithecus* that lived around 4.4 million years ago, with forested or at least semi-forested environments (e.g., White et al. 2009; Woldegabriel et al. 2009). Thus, while cooling and drying occurred, forests receded, and grasslands came to dominate more of the African environment in the Pleistocene, organisms are arguably more influenced by local environmental conditions than global trends, and these environments were more complicated than a savanna hypothesis suggests. Thus we must view the rise of grasslands not as a wholesale shift in environments but as an opening of additional environmental mosaics of opportunities for adaptive radiation.

Both human and nonhuman mammalian communities exploited the opportunities of these mosaics, and current models of the climatic influence on the origin, evolution, and dispersal of *Homo* focus more explicitly on the selective pressures of local environments (see Potts 2013). Many of the main models focus on the correlation between certain kinds of climatic changes and the occurrence of key events in the origin of *Homo*. Some suggest that a particular kind of environment fostered/drove the cyclical isolation and separation of *Homo* species by, for example, the presence or absence of deep lakes (e.g., Trauth et al. 2005) or high temperature (Passey et al. 2010). Still others focus on various aspects related to grassland expansion, such as aridity (deMenocal 2011) or cooling (Vrba 1995), as driving pulses of climate change that provide environmental mosaics resulting in opportunities for adaptive radiation of hominins and other mammals. In many instances these reflect opposite sides of the same coin: moisture-aridity, warm-cool. Hence, this focus on different environmental proxies and the differing conclusions about which of these influences the origin of *Homo* may, alternatively, signal that all of them are important. The variability hypothesis argues just this, that the underlying variability in environment, particularly after 2.5 million years ago, favored a particular kind of flexibility that first allowed the radiation of several species of *Homo*, and then the flourishing of a particularly flexible member of that genus, *H. erectus* (e.g., Potts 1998a, 1998b, 2013; see Potts, chapter 11 in this volume for a perspective on how climate further influenced the origin of *H. sapiens*).

Starting about 2.5 million years ago, cyclical changes in climate occurred more quickly and became more severe in amplitude. This cyclicity is related to the onset of northern hemispheric glaciation and resulted in deeper swings in temperature than were recorded in the oxygen isotopes of benthic foramina, yielding the so-called oxygen-isotope curves. When smoothed, these curves show the progressive cooling and drying trend that we have discussed, and when unsmoothed show shorter-term oscillations in temperature within that overall trend. The cycles that run on ~100,000-year, ~41,000-year, and ~21,000-year periodicities relate to the variation in solar radiation received due to changes in the direction and wobbles in the Earth's orbit (Milankovitch 1941; see primer in Potts 1998a and deMenocal 2004 for cogent reviews). This resulted in more severe swings in amplitude (much colder/much warmer), coupled with abrupt transitions during some periods. So within this overall cooling and drying trend, climate cycles are packed more tightly, implying that instead of pure directional selection tied to the overall trend, uncertainty about climate and environment may have driven organismal responses. Recent work suggests that periods of greater variability may correlate with periods of increased speciation in hominins and other mammals (e.g., Bobe and Behrensmeyer 2004; Potts 2013), and theoretical models support variability selection as a strong influence on both morphological and behavioral phenotypic variation (Grove 2011).

At the point of initial climate deterioration, we also see the rise of genus *Homo*. Potts argues that from a radiation of *Homo* species, environmental variability and uncertainty favored organisms that were more behaviorally and biologically flexible, and thus laid the foundation for the dispersibility of *Homo* (see Antón, Potts, and Aiello 2014). There is some indication that the correlation of more variable periods with increased species diversity may reflect sampling bias rather than a true causative connection (see Maxwell et al. 2018). However, while this may be critical to the link between speciation and variability, it is a moot point for the argument concerning whether the evolution of *H. erectus* was influenced by variable climate. I do not mean to imply by this that the variability selection hypothesis is *the* antidote to the savanna hypothesis, only that it is a current and comprehensive one that provides a critical balancing point from which to explore and acknowledge that there are a number of interrelated processes and factors whose relationships we need to plumb more fully. Certainly, *H. erectus* evolved during—and appears to be the only one of several hominin species and genera to persist through—the variable periods (all others apparently going extinct). It is also the only species to expand (at least visibly) out of the African continent into more temperate climates.

3. Evidence of Initial Hominin Dispersal

From ~7 million years ago to ~2 million years ago, our bipedal ancestors were entirely restricted to the continent of Africa. Although our ancestors lived in eastern Africa when there was no evidence of a geographic barrier to dispersal from the continent (PRISM Project Members 1995), hominin dispersal/migration did not happen until shortly after two million years ago. The earliest hominin fossils known outside of Africa are members of *H. erectus* from Dmanisi, Georgia (~1.8 Ma; Gabunia et al. 2000, 2001; Vekua et al. 2002; Ferring et al. 2011) and Sangiran, Indonesia (~1.6–1.8 Ma; Swisher et al. 1994; Antón and Swisher 2004). There are also somewhat younger archaeological sites in China[1] (~1.6–1.7 Ma; Zhu et al., 2004, 2008) bearing core and flake technologies. The first appearance date (FAD) of *H. erectus* in the fossil record of East Africa is about 1.89 million years ago in the Turkana Basin of current-day Kenya (Feibel, Brown, and McDougall 1989) and about two million years ago in South Africa (Herries et al. 2020). Thus all of these sites outside of Africa attest to an early, (geologically) quick, and likely extensive initial dispersal of hominins from Africa. Yet the origin and cause of the sustained dispersal(s) are not well understood.

Homo erectus is the fossil species most often considered ancestral to *H. sapiens*, and its dispersal is frequently attributed to some simple humanlike mechanism or feature: a new technology or an increased reproductive rate, for example (e.g., Klein 1999). But one of these mechanisms occurs much earlier and one much later than the appearance of *H. erectus*. Because living humans are broadly dispersed, we often look to our own attributes to answer the *why* questions, working backward into the record. We are biocultural organisms that outsource the solution of problems to a series of technologies, such as the use of knives rather than teeth to process meat. And we often see our biocultural, particularly our technological, savvy as being the trigger for dispersal (Klein 2009, 325). But the first hominin dispersal from Africa postdates the earliest stone tools by over a million years (~3.3 Ma; Harmand et al. 2015), the oldest Oldowan (core and flake technologies) assemblages by more than half a million years (~2.6 Ma; Semaw et al. 1997; de Heinzelin et al. 1999; Roche et al. 1999), and predates the earliest occurrence of the large cutting tools of the Acheulean tradition[2] (~1.78 Ma; Lepre et al. 2011) by tens of thousands of years. Thus, at least the gross technological innovations currently visible in the archaeological record do not correlate temporally with the initial dispersal. Hence, dispersal can hardly be attributed *solely* to a technological innovation.[3] Neither is a change in reproductive rate relative to earlier hominins likely to underlie dispersal. *H. erectus* individuals grew slightly more slowly than *Australopithecus* (Dean et al. 2001; Schwartz 2012). Slower development should correlate with slightly longer interbirth intervals and fewer offspring, and thus suggests that *H. erectus* should have had a lesser biological ability to populate new regions compared to their hominin ancestors.[4] The difference in growth rate between *H. sapiens* and *Australopithecus* is even larger; humans grow very slowly in comparison to any other hominin. However, in our species this slow growth rate is countered by the phenomenon known as "stacking offspring" (the ability to support multiple dependent offspring at one time), which results in a decreased interbirth interval. However, the substantially decreased interbirth intervals of industrialized societies (Wood 1994) likely occurred only late in our lineage. That is, human hunter-gatherer reproductive ecology (Howell 1979; Blurton-Jones et al. 1992) does not match that of industrial societies, suggesting that the ability to fuel population growth and dispersal by substantially stacking offspring significantly postdates even the origin of *H. sapiens* and certainly the initial dispersal of *Homo* from Africa. If simple prime-movers offer insufficient explanations for dispersal, more complex reasons must be considered.

In the effort to address the potential causes of dispersal, there are two big questions to disentangle: What provided the ability to disperse? What was the impetus/cause for dispersal? That is, *can you* disperse, and if you can, *do you* disperse? It is critical to keep from conflating these two. The *ability* to disperse revolves around capabilities and barriers. Are there geographic routes or barriers (deserts, bodies of water, etc.) that allow or constrain your movement? Are you biologically capable of moving from your place of origin (factors include body size, home range, suitability of diet, ability to deal with predators, and climate)? Are you behaviorally capable of dispersing (factors include flexibility in the face of changing resources and stress response)? If you can, whether you actually disperse may be due to other variables entirely. You might be pushed from an area by overpopulation, resource scarcity, predator abundance, or climate change. Or you might be pulled into a new area following resources like food or water, by climatic variables, or by following mates. Or you might for no particular reason wander farther, out of curiosity, out of disorientation, out of fear or even wonder. Evidence from extant mammals suggests that all of these alternatives and more may lie behind an organism's dispersal into new territories (see Hamilton and May 1977; Grant 1978; Arnold 1990; Lidicker and Stenseth 1992). Thus, trying to get to the answer to this second question of *why* an organism disperses, when it does (or doesn't) is the more challenging and interesting of the two—but requires that we first answer the question of whether dispersal is possible.

4. Can You Disperse?

We know some hominin species could disperse, because we see the evidence that they did. The first fossil hominins found out of Africa are *H. erectus*—and traditionally these are considered the first to be able to disperse. But this is largely due to a conflation of *can* and *do*. In fact, there were multiple hominin species in Africa from 2.5 million to 1.5 million years ago, including multiple members of genus *Homo*. Was only one of these able to disperse? If so, why? If not, then why is *H. erectus* the only one to have left a record of dispersal?

Since dispersal into Asia likely happened overland from eastern Africa, let us consider the diversity of East African hominin species between 2.5 and 1.5 million years ago. There is a lot, and our knowledge of that diversity has increased greatly in the past decade. About 2.8 million years ago we see what is likely the first member of genus *Homo* (Villmoare et al. 2015). And between 2.3 and 1.89 million years ago we see the FADs of, conservatively, three

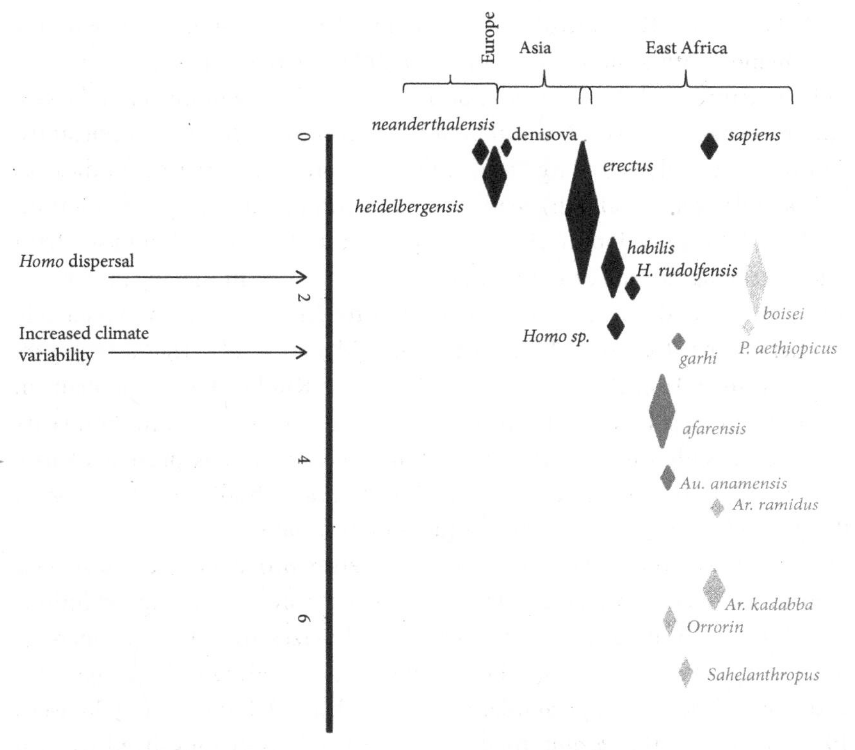

Figure 6.1 Distribution of fossil hominins in eastern Africa and Eurasia, mapped according to their persistence over time. Timescale on left is in millions of years. Diamond length reflects persistence of taxon. Diamond breadth has no meaning. *Homo* species are black diamonds (starting with *Homo* sp., the hypothesized first *Homo* species). *Paranthropus* species are gray hatched diamonds. *Australopithecus* species are solid gray diamonds. Taxa earlier than four million years (including *Ardipithecus*) that are stippled diamonds are from various genera. *H. sapiens* are shown in East Africa as their presumed region of origin.

species of early *Homo* (*H. habilis*, *H. rudolfensis*, and *H. erectus*; see Figure 6.1). Although they each have slightly different FADs, they all are present at the time of dispersal, and they are not progressively (unilineally) related to one another in any clear way. These species appear, instead, to be three separate ways of "being *Homo*" (Antón 2018). For example, these three taxa show several different kinds of faces that are based on different ways of assembling the tooth rows (Leakey et al. 2012; Spoor et al. 2015), differences that would likely have made their faces visually different from one another, which may have helped with mate recognition. *H. habilis* has the more "primitive" arcade structure, with *H. rudolfensis* and *H. erectus* having more derived faces, but faces that are derived in different ways from one another. Importantly, all three have relatively large brains (and we presume large-ish bodies) as well as relatively small(ish) teeth and jaws compared with their hominin forebears, the australopithecines. Also present in this interval in East Africa is the evolving lineage of *Paranthropus* with an apparent dietary preference for C_4 grasses and corms (compared to the three *Homo* species) and relatively smaller brains (Cerling, Mbua, et al. 2011; Macho 2014). The last appearance datum of *Australopithecus afarensis* (~3 Ma; Kimbel, Rak, and Johanson 2004) occurs just before the advent of *Homo,* and at about 2.5 million years ago the possibly distinct species *Australopithecus garhi* is present (Asfaw et al. 1999). Both have larger teeth and mostly smaller bodies and brains than the three *Homo* species. But there is individual variation.

So what does that mean for ability to disperse? Broadly dispersed species tend to share certain features, including being polytypic, being evolutionarily long-lived, having relatively larger body sizes than their congeners, possessing correlatively larger home range sizes, tending to be gregarious, and having broad ecological tolerance (Ehrlich 1989; Lidicker and Stenseth 1992). Although the *Paranthropus* lineage was relatively long-lived with an apparently broad preference for C_4 resources, it also seems to have been tied to wooded areas near water sources. The specialized cranio-dental anatomy of the lineage as well as its isotopic and environmental signatures are consistent with a predilection for corms—starchy C_4 resources that are prevalent in wooded, well-watered areas and require persistent mastication to break down (Macho 2014). Energetic models suggest that a *Paranthropus* individual could meet its dietary needs by spending about six hours a day processing corms. If this is accurate, this dietary predilection would limit the range of individuals to within a circumscribed radius of the food source and would make them vulnerable to increasing aridity. Both potentially limit the dispersibility of the group.

The *Homo* taxa, on the other hand, raise an interesting question. In the past, *H. erectus* has been thought to differ from the other two *Homo* species on the basis of its larger brain and body and its longer hind limbs (see Antón 2003). These differences were considered likely to indicate larger home range sizes and greater dispersibility. The causal links for these inferences are the positive allometric relationship between body size and home range size (McNab 1963) and a similar relationship that relates hind limb elongation and stride length, which increases locomotor efficiency and the ability to range over larger areas (Pontzer 2007). The inference of higher-quality diets is based on the energetic requirements of building large brains (e.g., Leonard and Robertson 1994, 1997; Aiello and Wheeler 1995; Navarrete et al. 2011) and the positive allometric relationship between diet quality and home range (Leonard and Robertson 1994). Together, these attributes suggested greater ability to disperse in *H. erectus* than other early *Homo* species (e.g., Antón, Robertson, and Leonard 2002). However, new evidence suggests instead that all three species possess relatively large brains (Spoor et al. 2015)—and we assume bodies. By this logic, each likely had higher-quality diets and large home ranges. And as stone tool manufacture is ubiquitous after two million years ago (Grove 2011; Antón, Potts, and Aiello 2014), they may well have been biocultural critters.

In short, with respect to the *Can you disperse?* question, there seem to be at least three taxa that meet the established dispersibility criteria. All three are present at the time of mosaic habitats with increasing C_4 grassland and expanding herbivore biomass. All three have large brains that indirectly imply a higher-quality diet. Small faces and teeth[5] speak to reduced dietary forces, and dental wear and isotopic values suggest that all three had great dietary breadth that is consistent with a higher-quality diet (Ungar 2012). Additionally, all three could have used tools to access meat and marrow. Such abilities would have facilitated a foraging change that increased the ability to leave particular kinds of habitats when no longer tied to particular plant products. It thus appears that all three of these were likely able to disperse. Yet, they do not seem to have done so—*or* if the others did, only *H. erectus* was able to survive long enough to leave traces in the archaeological record.

5. Do You Disperse? (Why *Homo erectus*?)

What was different about *H. erectus* that led it to disperse when the other species of *Homo* apparently did not? Because all three taxa overlap in time and space, often seemingly sharing the same environmental opportunities, external factors do not appear to provide an answer as to why *H. erectus* did and the others did not disperse. Similarly, all three appear to have been biocultural organisms, at least to the extent to which stone tool manufacture can be considered evidence of this, and the extent to which we can put tools in their hands. We know that the traditional tropes—of a really big body, a differentially larger brain, or a different type of tool for *H. erectus*—are not supported by the record. That is, so far we have been unable to identify something that is fundamentally or essentially different about all *H. erectus* that does not also apply (as far as we can tell) to other early *Homo*.

We do know, however, that the initial dispersal pattern of *H. erectus* does not look like the dispersal pattern of a typical primate; it is much quicker (Table 6.1). Based on the study of extant mammalian dispersals, the speed and success of a dispersal are related to many factors, including reproductive rate, home range size, competition with preexisting ecological communities, predator load in the new territory, and other aspects of the goodness of ecological fit to the new area. And while most dispersals are not successful (only about 10% of invasive species become established, with only 10% of these becoming pests), strength of ecological fit can ensure success (see Williamson 1996).

The diffusion coefficient (D) is used in the study of extant invasive species to provide an estimate of the efficiency and speed of an invasion based on the area invaded, time to spread, and the intrinsic rate of natural increase of the organism. In Table 6.1 I have estimated Ds for fossil macaques and two species of *Theropithecus* as they disperse from East Africa to India and from East to South Africa, respectively. I have similarly estimated Ds for the dispersal of *H. erectus* into Georgia and then subsequently from Georgia to Java. The cercopithecine monkey dispersals are all largely consistent with one another despite their different trajectories. However, the monkey D value estimates are substantially lower and less efficient than those for *H. erectus* dispersal. To achieve similar values for *H. erectus* dispersals, we would need to add more than a million years to their dispersal time.[6] That is, the currently known FADs in East Africa would have to underestimate the origin of *H. erectus* by more than a million years. From this we can conclude that at the point of this dispersal, *H. erectus* had stepped out of a typical primate pattern and was now filling a different ecological niche. And this niche was different not only from that of other primates but also from that of earlier hominins.

We should remember that the dispersal of *H. erectus* entailed a shift from the East African equatorial moist-dry seasonality—in which all hominins thrived—into both a temperate cool-warm seasonality (e.g., as witnessed by Dmanisi in the Republic of Georgia and sites in northeastern China) and into the monsoonal moist-dry seasonality of Southeast Asia (e.g., sites such as are found in the Sangiran Dome region of Indonesia). We presume this means

Table 6.1 Estimates of diffusion coefficients for ancient dispersals

Ancient Dispersals[1]	Intrinsic rate of natural increase (r)[2]	Time to occupy (t)	Area occupied (z)[3]	Diffusion coefficient (D)
Macaca sp. (Europe to Asia)	0.05	1.5 Ma	a) 2220 b) 3135	0.00001 0.00002
Theropithecus darti (to South Africa)	0.05	0.7 Ma	a) 1555 b) 2200	0.00002 0.00004
Theropithecus oswaldi (Kenya to India)	0.05	0.9 Ma	a) 2260 b) 3200	0.00003 0.00006
Homo erectus s.l. (to Georgia)	0.015	90 ka	a) 1885 b) 2665	0.007 0.015
Homo erectus s.l. (to Georgia)	0.01	90 ka	a) 1885 b) 2665	0.01 0.02
Homo erectus s.l. (Georgia to Java)	0.015	140 ka	a) 2250 b) 3100	0.004 0.008
Homo erectus s.l. (Georgia to Java)	0.01	140 ka	a) 2250 b) 3100	0.007 0.01

[1] Fossil identification and chronology references for *Macaca* (Andrews et al. 1996; Köhler, Moyà-Solà, and Alba 2000; Szalay and Delson 1979); *Theropithecus darti* (Eck 1993; Vrba 1995); *Theropithecus oswaldi* (Delson et al. 1993; Leakey 1993; Hughes, Elton, and O'Regan 2007); *Homo erectus* (Antón and Swisher 2004; Feibel, Brown, and McDougall 1989; Ferring et al. 2011). Issues of temporal resolution loom large in the fossil record. For this reason, all t values are calculated from first appearance dates at the origin and destination. For example, the t from Georgia to Java assumes dispersal begins as soon as hominins reached Dmanisi. Because FADs may not represent historical entry points, but more likely represent last possible dates for the first entry of an organism into a region, D values must be considered rough estimates of diffusion. Since the values at both departure and arrival ends may vary by unknown amounts, relative orders of magnitude rather than specific absolute values of D are the appropriate focus of the exercise.

[2] Malthusian parameter (r) values for fossil taxa use those from extant macaques (Sade et al. 1977) for the cercopithecine monkeys, and a range for human foragers (Blurton-Jones et al. 1992) for *H. erectus*. The human forager values are at the high end of those for extant *Pan* and should, therefore, represent conservative estimates for early *Homo* (see Antón, Robertson, and Leonard 2002).

[3] Z-values are the square root of the area defined by the shortest overland, linear distance between localities, multiplied by either (a) a transect 600 km wide or (b) a transect 1,200 km wide (see Antón, Robertson, and Leonard 2002 for details). The equation for D is $D^{1/2} = z/(t)(2r^{1/2})$, where z is the square root of the area invaded, t is the time in years over which invasion occurred, and r is the intrinsic rate of natural increase of the species (see Drake et al. 1989 for equation details).

that *H. erectus* had found a way—through biological or biocultural means—to improve their environmental fit and survivability in these new contexts. While we do not know how *H. erectus* accomplished this, we do know that by at least 1.8 million years ago this shift had occurred. And since contemporaneous hominin species did not disperse in this way, we likewise presume they had not found a means of adapting to such a diversity of environments.

How might we investigate via the fossil record whether *H. erectus* differed from the other taxa in some (biological or biocultural) way that allowed it to improve its fit to the environment, thus enhancing its dispersibility *and* ensuring the success of its dispersal? What if instead of looking across the entire species, we closely examine the fossil individuals that we have, and their environment outside of Africa, in order to consider what this might tell us about how *individuals* survived in their new environments? The success of these individuals is the foundation for their populations surviving and leaving strong archaeological traces of their species' dispersal into new regions. That is, let us not think just about long-term genetic adaptations or single essential features as triggers—but the messier stuff of short-term survival. Part of this path is genetic, but a large part is not; it's a complex interplay between individuals and the local environment, and humans are real masters of this trade-off.

Broadly dispersed organisms tend to have broad ecological tolerance (Baker 1974; Ehrlich 1989; Lidicker and Stenseth 1992; Cassey et al. 2004; Vasquez 2006) and to be behaviorally flexible in, for example, being able to solve problems, breed in multiple contexts, and utilize a variety of food resources (see, e.g., McLain, Moulton, and Sanderson 1999; Sol, Timmermans, and Lefebvre 2002; Chow, Lurz, and Lea 2018). Recent humans are such consummate dispersers because of our ability to make short-term accommodations (physiologically and culturally) to new situations over the life of an individual. That is, we exhibit a great deal of developmental plasticity that accommodates environments during growth, helping us to survive but also bearing long-term consequences for adult morphology. For example, we know that bodies strongly reflect certain environmental signals during an individual's lifetime. Nutritional sufficiency—how much food, the quality available, and the timing of that availability during growth—is a key driver of body size. Especially critical are conditions during early, insulin-dependent growth (<2 years of age) when resource inadequacy can permanently influence growth outcomes (Billewicz and McGregor 1982; Eveleth and Tanner 1990). Similarly, extrinsic mortality—how likely you are to die by, for example, predation, being killed in some other way such as by parasites, disease, or homicide—is also linked to adult outcomes in size, age at first reproduction, and neonate size (Belsky, Steinberg, and Draper 1991; Ellis et al. 2009). Could these new local environments, which differ in resource sufficiency and extrinsic mortality, be more survivable if you have greater developmental plasticity? And how would we assess this from skeletal and archaeological records?

In living humans, key biological indicators such as stature, trunk length, limb length, and body size and shape (but not brain size) manifest a great deal of plasticity across environments. And while short-term accommodations are not genetically encoded, there is evidence that they do influence offspring size and shape (Kuzawa 2005; Chung and Kuzawa 2014). That is, maternal and paternal environments influence physical outcomes for offspring. For example, Chung and Kuzawa (2014) showed that maternal lower limb length but not trunk length is a strong predictor of neonate birthweight, even more so than gestational age! And because leg length is also known to be more strongly linked to the mother's own early nutritional environments, this suggests that the mother's early growth environment influences her child's. This is one of the reasons that structural racism—while not based on biological racial differences—results in very real biological consequences over generations. We do not know for how many generations the influence lasts, but it is a strong effect. Typically, this change in variation over the short term has not been seen as very important to longer-term evolution—at least not under the modern synthesis version of evolution (Bateson, Gluckman, and Hanson 2014). But it may provide means for survival in the short term—and that is necessary for any long-term evolution.

If we reflect on *H. erectus* not as a giant homogeneous package of traits but as a set of individuals in particular contexts, and compare them to their contemporaries, can we tell whether *H. erectus* appears to respond to environments in ways similar to *H. sapiens*? We know that *H. erectus* is quite morphologically variable. Indeed, the early discoveries of these fossils from different geographic regions were not even recognized as belonging to a single genus, let alone one species (von Koenigswald and Weidenreich 1939; Antón 2003). It makes sense that *H. erectus* would be polytypic given its broad dispersion (see above). But does the structure of this variation between paleodemes[7] of *H. erectus* form patterns in the ways we would expect if it were reflecting developmental plasticity in the face of new environments? That is, is there a hint that it is their plasticity in multiple environments (presumably both physical and behavioral/technological) that allows *H. erectus* to persist?

Using a model developed by Kuzawa and Bragg (2012), we can set expectations about what kinds of skeletal accommodations we are likely to see in different environments. And pairing these expectations with the environments of *H. erectus* paleodemes, we can begin to assess whether *H. erectus* plasticity looks like ours (Antón and Kuzawa 2017). Following this model, in the best of all possible worlds—an abundance of food and few predators—you are born at an ideal time and are fat and sassy and grow up to an ideal size, and you are well set to have ideal offspring. Increase the predator load, and you mature and reproduce earlier, and that means smaller bodies at maturity and small babies (remember Chung and Kuzawa's work). A similar effect is achieved if you reduce food quantity or quality but keep predator load low: you may be able to grow a little longer because you are not penalized by being eaten by a predator prior to reproduction, but you are undernourished, so you will not reach full-size potential at maturity, nor will your neonates. And in the worst of all possible worlds of heavy extrinsic mortality risk and low nutritional sufficiency, you will be the smallest both at birth and at adulthood. Turning back to *H. erectus,* in high-predator-load, low-resource environments, such as at Dmanisi, we would expect small bodies, particularly small legs—but if a human pattern were in play, we would also expect to see brain sparing (Antón and Kuzawa 2017). In terms of bodies, the Dmanisi fossils meet these expectations; however, we do not seem to see brain sparing. This may suggest that brain size is not yet such an intrinsic part of our survival kit that *H. erectus* could not survive without *absolutely* great capacity. Or because there is also a signal of greater brain size variation (larger coefficients of variation) within *H. erectus* paleodemes than in humans or later fossil *Homo,* it may also suggest that at this particular moment, brain size is particularly labile, and there is an opportunity to change the integration between brain and body size. Whether this pattern was different from what another member of early *Homo* might have shown had they dispersed is unknowable, since none of the others appears to have left Africa.

These *H. erectus* findings suggest that the species may have benefited from greater plasticity that served to improve their ecological fit and their ability to successfully disperse and persist in different environments than did other hominins. In comparison with other, more localized species (such as Neanderthals), *H. erectus* does seem to vary more across its localities (Antón et al. 2016). However, for this to be a strongly supported reason for why *H. erectus* rather than other early *Homo* species dispersed, we would need more data from earlier species. Precisely because other early *Homo* species are found in only a few locales, are not abundant, and are apparently not polytypic, such comparisons are currently impossible to make.

Yet this pattern in itself may provide a clue. On the African continent, for example, *H. erectus* is found in both East African and South African localities, with nearly identical FADs (Herries et al. 2020). This suggests the early dispersibility of *H. erectus* (Antón 2020). Alternatively, despite having earlier FADs than *H. erectus*, no reliable record of either *H. habilis* or *H. rudolfensis* is known from South Africa, perhaps hinting at the lesser dispersibility of these species.

These differences in their pattern of distribution are suggestive, but the fossil record is very limited, and what we know from the extant record about how local environments and experiences influence individual skeletal outcomes is nearly as limited. Additional work needs to be done to lay the foundation of expectations that might help us to assess new fossils as they are found. Work like this requires very particular kinds of skeletal samples, ideally individuals whose behavior during their life was known and who were then skeletonized upon death. The forensic community is a bit ahead in this undertaking, and the newer body donation programs privilege self-donation and extensive premortem information (e.g., Wescott 2018). But, of course, we cannot answer this only with reference to *H. sapiens*. As the importance of local context becomes more appreciated, there are research groups that are working to retain skeletal remains of known *nonhuman* primate individuals and to tie specific life experience to skeletal outcomes, such as with mountain gorillas (McFarlin et al. 2014) and owl monkeys (Fernandez-Duque 2011). Additionally, working with the Cayo Santiago macaques, we are developing protocols for integrating data across fleshed and skeletal studies of short-term and long-term adaptation (see bonesandbehavior.org) in order to test these inferred relationships by establishing tight linkages between skeletal measures, somatometric measures, and environmental cues (Taboada et al. 2018). Yet much remains to be done to get to the point at which we can robustly interrogate how well the *H. erectus* pattern of variation matches that of *H. sapiens* across environments, or how greatly either differs from other primates.

6. Assessing the Success of *H. sapiens*

Regardless of how they succeeded in dispersing, *H. erectus* laid an important foundation upon which *H. sapiens* expanded and from which our success should be judged. By nearly two million years ago our ancestors had moved into ecological zones not previously inhabited by other hominins. These represented many of the kinds of zones into which *H. sapiens* would one day move. *H. erectus* then thrived for nearly two million years, with their last known appearance in the later Pleistocene (circa 120–250 ka; Indriati et al. 2011; Rizal et al. 2019). Their existence was not static, with arguably a great deal of local, climatic, and temporal change taking place over this period (e.g., Antón 2003; Kaifu et al. 2008). And their success was not guaranteed, as during this time other species of hominins—likely descended from *H. erectus* (such as *H. naledi* and *H. neanderthalensis*)—arose and went extinct in relatively circumscribed geographic areas (Klein 2009; Dirks et al. 2017). *H. sapiens*, building from this remarkable biocultural foundation, has persisted for no more than a quarter of the time that *H. erectus* did; the FAD for *H. sapiens* is maximally ~315,000 years ago (Hublin et al. 2017) or ~260,000 years ago (Grün et al. 1996), depending on how the species is defined. While we have survived longer and dispersed more broadly than Neanderthals or *H. naledi*, we are still closer to their short-term levels of persistence than we are to the longer persistence of *H. erectus*, or even the species *Paranthropus*.

Many of our current definitions of human evolutionary success rely on the fact that we defy short-term ecological and biological expectations for a primate of our size. I have shown here, however, that our lineage moved outside a typical primate niche—and even beyond early hominin patterns—by nearly two million years ago. This "success" happened in the distant past and is rightly attributed to our ancestor, *H. erectus*. We should thus be careful to distinguish the success of *H. sapiens* from the earlier success of ancestral *Homo* species.

Framed in an evolutionary context, measures of success include reproductive success (number of surviving offspring for an individual; persistence through time for a species or genus), reducing mortality risks (adaptations that reduce predator load or other extrinsic mortality factors for an individual; means of avoiding extinction for a species), success in occupying various ecological niches (survival in a variety of contexts for an individual; colonizing vast landscapes for a species), or success in acquiring resources (mates and nutritional sufficiency for an individual; population size for a species or genus).

Of these, we cannot compete with *H. erectus* based on persistence and dispersal. *H. erectus* spent nearly two million years in Africa and Asia. We have persisted a few hundred thousand years. For all our advances at reducing mortality risk, *H. erectus* still avoided extinction far longer than we have so far—and there is no guarantee we shall persist into the future. Similarly, *H. erectus* successfully colonized many of the habitats where we also now thrive, including both temperate and monsoonal zones. They did not, of course, inhabit (so far as we are aware) the Arctic.

In contrast, we may argue for our greater relative success based on our ability to acquire and distribute resources in a manner that allows greater population size. The stacking of offspring, which seems to be a relatively recent phenomenon of industrialized humans (see above; Grove, chapter 7, this volume) allows us to increase our rate of intrinsic population increase well beyond that of the great apes, and probably all other hominins, allowing us to fill (or overfill) space more quickly. Our exception to the generation-time law (Ginzburg and Colyvan 2004) by stacking offspring—thereby creating much shorter generation times than should be present in a large-brained primate of our size—does not coincide with the origin of our species. Nonetheless, it seems likely that we alone among hominins possess the ability to stack offspring, and this would mark us as uniquely successful relative to our forebears. Likewise, while the fossil record does not provide any good means to assess this, it seems equally likely that our ability to break Damuth's (1981) law by living at much higher population densities than should be feasible for a mammal of our size is also a recent feature, as it seems to occur relatively late in our own species. Of course, increased population size and density in itself also makes us more susceptible to transmissible disease, as we see in our current moment, and so may at some point work against our persistence. See Fuentes (chapter 10, this volume) for an exploration of the argument for the need to modify our definitions of success for *H. sapiens* beyond those of demographic expansion.

We might also simply argue that we are the last to survive, and therefore we win. This is a kind of single-year, World Series measure of success (to use a U.S. baseball analogy). Still, the *H. erectus* franchise won persistently for millions of years but now is gone. That is, just being here now may also be a short-sighted version of success. In some instances, and at certain tipping points, the good of the individual (breeding more, getting more) will sit in opposition to the good of the species (broad temporal and geographic persistence). We are likely at the point where an organism with such great technological and cognitive savvy can be considered a success only if we can minimally match our fossil forebears in persistence. Given our ability to outstrip ecological laws—and the resulting consequences of this on our ecosystems and social inequalities—it will take all our evolved cognitive and empathetic capacities to succeed.

Notes

1. There is some suggestion of archaeological sites possibly older than 2 Ma in China (Zhu et al. 2018). However, these dates are dependent on corroboration with the paleomagnetic timescale, which is not a calibrated timescale. While these may hint at an earlier hominin expansion, if correct, it would seem to be an expansion that did not persist.
2. The Acheulean Industry, although historically considered important for dispersal, may be further irrelevant to dispersal since the earliest sites outside of Africa (at Dmanisi, Georgia) lack the large cutting tools of the Acheulean but contain Mode I core and flake technologies broadly similar to the Oldowan.
3. It is worth noting, however, that the appearance of Mode 1 technologies and their ubiquity occur during high variability times, and this extension of plasticity, while not a sole driver of dispersal, certainly is an important piece of a package of behavioral flexibility, one that is likely to underlie dispersal (Grove 2011).
4. Maturation in *H. erectus* is at best only slightly slower than in *Australopithecus,* and there is some disagreement on this point. Despite this disagreement, however, no one argues that *H. erectus* matured faster than *Australopithecus.* And only faster growth and reproduction would increase dispersibility over the condition in *Australopithecus.*
5. The teeth and jaws of *H. erectus* are the most gracile of the three taxa, suggesting that *H. erectus* may have the greatest dietary shift from ancestral hominins and, by inference, perhaps the greatest reliance on meat and marrow (or at least less tough food items). However, so far there is insufficient microwear and isotopic evidence to fully support this suggestion.
6. Given similar FADs for all early *Homo,* it would make very little difference for D values were we to consider the Dmanisi fossils to represent *Homo habilis* (as some authors do; see Ferring et al. 2011). The conclusion that *Homo* had acquired a nontypical primate niche by 2 Ma would still stand.
7. Paleodemes are local "populations" of polytypic fossil taxa that are inferred to have shared a closer gene pool than their geographically and temporally more distant relatives (Howell 1999). Paleodemes are thus temporally and geographically restricted fossil groupings that are formed in an attempt to replicate the deme structure of extant populations—that is, the smallest reproductive population of a species (Simpson 1961).

References

Andrews, P., T. Harrison, E. Delson, R. L. Bernor, and L. Martin. 1996. "Distribution and Biochronology of European and Southwest Asian Miocene Catarrhines." In *The Evolution of Western Eurasian Neogene Mammal Faunas,* edited by R. L. Bernor, V. Fahlbusch, and H.-W. Mittman, 168–207. New York: Columbia University Press.

Aiello, L., and P. Wheeler. 1995. "The Expensive Tissue Hypothesis: The Brain and Digestive System in Human and Primate Evolution." *Current Anthropology* 36: 199–221.

Antón, S. C. 2003. "Natural History of *Homo erectus*." *Yearbook Physical Anthropology* 46: 126–170.

Antón, S. C. 2018. "The Many Faces of Early *Homo*." *General Anthropology* 25: 1–5. https://doi.org/10.1111/gena.12035.

Antón, S. C. 2020. "All Who Wander Are Not Lost." *Science* 368: 34–35.

Antón, S. C., F. Aziz, and Y. Zaim. 2001. "Dispersal and Migration in Plio-Pleistocene *Homo*." In *Humanity from African Naissance to Coming Millennia—Colloquia in Human Biology and Palaeoanthropology*, edited by P. V. Tobias, M. A. Raath, J. Moggi-Cecchi, and G. A. Doyle, 97–108. Florence: Florence University Press.

Antón, S. C., and C. Kuzawa. 2017. "Early *Homo* Plasticity and the Extended Synthesis." *Royal Society Interface Focus* 7: 20170004. doi: 10.1098/rsfs.2017.0004.

Antón, S. C., R. Potts, and L. C. Aiello. 2014. "Evolution of Early *Homo*: An Integrated Biological Perspective." *Science* 345 (6192): 1236828.

Antón, S. C., M. Robertson, and W. R. Leonard. 2002. "An Ecomorphological Model of the Initial Hominid Dispersal from Africa." *Journal Human Evolution* 43: 773–785.

Antón, S. C., and C. C. Swisher III. 2004. "Early Dispersals of *Homo* from Africa." *Annual Review of Anthropology* 33: 271–296.

Antón, S. C., H. G. Taboada, E. R. Middleton, C. W. Rainwater, A. B. Taylor, T. R. Turner, J. E. Turnquist, et al. 2016. "Morphological Variation in *Homo erectus* and the Origins of Developmental Plasticity." *Philosophical Transactions of the Royal Society B* 371: 20150236. https://doi.org/10.1098/rstb.2015.0236.

Arnold, Walter. 1990. "The Evolution of Marmot Sociality: I—Why Disperse Late?" *Behavioral Ecology and Sociobiology* 27: 229–237.

Asfaw, B., T. White, O. Lovejoy, B. Latimer, S. Simpson, and G. Suwa. 1999. "*Australopithecus garhi*: A New Species of Early Hominid from Ethiopia." *Science* 284 (5414): 629–635. doi:10.1126/science.284.5414.629.

Baker, H. G. 1974. "The Evolution of Weeds." *Annual Review Ecological System* 5: 1–24.

Bateson, P., P. Gluckman, and M. Hanson. 2014. "The Biology of Developmental Plasticity and the Predictive Adaptive Response Hypothesis." *Journal of Physiology* 592: 2357–2368. doi:10.1113/jphysiol.2014.271460.

Belsky, J., L. Steinberg, and P. Draper. 1991. "Childhood Experience, Interpersonal Development, and Reproductive Strategy: An Evolutionary Theory of Socialization. *Child Development* 62: 647–670.

Billewicz, W., and I. McGregor. 1982. "A Birth-to-Maturity Longitudinal Study of Heights and Weights in Two West African (Gambian) Villages, 1951–75." *Annals of Human Biology* 9: 309–320.

Blurton-Jones, N. G., L. C. Smith, J. F. O'Connell, K. Hawkes, and C. L. Kamuzora. 1992. "Demography of the Hadza, an Increasing and High Density Population of Savanna Foragers." *American Journal of Physical Anthropology* 89: 159–181.

Bobe, R., and A. K. Behrensmeyer. 2004. "The Expansion of Grassland Ecosystems in Africa in Relation to Mammalian Evolution and the Origin of Genus *Homo*." *Palaeogeography, Palaeoclimatology, Palaeoecology* 207: 399–420. doi:10.1016/j.palaeo.2003.09.033.

Cassey, P., T. M. Blackburn, D. Sol, R. P. Duncan, and J. L. Lockwood. 2004. "Global Patterns of Introduction Effort and Establishment Success in Birds." *Proceedings of Royal Society B* 271: S405–S408.

Cerling, T. E. 1992. "Development of Grasslands and Savannas in East Africa during the Neogene." *Palaeogeography, Palaeoclimatology, Palaeoecology* 97: 241–247.

Cerling, T. E., E. Mbua, F. M. Kirera, F. K. Manthi, F. E. Grine, M. G. Leakey, M. Sponheimer, and K. T. Uno. 2011. "Diet of *Paranthropus boisei* in the Early Pleistocene of East Africa." *Proceedings of the National Academy of Sciences* 108 (23): 9337–9341.

Cerling, T. E., J. G. Wynn, S. A. Andanje, M. I. Bird, D. K. Korir, N. E. Levin, W. Mace, et al. 2011. "Woody Cover and Hominin Environments in the Past 6 Million Years." *Nature* 475: 51–56.

Chow, P. K. Y., P. W. W. Lurz, and S. E. G. Lea. 2018. "A Battle of Wits? Problem-Solving Abilities in Invasive Eastern Grey Squirrels and Native Eurasian Red Squirrels." *Animal Behavior* 137: 11–20. doi:10.1016/j.anbehav.2017.12.022.

Chung, G. C., and C. W. Kuzawa. 2014. "Intergenerational Effects of Early Life Nutrition: Maternal Leg Length Predicts Offspring Placental Weight and Birth Weight among Women in Rural Luzon, Philippines." *American Journal of Human Biology* 26: 652–659. doi:10.1002/ajhb.22579.

Damuth, J. 1981. "Population Density and Body Size in Mammals." *Nature* 290 (5808): 699–700. doi:10.1038/290699ao.

Dart, R. 1925. "*Australopithecus africanus:* The Man-Ape of South Africa." *Nature* 364: 195–199.

Dean, C., M. G. Leakey, D. Reid, F. Schrenk, G. T. Schwartz, C. Stringer, and A. Walker. 2001. "Growth Processes in Teeth Distinguish Modern Humans from *Homo erectus* and Earlier Hominins." *Nature* 414: 628–631.

Dean, C., and B. H. Smith. 2009. "Growth and Development of the Nariokotome Youth, KNM-WT 15000." In *The First Humans: Origin and Early Evolution of the Genus Homo*, edited by F. E. Grine, J. G. Fleagle, and R. E. Leakey, 101–120. New York: Springer.

Delson, E. 1996. "The Oldest Monkeys in Asia." In *Abstracts, International Symposium: Evolution of Asian Primates*, edited by O. Takenaka, 40. Inuyama, Japan: Primate Research Institute.

Delson, E., G. G. Eck, M. B. Leakey, and N. G. Jablonski. 1993. "Appendix I: A Partial Catalogue of Fossil Remains *of Theropithecus*." In Theropithecus: *The Rise and Fall of a Primate Genus*, edited by N. G. Jablonski, 499–531. Cambridge: Cambridge University Press.

deMenocal, P. B. 2004. "African Climate Change and Faunal Evolution during the Pliocene-Pleistocene." *Earth and Planetary Science Letters* 220: 3–24.

deMenocal, P. B. 2011. "Climate and Human Evolution." *Science* 270: 53–59.

Dirks, P. H. G. M., E. M. Roberts, H. Hilbert-Wolf, J. D. Kramers, J. Hawks, A. Dosseto, et al. 2017. "The Age of *Homo naledi* and Associated Sediments in the Rising Star Cave, South Africa." *Elife* 6: e24231. doi:10.7554/eLife.24231.

Drake, J. A., H. A. Mooney, F. di Castri, R. H. Groves, F. J. Kruger, M. Rejmanek, and M. Williamson, eds. 1989. *Biological Invasions: A Global Perspective*. New York: John Wiley.

Eck, G. G. 1993. "*Theropithecus darti* from the Hadar Formation, Ethiopia." In Theropithecus: *The Rise and Fall of a Primate Genus*, edited by N. G. Jablonski, 15–84. Cambridge: Cambridge University Press.

Ehrlich, P. R. 1989. "Attributes of Invaders and the Invading Process: Vertebrates." In *Biological Invasions: A Global Perspective*, edited by J. A. Drake, H. A. Mooney, F. di Castri, R. H. Groves, F. J. Kruger, M. Rejmanek, and M. Williamson, 315–328. New York: John Wiley.

Ellis, B. J., A. J. Figueredo, B. H. Brumbach, and G. L. Schlomer. 2009. "Fundamental Dimensions of Environmental Risk: The Impact of Harsh versus Unpredictable Environments on the Evolution and Development of Life History Strategies." *Human Nature* 20: 204–268. doi:10.1007/s12110-009-9063-7.

Eveleth, P. B., and J. M. Tanner. 1990. *Worldwide Variation in Human Growth*. 2nd ed. New York: Cambridge University Press.

Feibel, C. S., F. H. Brown, and I. McDougall. 1989. "Stratigraphic Context of Fossil Hominids from the Omo Group Deposits: Northern Turkana Basin, Kenya and Ethiopia." *American Journal of Physical Anthropology* 78: 595–622.

Fernandez-Duque, E. 2011. "Rensch's Rule, Bergman's Effect and Adult Sexual Dimorphism in Wild Monogamous Owl Monkeys of Argentina." *American Journal of Physical Anthropology* 146: 38–48.

Ferring, R., O. Oms, J. Agusti, F. Berna, M. Nioradze, T. Shelia, M. Tappen, et al. 2011. "Earliest Human Occupations at Dmanisi (Georgian Caucasus) Dated to 1.85–1.78 Ma." *Proceedings of the National Academy of Sciences* 108 (26): 10432–10436. doi:10.1073/pnas.1106638108.

Gabunia, L., S. C. Antón, D. Lordkipanidze, A. Vekua, C. C. Swisher III, and A. Justus. 2001. "Dmanisi and Dispersal." *Evolutionary Anthropology* 10: 158–170.

Gabunia, L., A. Vekua, D. Lordkipanidze, C. C. Swisher III, R. Ferring, A. Justus, M. Nioradze, et al. 2000. "Earliest Pleistocene Cranial Remains from Dmanisi, Republic of Georgia: Taxonomy, Geological Setting, and Age." *Science* 288: 1019–1025.

Ginzburg L. R., and M. Colyvan. 2004. *Ecological Orbits: How Planets Move and Populations Grow*. New York: Oxford University Press.

Grant, P. R. 1978. "Dispersal in Relation to Carrying Capacity." *Proceedings of the National Academy of Sciences* 75: 2854–2858.

Grove, M. 2011. "Speciation, Diversity, and Mode 1 Technologies: The Impact of Variability Selection." *Journal of Human Evolution* 61: 306–319.

Grün, R., J. S. Brink, N. A. Spooner, L. Taylor, C. B. Stringer, R. G. Franciscus, and A. S. Murray. 1996. "Direct Dating of Florisbad Hominid." *Nature* 382 (6591): 500–501.

Hamilton, W. D., and R. M. May. 1977. "Dispersal in Stable Habitats." *Nature* 269: 578–581.

Harmand, S., J. E. Lewis, C. S. Feibel, C. J. Lepre, S. Prat, A. Lenoble, X. Boës, et al. 2015. "3.3-Million-Year-Old Stone Tools from *Lomekwi* 3, West Turkana, Kenya." *Nature* 521 (7552): 310–315.

Heinzelin, J. de, J. D. Clark, T. White, W. Hart, P. Renne, G. Woldegabriel, Y. Beyene, and E. Vrba. 1999. "Environment and Behavior of 2.5-Million-Year-Old Bouri Hominids." *Science* 284: 625–629.

Herries, A. I., J. M. Martin, A. B. Leece, J. W. Adams, G. Boschian, R. Joannes-Boyau, T. R. Edwards, et al. 2020. "Contemporaneity of Australopithecus, Paranthropus, and Early *Homo erectus* in South Africa." *Science* 368: eaaw7293. doi:10.1126/science.aaw7293.

Howell, F. C. 1999. Paleo-Demes, Species, Clades and Extinctions in the Pleistocene Hominid Record." *Journal of Anthropological Research* 55: 191–243.

Howell, N. 1979. *Demography of the Dobe! Kung*. New York: Academic Press.

Huang, W., R. Ciochon, G. Yumin, R. Larick, F. Qiren, H. Schwarcz, C. Yonge, J. de Vos, and W. Rink. 1995. "Early *Homo* and Associated Artefacts from Asia." *Nature* 378: 275–278.

Hublin, J.-J., A. Ben-Ncer, S. Bailey, S. Friedline, S. Neubauer, M. M. Skinner, I. Bergmann, et al. 2017. "New Fossils from Jebel Irhoud, Morocco and the Pan-African Origin of *Homo sapiens*." *Nature* 546: 289–292. doi:10.1038/nature22336.

Hughes, J. K., S. Elton, and H. J. O'Regan. 2007. "*Theropithecus* and 'Out of Africa' Dispersal in the Plio-Pleistocene." *Journal of Human Evolution* 54: 43–77.

Indriati, E., C. C. Swisher III, C. Lepre, R. L. Quinn, R. A. Suriyanto, T. H. Agus, R. Grün, et al 2011. "The Age of the 20 Meter Solo River Terrace, Java, Indonesia and the Survival of *Homo erectus* in Asia." *PLoS ONE* 6 (6): e21562. doi:10.1371/journal.pone.0021562.

Kaifu, Y., F. Aziz, E. Indriati, T. Jacob, I. Kurniawan, and H. Baba 2008. "Cranial Morphology of Javanese *Homo erectus*: New Evidence for Continuous Evolution, Specialization, and Terminal Extinction." *Journal of Human Evolution* 55: 551–580.

Kimbel, W. R., Y. Rak, and D. C. Johanson. 2004. *The Skull of* Australopithecus afarensis. New York: Oxford University Press.

Klein, R. 1999. *The Human Career: Human Biological and Cultural Origins.* 2nd ed. Chicago: University of Chicago Press.

Klein, R. 2009. *The Human Career: Human Biological and Cultural Origins.* 3rd ed. Chicago: University of Chicago Press.

Köhler, M., S. Moyà-Solà, and D. M. Alba. 2000. "*Macaca* (Primates, Cercopithecidae) from the Late Miocene of Spain." *Journal of Human Evolution* 38: 447–452.

Kuzawa, C. W. 2005. "Fetal Origins of Developmental Plasticity: Are Fetal Cues Reliable Predictors of Future Nutritional Environments?" *American Journal of Human Biology* 17: 5–21.

Kuzawa, C. W., and J. M. Bragg. 2012. "Plasticity in Human Life History Strategy: Implications for Contemporary Human Variation and the Evolution of Genus *Homo*." *Current Anthropology* 53 (6): S369–S382.

Leakey, M. G. 1993. "Evolution of *Theropithecus* in the Turkana Basin." In Theropithecus: *The Rise and Fall of a Primate Genus*, edited by N. G. Jablonski, 85–124. Cambridge: Cambridge University Press.

Leakey, M. G., F. Spoor, M. C. Dean, C. S. Feibel, S. C. Antón, C. Kiarie, and L. N. Leakey. 2012. "New Fossils from Koobi Fora in Northern Kenya Confirm Taxonomic Diversity in Early *Homo*." *Nature* 488: 201–204. doi:10.1038/nature11322.

Leonard, W. R., and M. L. Robertson. 1994. "Evolutionary Perspectives on Human Nutrition: The Influence of Brain and Body Size on Diet and Metabolism." *American Journal of Human Biology* 6: 77–88.

Leonard, W. R., and M. L. Robertson. 1997. "Comparative Primate Energetics and Hominid Evolution." *American Journal of Physical Anthropology* 102: 265–281.

Lepre, C., H. Roche, D. Kent, S. Harmand, R. L. Quinn, J.-P. Brugal, P.-J. Texier, A. Lenoble, and C. S. Feibel. 2011. "An Earlier Origin for the Acheulian." *Nature* 477: 82–85. doi:10.1038/nature10372.

Lidicker, W. Z., and N. C. Stenseth. 1992. "To Disperse or Not to Disperse: Who Does It and Why?" In *Animal Dispersal: Small Mammals as a Model*, edited by W. Z. Lidicker and N. C. Stenseth, 233–261. New York: Chapman Hall.

Macho, G. A. 2014. "Baboon Feeding Ecology Informs the Dietary Niche of *Paranthropus boisei*." *PLoS One* 9 (1): e84942. doi:10.1371/journal.pone.0084942.

Maxwell, S. J., P. J. Hopley, P. Upchurch, and C. Soligo. 2018. "Sporadic Sampling, Not Climatic Forcing, Drives Observed Early Hominin Diversity." *Proceedings of the National Academy of Sciences* 115 (19): 4891–4896. doi:10.1073/pnas.1721538115.

McFarlin, S. C., D. J. Reid, K. Arbenz-Smith, M. R. Cranfield, F. Nutter, T. S. Stoinski, C. Whittier, T. G. Bromage, and A. Mudakikwa. 2014. "Histological Examination of Dental Development in a Juvenile Mountain Gorilla from Volcanoes National Park, Rwanda." *Bulletin of the International Association for Paleodontology* 8 (1): 149.

McLain, D. K., M. P. Moulton, and J. G. Sanderson. 1999. "Sexual Selection and Extinction: The Fate of Plumage-Dimorphic and Plumage-Monomorphic Birds Introduced onto Islands." *Evolution Ecology Research* 1: 549–565.

McNab, B. K. 1963. "Bioenergetics and the Determination of Home Range Size." *American Naturalist* 97: 133–140.

Milankovitch, M. M. 1941. *Canon of Insolation and the Ice-Age Problem.* Belgrade: Koniglich Serbische Academie.

Navarrete, A., C. P. van Schaik, and K. Isler. 2011. "Energetics and the Evolution of Human Brain Size. *Nature* 480: 91–94.

Passey, B. H., N. E. Levin, T. E. Cerling, F. H. Brown, and J. M. Eiler. 2010. "High-Temperature Environments of Human Evolution in East Africa Based on Bond Ordering in Paleosol Carbonates." *Proceedings of the National Academy of Sciences* 107: 11245–11249.

Pontzer, H. 2007. "Effective Limb Length and the Scaling of Locomotor Cost in Terrestrial Animals." *Journal of Experimental Biology* 210: 1752–1761. doi:10.1242/jeb.002246.

Potts, R. 1998a. "Environmental Hypotheses of Human Evolution." *Yearbook of Physical Anthropology* 41: 93–136.

Potts, R. 1998b. "Variability Selection in Hominid Evolution." *Evolutionary Anthropology* 7: 81–96.

Potts, R. 2013. "Hominin Evolution in Settings of Strong Environmental Variability." *Quaternary Science Reviews* 73: 1–13.

PRISM Project Members. 1995. "Middle Pliocene Paleoenvironments of the Northern Hemisphere." In *Paleoclimate and Evolution, with Emphasis on Human Origins*, edited by E. S. Vrba, G. H. Denton, T. C. Partridge, and L. H. Burkle, 197–212. New Haven, CT: Yale University Press.

Rizal, Y., K. E. Westaway, Y. Zaim, G. D. van den Bergh, E. A. Bettis, M. J. Morwood, O. F. Huffman, et al. 2019. "Last Appearance of *Homo erectus* at Ngandong, Java, 117,000–108,000 Years Ago." *Nature* 577: 381–385. doi:10.1038/s41586-019-1863-2.

Robinson, J. T. 1954. "The Genera and Species of the Australopithecinae." *American Journal of Physical Anthropology* 12: 181–200.

Robinson, J. T. 1966. "Adaptive Radiation in the Australopithecines and the Origin of Man." In *African Ecology and Human Evolution*, edited by F. C. Howell and F. Bourliere, 385–416. Chicago: Aldine.

Roche, H., A. Delagnes, J. P. Brugal, C. Feibel, M. Kibunjia, V. Mourre, and J.-P. Texier. 1999. "Early Hominid Stone Tool Production and Technical Skill 2.34 Myr Ago in West Turkana, Kenya." *Nature* 399: 57–60.

Sade, D. S., J. M. Schneider, A. Figueroa, J. R. Kaplan, B. F. Chepko-Sade, K. Cushing, P. Cushing et al. 1977. "Population Dynamics in Relation to Social Structure on Cayo Santiago." *Yearbook of Physical Anthropology* 20: 253–262.

Schwartz, G. T. 2012. "Growth, Development, and Life History throughout the Evolution of *Homo*." *Current Anthropology* 53 (S6): S395–408.

Semaw, S., P. Renne, J. W. K. Harris, C. S. Feibel, R. L. Bernor, N. Fesseha, and K. Mowbray. 1997. "2.5 Million Year Old Stone Tools from Gona, Ethiopia." *Nature* 385: 333–336.

Simpson, G. G. 1961. *Principles of Animal Taxonomy.* New York: Columbia University Press.

Sol, D., S. Timmermans, and L. Lefebvre. 2002. "Behavioural Flexibility and Invasion Success in Birds." *Animal Behaviour* 63: 495–502. doi:10.1006/anbe.2001.1953.

Spoor, F., P. Gunz, S. Neubauer, S. Stelzer, N. Scott, A. Kwekason, and M. C. Dean. 2015. "Reconstructed *Homo habilis* Type OH 7 Suggests Deep-Rooted Species Diversity in Early *Homo*." *Nature* 519: 83–86.

Swisher, C. C., III, G. H. Curtis, T. Jacob, A. G. Getty, A. Suprijo, and Widiasmoro. 1994. "Age of the Earliest Known Hominids in Java, Indonesia." *Science* 263: 1118–1121.

Szalay, F., and E. Delson. 1979. *Evolutionary History of the Primates*. New York: Academic Press.

Taboada, H. G., S. C. Antón, S. A. Williams, M. F. Laird, M. K. Stock, C. I. Villamil, S. E. Bauman Surrat, et al. 2018. "Intra-demic Morphological Variation: Using the Cayo Santiago Macaques to Model Plasticity in the Fossil Record." *American Journal of Physical Anthropology*, (Abstract) Supplement S66: 269. doi:10.1002/ajpa.23489.

Trauth, M. H., M. A. Maslin, A. L. Deino, and M. R. Strecker. 2005. "Late Cenozoic Moisture History of East Africa." *Science* 309: 2051–2053.

Ungar, P. S. 2012. "Dental Evidence for Diet in Early *Homo*." *Current Anthropology* 53 (S6): S318–S329. doi:org/10.1086/666700.

Uno, K. T., P. J. Polissar, K. E. Jackson, and P. B. deMenocal. 2016. "Neogene Biomarker Record of Vegetation Change in Eastern Africa." *Proceedings of National Academy of Sciences, U.S.A.* 113: 6355–6363. doi.org/10.1073/pnas.1521267113.

Vazquez, D. P. 2006. "Exploring the Relationship between Niche Breadth and Invasion Success." In *Conceptual Ecology and Invasion Biology: Reciprocal Approaches to Nature*, edited by M. W. Cadotte, S. M. Mcmahon, and T. Fukami, 307–322. Dordrecht: Springer.

Vekua, A., D. Lordkipanidze, G. P. Rightmire, J. Agusti, R. Ferring, G. Majsuradze, A. Mouskhelishvili, et al. 2002. "A New Skull of Early *Homo* from Dmanisi Georgia." *Science* 297: 85–89.

Villmoare, B., W. H. Kimbel, C. Seyoum, C. J. Campisano, E. N. Dimaggio, J. Rowan, D. R. Braun, J. R. Arrowsmith, and K. E. Reed. 2015. "Early *Homo* from 2.8 Ma from Ledi-Geraru, Afar, Ethiopia." *Science* 347: 1352–1355.

von Koenigswald, G. H. R., and F. Weidenreich. 1939. "The Relationship between *Pithecanthropus* and *Sinanthropus*." *Nature* 144: 926–929.

Vrba, E. S. 1995. "The Fossil Record of African Antelopes Relative to Human Evolution." In *Paleoclimate and Evolution, with Emphasis on Human Origins*, edited by E. S. Vrba, G. H. Denton, T. C. Partridge, and L. H. Burkle, 385–424. New Haven, CT: Yale University Press.

Wescott, D. J. 2018. "Recent Advances in Forensic Anthropology: Decomposition Research." *Forensic Science Research* 3: 278–293. doi:10.1080/20961790.2018.1488571.

White, T. D., S. H. Ambrose, G. Suwa, D. F. Su, D. DeGusta, R. L. Bernor, J.-R. Boisserie, et al. 2009. "Macrovertebrate Paleontology and the Pliocene Habitat of *Ardipithecus ramidus*." *Science* 326: 67–93. doi:10.1126/science.1175822.

Williams, Terrie M. 2019. "The Biology of Big." *Science* 366: 1316–1317. doi:10.1126/science.aba1128.

Williamson, Mark. 1996. *Biological Invasions*. New York: Chapman Hall.

Woldegabriel, G., S. H. Ambrose, D. Barboni, R. Bonnefille, L. Bremond, B. Currie, D. DeGusta, W. K. Hart, et al. 2009. "The Geological, Isotopic, Botanical, Invertebrate, and Lower Vertebrate Surroundings of *Ardipithecus ramidus*." *Science* 326: 65–65e5. doi:10.1126/science.1175817.

Wood, J. W. 1994. *Dynamics of Human Reproduction: Biology, Biometry and Reproduction*. New York: de Gruyter.

Zhu, Z., R. Dennell, W. Huang, Y. Wu, S. Qiu, S. Yang, Z. Rao, et al. 2018. "Hominin Occupation of the Chinese Loess Plateau since about 2.1 Million Years Ago." *Nature* 559 (7715): 608–612. doi:10.1038/s41586-018-0299-4.

Zhu, Z., R. Potts, Y. X. Pan, H. T. Yao, L. Q. Lu, X. Zhao, X. Gao, et al. 2008. "Early Evidence of the Genus *Homo* in East Asia." *Journal of Human Evolution* 55: 1075–1085.

Zhu, Z., R. Potts, F. Xie, K. A. Hoffman, C. L. Deng, C. D. Shi, and Y. X. Pan. 2004. "New Evidence on the Earliest Human Presence at High Northern Latitudes in Northeast Asia." *Nature* 431: 559–562.

7

Culture as a Life History Character

The Cognitive Continuum in Primates and Hominins

Matt Grove

1. Introduction

Humans are unique, but as Foley (1987) eloquently points out, so is every other species on the planet. Scholars of human evolution have often attempted to denote our uniqueness by highlighting specific characteristics or behaviors that are possessed by no other species, but one by one these dividing lines have been blurred or broken. Tool use is now understood to be widespread in the animal kingdom, some primates and cetaceans have rudimentary equivalents of language, and chimpanzees and New Caledonian crows are capable of mental time travel (e.g., Emery and Clayton 2004; McComb and Semple 2005; Mulcahy and Call 2006; Janik 2009; Wheeler 2009; Bentley-Condit and Smith 2010). Any differences in characteristics, therefore, are quantitative rather than qualitative. Following Foley (1987), human uniqueness stems from the fact that, while any given human trait can be found in similar form in another species, the specific constellation of traits we possess is, in its totality, unique to humans. The apparent success of humans must derive from this particular constellation of traits.

From an evolutionary perspective, success could reasonably be measured in terms of the overall population size, total biomass, geographical extent, chronological duration, or ecological dominance of a species. None of these measures is particularly satisfactory, and humans are uniquely successful only on the last—with devastating effects for the remainder of the Earth's biota (e.g., Vitousek et al. 1997). Hexapods such as ants and springtails are the perpetual winners of the population size and biomass games; rats, like humans, live on all continents except Antarctica (though this is at least partly due to their position as human commensal species); horseshoe crabs have

Matt Grove, *Culture as a Life History Character* In: *Human Success.* Edited by: Hugh Desmond and Grant Ramsey, Oxford University Press. © Oxford University Press 2023. DOI: 10.1093/oso/9780190096168.003.0007 roamed the Earth, with a largely unchanged body plan, for over 400 million years.

However, we are uniquely successful *as primates* in terms of numbers, biomass, and geographical extent. (In terms of chronological duration, we are over a million years from being even the most successful hominin.) In terms of both population size and biomass, we dwarf all other primates. According to IUCN Red List data, approximately 60% of primate species are now threatened with extinction, and approximately 75% have declining populations (Estrada et al. 2017). All other members of the family Hominidae are classified as either endangered or critically endangered—regrettably, in each case *Homo sapiens* is largely responsible for their decline. The only primate genus with a geographical extent approaching that of *Homo* is *Macaca*, and while there are currently 22 recognized species of macaques, there is of course only one extant species of the genus *Homo*. As Thierry (2007) argues, while the extant species of *Macaca* emerged through a process of evolutionary diversification, *Homo sapiens* has achieved its broad geographical distribution through a processes of cultural diversification. This reliance on cultural processes is certainly not unique but is undoubtedly greater in *Homo sapiens* than in any other species.

The fact that *Homo sapiens* as a species has achieved a greater geographical distribution than all 22 members of the genus *Macaca* combined is indicative of our behavioral diversity. It may also be an indicator of this behavioral diversity that—in stark contrast to the macaques—we are the last remaining member of our genus. In the only other cases where a catarrhine primate genus is monospecific, it either is represented by a number of subspecies (in siamangs, geladas, and patas monkeys) or has a very restricted geographic range (as with Allen's swamp monkey [Republic of the Congo and Democratic Republic of the Congo] and the proboscis monkey [Borneo]; Groves 2005). This chapter therefore views human success in the context of our primate heritage, with a focus on the extent to which we have amplified the primate tendency toward behavioral diversification via cultural as well as biological means.

The argument presented here is that the success of *Homo sapiens* relative to the other members of our order—whether measured by population size, biomass, or geographical extent—is due to a nexus of tightly interlocking traits that center on brain size and life history. In humans, these traits follow the primate pattern to a certain extent but are located at one extreme end point of the observable range of primate variation. In short, humans have very large brains and very slow life histories, with delayed development and long reproductive lives. This further implies that we have very slow rates of biological evolution. The analyses presented here suggest that large brains compensate for slow rates of biological evolution by facilitating an increased reliance on the production of cultural solutions to environmental challenges, and that this trade-off is central to human adaptation. In order to assemble the components of this argument, the following sections detail the relevant aspects of life-history theory, survey the history of genetic investigations into rates of evolution, and reconstruct life histories and evolutionary rates for a sample of hominin species, before returning once more to an appraisal of the nature of human success.

2. Life-History Theory

Perhaps the best-known characterization of life-history variation is Pianka's (1970) expansion of MacArthur and Wilson's (1967) *Theory of Island Biogeography*. Pianka's (1970, 593) famous treatment of *r*- and *K*-selected species has been cited over 2,000 times, and a modified formulation is reproduced here as Table 7.1. Individuals of *r*-selected species have small body sizes, develop rapidly, reproduce early, and have short lifespans. By contrast, individuals of *K*-selected species have large body sizes, develop slowly, reproduce later, and show greater longevity. It should be emphasized that these conditions cannot be absolute: one can refer to a species as *relatively r*- or *K*-selected only by comparison to another species. Although the theory of *r*- and *K*-selection has received considerable criticism (reviewed in Reznick, Bryant, and Bashey 2002), the basic essence of the theory—that animals can be arranged on a life-history continuum between those that have fast reproductive rates (*r*) and those that have slow reproductive rates (*K*)—is evidently true and continues to permeate the literature (e.g., Lande, Engen, and Saether 2017; Engen and Saether 2017; Wright et al. 2019).

Factor analyses conducted by Bielby and colleagues (2007) suggest that the fast-slow continuum in mammals could profitably be divided into two axes: the first is essentially that described above, while the second describes a trade-off between the number and the size of offspring. This is an important development in terms of mammalian life-history theory, but it should be noted that primates as an order are heavily skewed toward producing few, large offspring. For example, 84% of the primate species in the database

Table 7.1 Trait variation between species with relatively fast (*r*-selected) and slow (*K*-selected) life histories

		Life history	
	Trait	**"Fast" (*r*)**	**"Slow" (*K*)**
Pianka (1970)	Body mass	Lower	Higher
	Development	Faster	Slower
	Reproduction	Earlier	Later
	Fecundity	Higher	Lower
	Lifespan	Shorter	Longer
	Climate	Unpredictable	Predictable
Additional	Mutation rate	Higher	Lower
	Brain mass	Lower	Higher
	Reliance on learning	Lower	Higher

Note: The Additional section shows other traits now known to be associated with these life-history patterns.

Source: After Pianka 1970.

analyzed by Bielby and colleagues give birth to a single offspring in the majority of cases, with another 15% giving birth to twins. Of the species represented in this database, only the brown mouse lemur *Microcebus rufus* routinely gives birth to litters of more than two offspring. As such, analyses of hominin life history tend to focus on variables loading heavily on the first factor identified in the analyses of Bielby and colleagues. Since fossil hominin body mass can be estimated with relative accuracy (though rarely with great precision), scholars of human evolution have often used body mass estimates as the basis of life-history reconstructions for extinct taxa (e.g., Robson and Wood 2008; Lee 2012). Broadly speaking, hominins are considered to have late ages at sexual maturity, long interbirth intervals, low fecundity, long generation times, and long lifespans relative to most other primates.

This widely observed suite of intercorrelated traits in mammals allows Bromham (2011, 2503) to assert, "[I]f you give me a mammal in a box, but you do not tell me what the species is, I can make a fair guess at many of its characteristics based on the size of the box." Most of these characteristics relate to the mammal's life history, but crucially the inference can also be made that "a gene in a mammal in a small box is likely to evolve faster than the same gene in a larger mammal" (2503). Bromham (2009, 2011) has been at the forefront of a steadily accumulating body of literature that makes the inference that if reproductive rates vary between species, then rates of potential evolution must also vary. Although this link has recently come to prominence, studies concordant with it date back at least as far as the work of Goodman and colleagues (e.g., Goodman 1961, 1985). The link relates to the fact that mutation rates vary considerably between species (e.g., Bromham, Rambaut, and Harvey 1996; Lanfear, Welch, and Bromham 2010), with a substantial part of this variation correlated with body size and life-history variables (e.g., Nikolaev et al. 2007; Nabholz, Glemin, and Galtier 2008). Since mutation provides the raw material for evolutionary change, the potential for evolutionary change is greater in those species with higher mutation rates.

3. Rates of Mutation, Substitution, and Evolution

First, it is important to note the distinction between the terms "mutation rate" and "substitution rate." The former refers to the rate at which new changes spontaneously occur in individual genes, while the latter refers to the rate at which (some of) those changes accumulate in an evolving population or lineage over time. Substitution rates therefore depend both on mutation rates and on natural selection, sexual selection, or genetic drift, and are indicative of rates of DNA evolution. Following Kimura (1968, 1991), it is widely argued that the majority of mutations are selectively neutral; of those that have selective effects, the vast majority are deleterious, with only a small proportion being beneficial (e.g., Keightley and Eyre-Walker 2007). Only pedigree-based studies (those that measure genomic differences between parents and offspring) can directly estimate mutation rates. Phylogenetic studies necessarily measure substitution rates. However, the possibility that substitution rates may be a very good proxy for mutation rates emerges from the fundamental basis of the neutral theory. A diploid population of n individuals contains $2n$ copies of each gene, and with a mutation rate of m per gene per generation there are $2nm$ new mutations per generation. If these new mutations are all selectively neutral, then each has an equal chance— $1/2n$ —of becoming a substitution. The number of mutations that become substitutions is therefore $2nm \times \frac{1}{2n} = m$, and the substitution rate is equal to the mutation rate. The neutral theory not only relates mutation and substitution rates but also provides a useful null model, departures from which indicate the action of selection. Not all mutations are neutral, but synonymous mutations (those that change a DNA sequence without changing the amino acid that it codes for) necessarily are, so the ratio of non-synonymous to synonymous mutations (denoted ω) permits assessment of the relative effects of selection and drift.

Although there are myriad ways in which mutations can arise (e.g., Loewe and Hill 2010), these can crudely be separated into those that arise during the life of an individual (in vivo mutations) and those that arise via DNA replication errors during recombination. In terms of the latter, it is also important to note that recombination itself generates new genomes, and would do so even in the absence of mutation, simply because the genome of the offspring is a combination of elements of the maternal and paternal genomes, and therefore differs from both. The distinction between the genetic variation that arises from in vivo mutations and that which arises from the recombination process corresponds to a distinction between two major competing hypotheses regarding the correlation between body mass and mutation rate. Focusing on recombination as the major generator of genetic variation leads to the hypothesis that small-bodied, rapidly reproducing species will have higher mutation rates because they undergo a greater number of recombination events per unit time (e.g., Laird, McConaughy, and McCarthy 1969; Wu et al. 1985; Li, Tanimura, and Sharp 1987; Li et al. 1996; Ohta 1993; Nikolaev et al. 2007; Nabholz, Glemin, and Galtier 2008). As well as correlating negatively with life-history variables such as age at sexual maturity and longevity, however, body mass also correlates negatively with metabolic rate. Since in vivo mutations are largely the result of metabolic by-products such as free radicals, the hypothesis emerges that small-bodied species should produce more mutations per unit time as a result of their higher metabolic rates (e.g., Martin et al. 1992; Martin and Palumbi 1993; Bleiweiss 1998; Gillooly et al. 2005).

Although either of these hypotheses can account for the negative correlation between body mass and mutation rate, studies that have evaluated them simultaneously have tended to favor explanations based on life history and recombination, finding little support for the influence of metabolic rate once life-history variables are factored out (e.g., Bromham, Rambaut, and Harvey 1996; Rowe and Honeycutt 2002; Lanfear et al. 2007). Lanfear and colleagues (2007), for example, demonstrate that variation in mutation rate across a phylogenetically broad sample of more than 300 metazoan taxa correlates significantly with body mass itself, but not with basal metabolic rate.

As well as these two major hypotheses, there are other potential confounds that could affect mutation rates and, ultimately, rates of evolution. One of these that may be of particular importance to primate and human evolution is effective population size (N_e). Effective population size plays an important role in determining the relative importance of selection and drift. A slightly deleterious mutation that will be removed by purifying selection in a large population may instead become fixed via drift in a small population. Similarly, a slightly beneficial mutation that will be fixed via directional selection in a large population may be lost via drift in a small population. In smaller populations, therefore, a greater percentage of all mutations behave as if they were neutral, leading to the prediction that substitution rates should be higher in smaller populations. In addition, as most studies suggest that beneficial mutations are exceptionally rare relative to deleterious mutations (e.g., Keightley and Eyre-Walker 2007), the ratio of non-synonymous to synonymous mutations (ω) should also be higher in smaller populations (e.g., Ohta 1992; Woolfit 2009).

Population size in studies of human evolution becomes an important confound because it is widely suggested that larger-bodied mammals will have smaller N_e. Damuth (1981a, 1987) has shown that population *density* is negatively correlated with body mass in mammals, though whether population density can be taken as a proxy for N_e is questionable. Since larger animals also tend to have larger home ranges and greater home range overlap between individuals (Damuth 1981b), N_e does not necessarily scale directly with population density. Furthermore, N_e is affected by factors such as dominance hierarchies (which skew mating probabilities and are prevalent in primate societies) and population structure (which is likely to have been of considerable importance in human evolution; e.g., Scerri et al. 2018). Nonetheless, Kosiol and colleagues (2008) show that the great apes have higher ω values than other primates and interpret this finding as reflecting the larger effective population sizes of the latter. Similar conclusions are reached by Nielsen (2005) and Keightley et al. (2005). Thus larger-bodied animals should have lower *mutation* rates due to their slower life histories but could have higher *substitution* rates if they have lower effective population sizes.

3.1 Rate Variation and the Molecular Clock

Over the past 50 years, the number of studies demonstrating mutation rate variation in mammals and other animals has steadily increased. These studies have necessarily proceeded in parallel with developments in the use of "molecular clocks" to study divergence dates between species. Seven years after Zuckerkandl and Pauling (1962; see also 1965) established the mechanistic basis of the protein molecular clock, Laird, McConaughy, and McCarthy (1969) produced the first demonstration that substitution rates differ substantially between taxa when measured in units of years, but are much more similar when measured in units of generations. Kohne (1970) subsequently demonstrated that mutation rates in humans were 20% lower than those in the other great apes, suggesting explicitly that this difference was due to the reduced number of germline cell replications per year experienced by species with longer generation times. Despite objections to these and similar conclusions (e.g., Sarich and Wilson 1973), they were supported in a more comprehensive analysis by Wu and Li (1985), who found considerably faster rates of evolution in rodents than in humans. In supporting the "generation time hypothesis" (GTH), these authors cautioned against using average mammalian mutation rate estimates to reconstruct divergence times between species, noting that apes "may have acquired their long generation times only in the relatively recent past" (Wu and Li 1985, 1745).

The roots of the modern GTH can be traced to the work of Goodman and colleagues in the early 1960s (e.g., Goodman 1961, 1962, 1963). Goodman's analyses of serum proteins demonstrated that there were relatively minor immunological differences between humans and the other great apes, greater differences between the great apes and Old World monkeys, and still greater differences between other relatively closely related mammalian groups. Further analyses by this group demonstrated that (1) the assumption of a constant molecular clock substantially underestimates primate divergence times (Goodman 1981); (2) rates of sequence evolution in primates are considerably less than the mammalian average, with the fastest primate rates shown by primitive strepsirrhines such as lemurs and the slowest by humans and orangutans (Koop et al. 1986); and (3) that the slowing of rates occurred in a "graded" fashion, such that galagos, tarsiers, spider monkeys, gibbons, nonhuman great apes, and humans show successively slower rates of evolution (Bailey et al. 1991). By the early 1980s, Goodman and colleagues were attributing these differences to differences in generation times among primate species (e.g., Goodman 1985; Bailey et al. 1991). The resultant hypothesis, which came to be known as the "hominoid slowdown" (e.g., Goodman 1985), proved controversial at the time due to its incompatibility with the constant rate assumption of molecular clock analyses (see Sarich and Wilson 1967a, 1967b; Wilson and Sarich 1969; Herbert and Easteal 1996), but modern genome sequencing techniques and the vast quantities of genetic data now available have provided substantial evidence in its favor.

Other landmark studies included Ohta's (1993) prediction, consistent with the nearly neutral theory, that generation time should have a greater effect on synonymous than non-synonymous substitution rates, and her presentation of preliminary data in support of this prediction. Li, Tanimura, and Sharp (1996) expanded the work of Wu and Li (1985) to show that both protein and DNA sequences evolve at significantly higher rates in the Old World monkey lineage than in the human lineage, and further interpreted their results as implying that mutation arises primarily from DNA replication (see also Elango et al. 2009). Bromham, Rambaut, and Harvey (1996) introduced modern comparative methods (*sensu* Harvey and Pagel 1991) to the study of mutation rate variation and found a strong generation time effect on rates of protein sequence evolution. Nikolaev et al. (2007) demonstrated slower rates of evolution in primates relative to other mammals, a finding that they interpret as being linked to primate generation times.

Elango et al. (2006) demonstrated that the molecular clock ticks around 2% faster in chimpanzees than in humans, and around 11% faster in gorillas than in humans. Using the approximately equal dates of the common ancestors of the baboon-macaque clade and the human-chimpanzee clade, the analyses of Kim et al. (2006) showed that rates of evolution have been around 30% slower in the latter (see also Yi, Ellsworth, and Li 2002; Steiper, Young, and Sukarna 2004). Elango et al. (2009) demonstrated substantial rate variation within the Old World monkey clade, variation which was shown to be negatively correlated with generation time by Yi (2013). In the most comprehensive study to date, Moorjani et al. (2016) employed whole genome sequences of 10 primate species, demonstrating that Old World monkey lineages have evolved around 33% faster and New World monkey lineages around 64% faster than hominoid lineages. They also confirmed the result of Elango et al. (2006) of a 2% faster rate in chimpanzees than in humans, and revised the gorilla rate to around 7% faster than that in humans. A number of these studies (Kim et al. 2006; Elango et al. 2009; Moorjani et al. 2016) also show that while there is substantial variation in non-CpG sites (at which mutations arise primarily through recombination), there is very little at CpG sites (at which mutations arise primarily through methylation), providing further support for the GTH over the metabolic rate hypothesis.

Finally, findings supporting the GTH are not confined solely to mammalian lineages. Bromham (2002) reported similar findings for a wide range of reptilian species, and Saclier et al. (2018) demonstrated that extended generation times and reduced nuclear substitution rates have evolved in correlated fashion on multiple, independent occasions in isopoda. Thomas et al. (2010) extended the analysis of the GTH to a broader sample of invertebrates, using 15 genes from 143 species, including all major superphyla, and finding highly significant negative correlations between generation time and mutation rate for mitochondrial proteins and RNAs as well as nuclear RNAs.

The oft-repeated finding of substantial mutation rate variation per unit time in mammals of course contradicts the basic assumption of the molecular clock, as originally conceived. This tension may have been one reason why early findings of mutation rate heterogeneity were regarded with suspicion (Yi 2013). Yet more recently researchers have suggested that the systematic bias created by the generation time effect could be profitably incorporated into genetic clock analyses (e.g., Welch, Bininda-Emonds, and Bromham 2008; Thomas et al. 2010). A pioneering early attempt at such a correction was provided by Lovejoy, Burstein, and Heiple (1972), who used neocortical sizes estimated from fossil endocasts as proxies for generation times in extinct primate taxa. Using this method, these authors calculated the date of human-chimpanzee divergence as 12–14 Ma. Although their analyses rested on a number of simplifying assumptions and produced estimates that we now know to be too old, their study was an important example of how information on mutation rate variation could be used to inform molecular clock divergence estimates.

The overwhelming evidence of life-history effects on rates of evolution has led to a gradual shift in approaches; for example, Lepage et al. (2007, 2669) caution that "the strict molecular clock hypothesis is not biologically realistic," and Thomas et al. (2010, 1179) are able to conclude that "assumptions of rate constancy in molecular dating are no longer prevalent." As well as a proliferation of "relaxed molecular clock" methods (e.g., Drummond et al. 2006; Huelsenbeck, Larget, and Swofford 2000; Smaers, Mongle, and Kandler 2016; Yoder and Yang 2000) that do not require the strict assumption of an invariant evolutionary rate across the tree, a number of studies have followed the lead of Lovejoy, Burstein, and Heiple (1972) in attempting to recalibrate divergence dates using information on life history and other ecological variables. The most comprehensive of these studies to date is that of Steiper and Seiffert (2012), who used data from both extant and fossil primates to reconstruct body sizes, endocranial volumes, and relative endocranial volumes at various ancestral nodes in the primate phylogeny. Using significant negative correlations between each of these three variables and mutation rates, they then used each to reconstruct ancestral mutation rates, which were in turn used to recalibrate molecular clock estimates of several key events in primate evolution. Their recalibrated clock produced estimated divergence times that are far more consistent with the fossil record than are standard molecular clock estimates.

Critically, Steiper and Seiffert (2012) justify their choice of phenotypic traits by noting that body size, brain size, and relative brain size are correlated with life-history variables, which are in turn correlated with molecular evolutionary rates. As well as providing more realistic divergence time estimates than traditional molecular clock analyses, their reconstructions also suggest that the first crown primates were small-bodied and small-brained, and that all primate lineages have since evolved larger bodies, larger brains, and slower rates of molecular evolution. These results therefore support both the original "hominoid slowdown" of Goodman and colleagues (e.g., Goodman 1961, 1962, 1985) and the more comprehensive recent work of Bromham and colleagues (e.g., Bromham 2009, 2011; Welch, Bininda-Emonds, and Bromham 2008; Thomas et al. 2010). Steiper and Seiffert (2012) also take the important step of incorporating brain size as a potential determinant of evolutionary rates; the relationship between brain size and life history more generally has also been a recent focus of anthropological investigations.

3.2 Bodies or Brains?

Standard life-history theory, as embodied by the *r*/*K* continuum, is silent on the relationship between encephalization and life-history variables. Yet brain size and body size are tightly correlated in mammals (Jerison 1973; Herculano-Houzel et al. 2015; Tsuboi et al. 2018), and a number of recent studies suggest that brain size is in fact the more likely determinant of life history variation (e.g., Deaner, Barton, and van Schaik 2003; Barrickman et al. 2008; Isler and van Schaik 2009a, 2009b). Revising and extending the expensive tissue hypothesis (Aiello and Wheeler 1995), Isler and van Schaik (2009a) suggest that the energetic cost of developing and maintaining a large brain could lead to a reduction in the energy allocated to reproduction per unit time. Their analyses demonstrate that relatively large-brain mammals show reduced annual fertility rates (Isler and van Schik 2009a), and that r_{max}, a composite measure of reproductive rate, correlates more strongly with brain size than with body size across broad phylogenetic samples of both mammals and birds (Isler and van Schaik 2009b). Analyses specific to primates (Grove 2017) further consolidate this perspective, yet they also raise questions about the wider generality of the result.

Life-history theory has long revolved around two fundamental trade-offs: between somatic growth and reproductive output, and between quantity and quality of offspring (e.g., Lack 1954; Bielby et al. 2007). Phrased in terms of energetics, the cost of attaining sexual maturity determines the age at which reproduction commences, and the cost of raising offspring to independence determines the rate at which offspring can be produced. In highly encephalized species, a greater proportion of these costs is due to the growth and maintenance of metabolically expensive brain tissue, and it is therefore reasonable to expect brain size to be a major determinant of life-history patterns in these species. In less encephalized species, however, the brain is less of an energetic burden and overall body size is the better predictor of life-history variables. The variation between species described by Isler and van Schaik (2009a) appears to follow this pattern, and since primates are often presented as anomalous in terms of both encephalization and life history, it should be noted that the patterns they demonstrate may not be widely applicable. As noted by Grove (2017), the tight correlation between body mass and brain mass, coupled with the relatively high metabolic cost of the latter, is what causes body mass to be excluded from multiple regression analyses seeking the correlates of life-history variables in primates. Such results are statistically correct but logically incomplete, as it is inconceivable that body mass has no effect on life-history variation.

Nonetheless, it is certainly the case that in primates, brain mass is a better predictor of life-history variables than is body mass (Isler and van Schaik 2009a, 2009b; Grove 2017), and when combined with the relationship between life-history variables and mutation rates, this suggests an intriguing extension to the *r/K* continuum that encompasses a trade-off between rates of biological evolution and the reliance on cultural solutions to environmental challenges (Grove 2017). To recap, primates with larger brains have

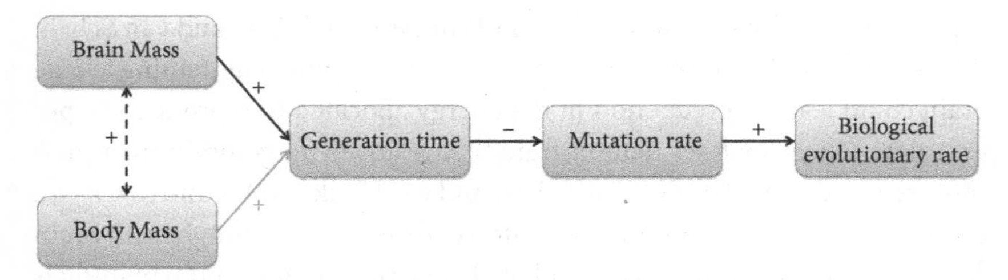

Figure 7.1 The basic scenario in which greater brain mass (and, to a lesser degree, greater body mass) leads to longer generation times, lower mutation rates, and a reduced rate of biological evolution. Plus and minus signs indicate positive and negative correlations between pairs of variables.

longer generation times and equivalently "slower" life histories. As this involves fewer instances of recombination per unit time, they also have lower mutation rates. Since mutation provides the raw material on which all evolutionary change is based, they also have slower rates of potential biological evolution. This overall scenario is depicted in Figure 7.1.

3.3 The Cognitive Continuum

This trajectory necessarily creates a deficit in the ability to deal with both biotic factors, such as arms races with predators and prey, and abiotic factors, such as changing environments. In terms of the latter, life-history theory has traditionally suggested that highly variable environments will be associated with faster life histories (as per Table 7.1), yet as climate has become more variable over the past three million years, the lineage leading to *Homo sapiens* has apparently evolved in the opposite direction (Grove 2017). It is hard to imagine natural selection favoring this trajectory unless its root cause—encephalization—is capable of recovering the deficit caused by a reduction in biological evolutionary rates. The clearest route by which encephalization could recover this deficit is by facilitating an increasing reliance on cultural evolution.

The term "cultural evolution" requires careful dissection, and it is perhaps more accurate to suggest that in encephalized species the burden of dealing with novel circumstances is shifted toward a greater reliance on learning than on the genetic system. Yet individual and social learning combine to create cultural variation, and by analogy with the arguments regarding mutation above, cultural variation is the raw material from which cultural change arises. Considerable evidence exists for increased rates of both individual and social learning in larger-brain animals (e.g., Reader and Laland 2002; Emery 2006; Rendell and Whitehead 2001; Street et al. 2017), and thus for the potential for an increased reliance on culture in these species. In a seminal study on the subject, Reader and Laland (2002) gathered hundreds of documented instances of social learning and innovation (≈ individual learning) from the primatological literature and demonstrated that the frequencies of both correlated positively and significantly with executive brain volume (the sum of the neocortex and striatum volumes) across 47 primate species. Importantly, instances of individual and social learning were themselves positively correlated, suggesting a generic increase in cultural capabilities in larger-brain species. Using an expanded data set, Street and colleagues (2017) found a positive correlation between endocranial volume and social learning richness across 150 primate species. (This latter study did not evaluate innovation rates.) These authors also found that measures of longevity are stronger predictors of social learning frequency than is endocranial volume itself, and that they remain so even when maternal investment (measured as the sum of gestation and lactation periods) is included as an additional predictor. The latter is an important finding, as it has sometimes been asserted that encephalization is better explained by maternal investment than by measures of post-weaning life history (e.g., Martin 1996; Barton and Capellini 2011); Street and colleagues (2017) demonstrate that this is not the case. From an archaeological perspective, it is also of interest that many instances of both individual and social learning relate to tool use, and that frequencies of tool use itself are also positively correlated with brain size (e.g., Reader and Laland 2002; Overington et al. 2009; Reader, Hager, and Laland 2011; Navarrete et al. 2016).

There is therefore an increasing body of evidence that encephalization could increase the rate at which cultural variation is produced and could be central to recovering the deficit caused by reduced rates of biological evolution in larger-brain animals. This suggests that the *r*/*K* relationship should be reframed as a cognitive continuum, with the position on the continuum defined as the relative reliance on biological (*r*) as opposed to cultural (*K*) solutions to challenges caused by biotic and abiotic factors. If the costs of encephalization (in terms of the reduced rates of biological evolution caused by slower life histories) are recovered *exactly* via enhanced cultural abilities, a given species might have access to numerous positions on the continuum.

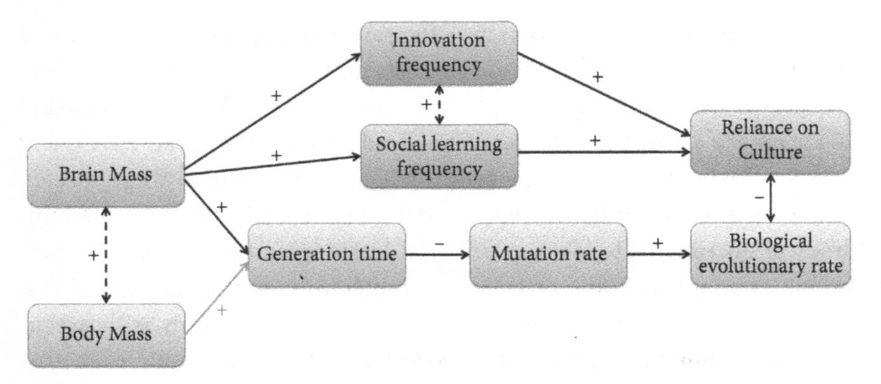

Figure 7.2 A more complex model in which greater brain mass also leads to greater frequencies of innovation and social learning, thus increasing the reliance on cultural solutions as rates of biological evolution decline. Plus and minus signs indicate positive and negative correlations between pairs of variables.

It is more likely, however, that unique sets of circumstances and challenges faced by different species lead to the "cultural route" being either more or less efficacious than the "biological route." Further to this, there is no a priori reason to believe that the cultural route is superior to the biological route; they are simply different ways of accommodating environmental challenges. Figure 7.2 extends Figure 7.1 to map both the biological and cultural routes, suggesting that both may ultimately be driven by the correlated evolution between brain size and life-history characteristics.

4. The Hominin Slowdown

To establish estimates of life history and substitution rate variation in the hominins, regression equations were first established via extant primate samples, and were then used to predict values for extinct hominin specimens. Data on primate body mass and endocranial volume were obtained from Isler and colleagues (2008). Endocranial volumes were converted to brain masses using the standard density of brain tissue at 1.036kg/liter (Snyder et al. 1975). This conversion assumes that endocranial volume is equivalent to brain volume, but Aiello and Dunbar (1993) demonstrate that this is very nearly the case. (Their regression of brain mass on endocranial capacity yielded an

R^2 value of 0.995 and a near-isometric slope.) (Data on numerous life-history variables are available from the Pantheria database: Jones et al. 2009). Given the focus on generation time in previous studies of mutation rate (e.g., Martin and Palumbi 1993; Bromham, Rambaut, and Harvey 1996; Li et al. 1996; Lehtonen and Lanfear 2014), a suitable proxy for this was sought. Following Bromham, Rambaut, and Harvey (1996), age at first birth was identified as a suitable proxy, since this accurately indexes the time elapsed between a birth in one generation and a birth in the next. Matching data from Isler and colleagues (2008) and Jones and colleagues (2009) and applying appropriate conversions yielded values of brain mass (BrM), body mass (BM), and age at first birth (AFB) for 94 primate species. One of these species, the spectral tarsier (*Tarsius tarsier*), is not represented in the GenBank taxonomy and was therefore removed from all analyses to ensure that the raw and phylogenetically controlled analyses used exactly the same data.

Data on substitution rates (SR) are harder to obtain for large numbers of primate species, but Nabholz, Glemin, and Galtier (2008) estimated substitutions per site per million years using GenBank cytochrome *b* third codon position data for a number of primate species, including 40 of those for which data on both endocranial volume and AFB were available. Mitochondrial substitution rates may be somewhat higher than those for nuclear loci (e.g., Allio et al. 2017), but these data can nonetheless be considered indicative of general patterns—if not magnitudes—of variation. Data on all four variables were natural-log transformed prior to analysis.

All regression analyses were carried out using four separate methods:

1. Type I (ordinary least squares [OLS]) regression, carried out in IBM SPSS 25 (IBM Corp., Armonk, NY).
2. Type II (reduced major axis [RMA]) regression, carried out in Matlab R.2019b (The Mathworks, Natick, MA).
3. Phylogenetic generalized least squares (PGLS) regression, (a phylogenetically corrected version of OLS), carried out using the pgls function of the caper package (Orme et al. 2018) in R v.3.6.3 (R Core Team 2020) with lambda (λ) values estimated via maximum likelihood.
4. Phylogenetic RMA (PGRMA) regression, carried out using the phyl.RMA function of the phytools package (Revell 2012) in R v.3.6.3, with λ values fixed to those produced in the equivalent PGLS analyses for comparative purposes.

Methods (3) and (4) employed phylogenetic trees downloaded from the 10k Trees Project Version 3 (Arnold, Matthews, and Nunn 2010; https://10ktrees.nunn-lab.org/Primates/downloadTrees.php) using the GenBank taxonomy with branch lengths proportional to the number of substitutions per site.

RMA and PGRMA regressions are the preferred techniques for analysis of this data set, as both dependent and independent variables are subject to measurement error (see, e.g., Aiello 1992; Sokal and Rohlf 1995, 541ff.); however, full model statistics can be reliably produced only for OLS and PGLS models. Accordingly, full statistics are given for OLS and PGLS models, with RMA PGRMA models represented only by their regression equations.

Initial examination of the data via partial correlation revealed that the relationship between BrM and AFB was positive and significant even when controlling for BM (partial $r = .429$, $p < .001$); conversely, the relationship between BM and AFB was not significant when controlling for BrM, and was actually slightly negative (partial $r = -.093$, $p = .379$). Accordingly, BM was excluded from the analyses and AFB was regressed on BrM using the four methods outlined above. Using OLS, BrM is a significant predictor of AFB ($\beta = .873$, $t(91) = 17.058$, $p < .001$), accounting for ≈76% of the variance in the latter (Adj. $R^2 = .759$, $F(1,91) = 290.975$, $p < .001$). The OLS equation is

$$\log(\text{AFB}) = 0.390 \times \log(\text{BrM}) + 5.741 \quad [1]$$

The equivalent RMA equation is

$$\log(\text{AFB}) = 0.446 \times \log(\text{BrM}) + 5.539 \quad [2]$$

The use of PGLS resulted in a slight loss of statistical power, but BrM remains a significant predictor of AFB ($\beta = .762$, $t(91) = 8.126$, $p < .001$), accounting for ≈41% of the variance in the latter (Adj. $R^2 = .414$, $F(1,91) = 66.030$, $p < .001$). The PGLS equation is

$$\log(\text{AFB}) = 0.340 \times \log(\text{BrM}) + 5.975 \quad [3]$$

The PGRMA equation is

$$\log(\text{AFB}) = 0.524 \times \log(\text{BrM}) + 5.467 \qquad [4]$$

The maximum likelihood (ML) estimate of Pagel's λ (Pagel 1999a, 1999b) is 0.759, indicating a relatively strong phylogenetic signal in the data for the relationship between these variables.

AFB is, in turn, a significant predictor of SR ($\beta = -.523$, $t(38) = -3.784$, $p = .001$), accounting for ≈26% of the variance in the latter (Adj. $R^2 = .255$, $F(1,38) = 14.321, p = .001$). The OLS equation is

$$\log(\text{SR}) = -0.501 \times \log(\text{AFB}) + 0.158 \qquad [5]$$

The equivalent RMA equation is

$$\log(\text{SR}) = -0.958 \times \log(\text{AFB}) + 3.367 \qquad [6]$$

The use of PGLS resulted in only a very slight loss of statistical power, with AFB remaining a significant predictor of SR ($\beta = -.524$, $t(38) = -3.147$, $p = .003$), accounting for ≈19% of the variance in the latter (Adj. $R^2 = .186$, $F(1,38) = 9.903, p = .003$). The PGLS equation is

$$\log(\text{SR}) = -0.502 \times \log(\text{AFB}) + 0.121 \qquad [7]$$

The PGRMA equation is

$$\log(\text{SR}) = -1.104 \times \log(\text{AFB}) + 4.252 \qquad [8]$$

The ML estimate of λ is 0.213, indicating a relatively weak phylogenetic signal in the data for the relationship between these variables.

For completeness, it is also documented here that there is a direct relationship between BrM and SR, with BrM being a significant predictor of SR

($\beta = -.520$, $t(38) = -3.753$, $p = .001$), accounting for ≈25% of the variance in the latter (Adj. $R^2 = .251$, $F(1,38) = 14.089$, $p = .001$). The OLS equation is

$$\log(\text{SR}) = -0.237 \times \log(\text{BrM}) - 2.556 \qquad [9]$$

The equivalent RMA equation is

$$\log(\text{SR}) = -0.456 \times \log(\text{BrM}) - 1.811 \qquad [10]$$

The PGLS analysis also yields a significant result ($\beta = -.541$, $t(38) = -2.826$, $p = .007$), with BrM accounting for ≈15% of the variance in SR (Adj. $R^2 = .152$, $F(1,38) = 7.986$, $p = .007$). The PGLS equation is

$$\log(\text{SR}) = -0.247 \times \log(\text{BrM}) - 2.584 \qquad [11]$$

The PGRMA equation is

$$\log(\text{SR}) = -0.594 \times \log(\text{BrM}) - 1.558 \qquad [12]$$

The ML estimate of λ is 0.409, indicating a moderate phylogenetic signal in the data for the relationship between these variables. Plots of the BrM and AFB data and equations [1–4] are shown in Figure 7.3a, of the AFB and SR data and equations [5–8] in Figure 7.3b, and of the BrM and SR data and equations [9–12] in Figure 7.3c.

In order to predict changes in AFB and SR during hominin evolution, hominin endocranial volumes and estimated ages were obtained from Grove, Pearce, and Dunbar (2012), with additions from Falk et al. (2005; for *Homo floresiensis*), Berger et al. (2015; for *Homo naledi*), and Du et al. (2018). Where age estimates or taxonomic attributions differed between Grove, Pearce, and Dunbar (2012) and Du et al. (2018), preference was given to the latter study. As noted above, RMA and PGRMA methods are most suitable for the data analyzed here. However, predictions of AFB and SR from the PGRMA equations would require a fully resolved hominin phylogenetic tree; as no such tree is available, predictions were generated from the RMA

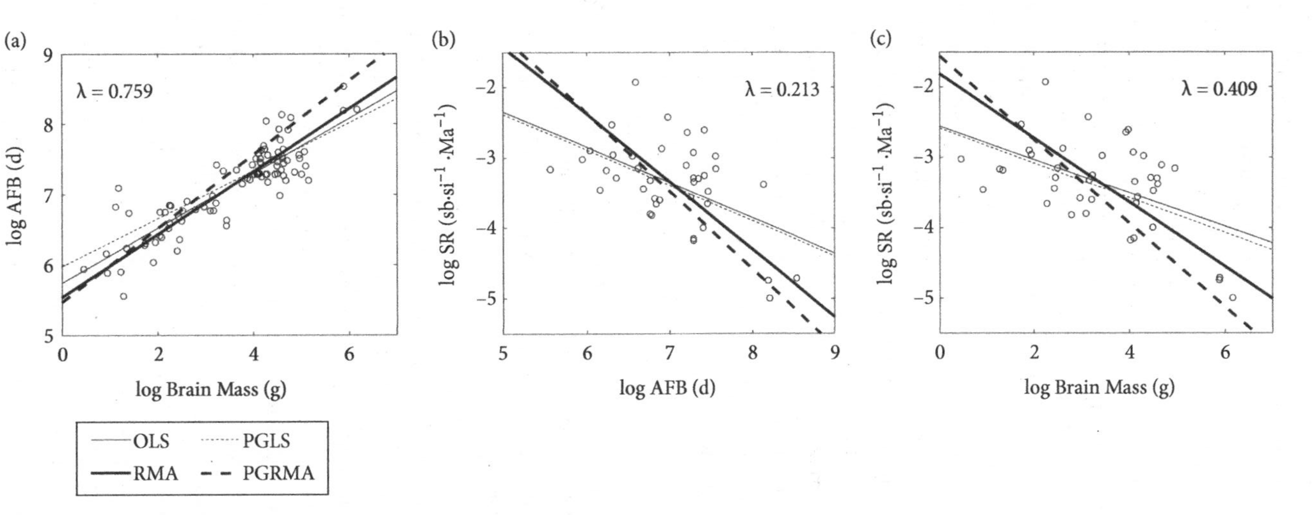

Figure 7.3 (A) regression of log age at first birth on log brain mass for 94 extant primate species; (B) regression of log substitution rate on log age at first birth for 40 extant primate species; (C) regression of log substitution rate on log brain mass for 40 extant primate species. Dashed lines indicate Model I (least squares) regressions; solid lines indicate Model II (reduced major axis) regressions. OLS = ordinary least squares; PGLS = phylogenetic least squares; RMA = reduced major axis; PGRMA = phylogenetic reduced major axis.

equations. Accordingly, equation [2] was used to predict AFB from BRM values for the hominins, and from these predictions, equation [6] was used to predict SR_1. As a separate analysis, equation [10] was used to predict SR_2 directly from BrM. Predicted AFB (from BrM) and SR_1 (from predicted AFB) through time for the hominin sample are plotted in Figures 7.4a and 7.4b respectively; Figure 7.4c shows SR_2 predicted directly from BrM, and Figure 7.4d demonstrates the relationship between predictions SR_1 and SR_2.

The potential problems of using serial regressions (i.e., employing the output of one regression equation as the input for a second) are well known (e.g., Smith 1996), and these results should be interpreted as indicating general trends rather than specific values. The analysis represented by Figure 7.4c circumvents this problem, however, and Figure 7.4d demonstrates that the estimates of SR based directly on BrM are very close to those produced via AFB as an intermediate variable. (The direct estimates are on average 0.00046 substitutions per site per million years lower than those produced via AFB.) Critically, the slope of the line passing through the points of Figure 7.4d is statistically indistinguishable from 1. By way of comparison, the estimates of AFB produced here for modern humans (95% CI: 16.49–18.55) fall toward the lower end of the range reported for 17 groups of extant hunter-gatherers (95% CI: 16.52–24.30) by Walker et al. (2006); the estimates of SR produced here for modern humans (SR_1 95% CI: 0.00620–0.00694; SR_2 95% CI: 0.00571–0.00644) fall toward the lower end of the range of those reported in a recent phylogenetic mtDNA analysis (95% CI: 0.00570–0.01930) by Fu et al. (2013).

To calculate predicted values for particular hominin species, all estimates of AFB and SR for each species were averaged. While this method fails to represent changes through time *within* species, it is useful in demonstrating differences *between* species over the period in question. Figure 7.5 shows estimated mean AFB for each species, following the splitting taxonomy and first and last appearance dates given by Wood and Boyle (2016). Figure 7.6 shows an equivalent plot for estimated SR_2. Figures 7.4–7.6 demonstrate the relative increase in generation time and the associated decrease in substitution rates experienced by hominin species over the past 4.5 million years. Note, however, that these features are not universals of hominin evolution; two species, *Homo naledi* and *Homo floresiensis*, persisted until relatively recently (~300 Ka and ~20 Ka, respectively) with small brains, and therefore with relatively fast predicted life histories and high predicted substitution rates. While results suggest a hominin slowdown overall, it appears that these two species followed an alternative trajectory.

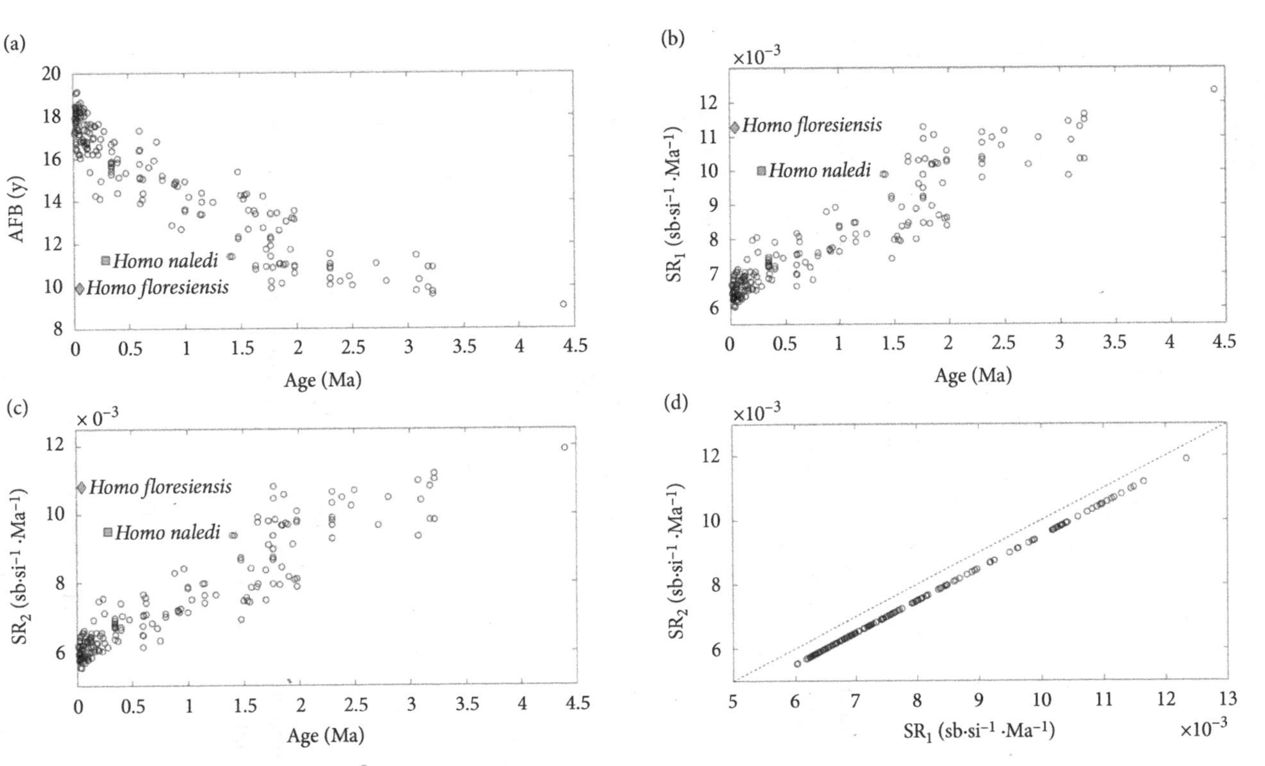

Figure 7.4 (A) predicted age at first birth for 196 hominin specimens, based on inputting estimated brain masses into equation [2]. (B) predicted substitution rate for 196 hominin specimens, based on inputting the output of equation [2] into equation [6]. (C) predicted substitution rate for 196 homimin specimens based on inputting estimated brain masses into equation [10]. (D) congruence of estimates of substitution rate between equations [6] and [10]. Outlying values for *Homo naledi* and *Homo floresiensis* are labeled.

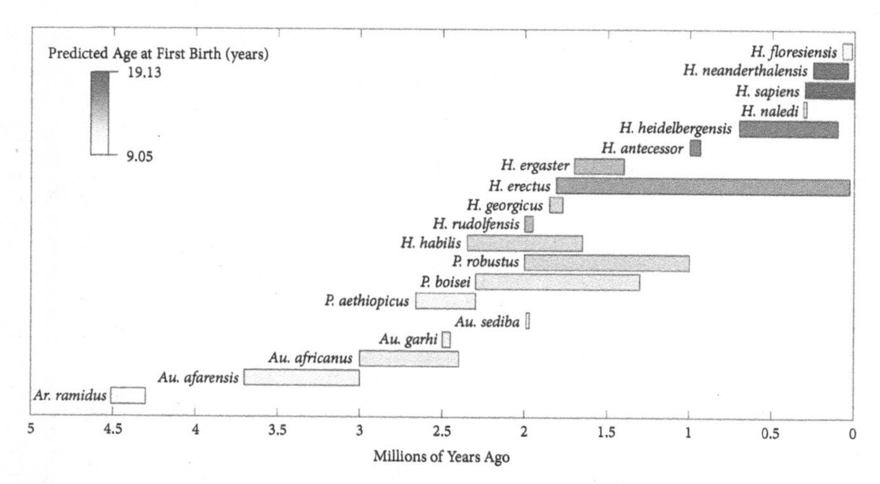

Figure 7.5 Predicted species age at first birth averages, plotted according to the first and last appearance dates of each species.

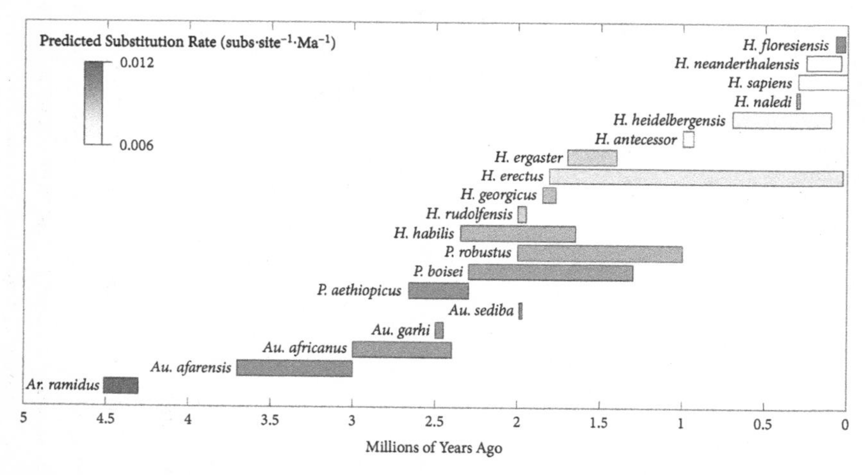

Figure 7.6 Predicted species substitution rate averages, plotted according to the first and last appearance dates of each species.

5. Discussion

5.1 The Expensive Brain

The positive relationship between brain size and AFB in primates suggests that a late switch from somatic growth to reproductive output is a primary cost of encephalization. This conclusion is supported by the particularly high metabolic cost of brain tissue relative to most other somatic components (e.g., Aiello and Wheeler 1995; Isler and van Schaik 2009a). For natural selection to favor encephalization, then, it must create benefits that outweigh the costs associated with a delayed onset of reproduction. Furthermore, these benefits must relate directly to reproduction itself, and can therefore be divided into two main areas. Larger-brain animals must either (1) produce the same number of offspring despite the late onset of reproduction, via increasing the reproductive lifespan or the number of offspring produced per unit time, or (2) produce an equal or lesser number of offspring of higher quality (where "quality" can be understood crudely as the probability that an offspring will survive to reproduce itself). In either case, if we let μ be the number of offspring per parent that survive to reproduce, then

$$\mu = \varphi \frac{(\delta - \beta)}{\gamma} \quad [13]$$

where φ is the probability of offspring survival to reproduction, δ is the age at reproductive senescence, β is the AFB, and γ is the period between births (i.e., the "interbirth interval"). Crucially, μ must be greater in larger-brain species for encephalization to be favored. Since larger brains are associated with increases in β, this must be balanced by increases in δ, γ, or both (via argument (1) above) or increases in φ (via argument (2) above).

Greater longevity is another standard component of a "slow" life history, and as such it appears that the deficit created by late reproductive onset is at least partly recovered via an extended reproductive period. However, Isler and van Schaik (2009b) found that, among extant birds and primates, increases in reproductive lifespan do not fully compensate for delayed reproduction in larger-brain species. As a result, these authors hypothesize that each lineage reaches an upper brain size threshold or "gray ceiling" beyond which further encephalization would depress reproductive output toward dangerously low levels. Their solution to this problem, particularly regarding human evolution, is to focus on the energetic requirements of reproduction and to note that allomaternal care could potentially both reduce the interbirth interval and increase the quality of offspring (Isler and van Schaik 2012b). In their analysis of 445 placental mammalian species, allomaternal care was positively correlated with brain size, suggesting that provisioning (and other helping activities) had direct benefits for offspring. In primates they found an additional, positive effect of allocare on rates of fertility, suggesting a complementary benefit for mothers that may be limited to the primate order (Isler and van Schaik 2012b). If allomaternal care in primates can reduce the energetic burden of reproduction for females, it may facilitate a reduction of the interbirth interval, enabling species of this clade to break through the gray ceiling toward further encephalization without a prohibitive decrease in reproductive rates (Isler and van Schaik 2012a, 2012b).

Applying this logic to human evolution, Isler and van Schaik (2012a) suggest that a brain size of around 600–700 cm^3 would be the maximum achievable without reliance on at least a degree of allomaternal care. They therefore align the emergence of allomaternal care with the appearance of *Homo erectus*, around 1.9 Ma. A crucial element of this argument is that extant humans have higher reproductive rates than would be expected for such an encephalized primate. If the higher reproductive rates of extant humans have deep ancestral roots, it would cast doubt on the idea that evolutionary rates have slowed over the course of hominin evolution. However, it is not at all clear when the extant human pattern first appeared. Based on the differences between the human and common chimpanzee genomes, coupled with the significant differences in the evolutionary rates of the two species, recent genetic studies suggest that the transition to modern life history patterns may have been relatively recent. Moorjani et al. (2016, 10611), for example, suggest that the life histories of hominins and chimpanzees "may have been fairly similar for much of their evolutionary history," while Elango et al. (2009) suggest a date for the emergence of modern human life history around 1 Ma. Besenbacher et al. (2019), employing pedigree-based estimates of chimpanzee, gorilla, and orangutan mutation rates and a recalibrated molecular clock, push this date even later, suggesting that the slowdown in human mutation rates must have occurred over the past 400 Ka.

Recent analyses by DeCasien and colleagues (2018) suggest that, in the primate order, the transition from small to large brain size generally precedes the transition from short to long lifespan, and that large brains may therefore have facilitated the evolution of long lifespans. If hominin evolution followed the general primate pattern, the modern human life-history pattern may not have emerged until after hominin brain masses were within the modern range of variation. This would imply an origin of humanlike life-history patterns only after around 600–700 Ka, with the appearance of *Homo heidelbergensis*, an estimate that is supported by the synthesis of hominin body size, brain size, and dental development data provided by Robson and Wood (2008). However, while broad trait reconstructions imply overall similarity between *Homo heidelbergensis*, *Homo neanderthalensis*, and *Homo sapiens*, the finer points of the trajectories of growth in these species may differ, with important consequences for reproductive energetics and cognitive abilities (e.g., Leigh 2006; Leigh and Blomquist 2007). For example, a number of studies have compared reconstructed ontogenetic trajectories of Neanderthals with those of modern humans (e.g., Coqueugniot et al. 2007; Guatelli-Steinberg 2009; Smith et al. 2007, 2010), demonstrating that growth rates in the former were significantly faster. Smith et al. (2010) analyzed dental ontogeny in 28 Neanderthal individuals, finding that Neanderthals matured significantly faster than modern humans; by contrast, Middle Paleolithic *Homo sapiens* (of similar chronological age to the Neanderthal specimens sampled) showed an essentially modern trajectory. The Engis Neanderthal infant, for example, died at approximately three years of age but had an endocranial capacity of ~1400cm^3, well within the adult Neanderthal range (Smith et al. 2010). In a similar vein, Coqueugniot et al. (2004) studied a *Homo erectus* infant from Mojokerto, revealing that it was less than 1.5 years old at death but had already attained 72–84% of adult brain size. This trajectory is considerably more like that of the common chimpanzee than that of *Homo sapiens*. Smith and colleagues (2010, 20926) conclude their study by postulating that "consistently prolonged dental development may have first appeared in *Homo sapiens*."

It is possible, therefore, that the anomalously short interbirth interval in *Homo sapiens* is a recent, species-specific phenomenon, and that the allomaternal care that may have supported it is also a recent development. There are also potential reasons for suggesting that the developmental pattern in contemporary humans was not present even in earlier members of our own species. The pace of life history is determined to a great extent by mortality risk; a high mortality risk during the reproductive phase prompts rapid development in an attempt to maximize reproductive output (e.g., Harvey, Promislow, and Read 1989). Walker and colleagues (2006, 308) therefore conclude that the apparently elevated reproductive rates of modern humans "may actually be a recently derived characteristic of some human groups in response to high mortality from parasites and infectious disease, probably exacerbated by higher population densities." If this is the case, then the short interbirth intervals of contemporary *Homo sapiens* may have little to do with ancestral patterns of encephalization. An intriguing extension is provided by Kuzawa and Bragg (2012, 379), who propose a "phenotype first" model in which behavioral decisions enabled by plasticity in life-history traits "preceded and ultimately facilitated genetic adaptations" that led to the life-history characteristics of *Homo sapiens*. Again, such decisions are predicated upon levels of nutrition and cues of mortality risk. The timing of these changes is unclear, but it is certainly true that modern humans retain significant life-history plasticity. In a study of modern English populations, Nettle (2010) demonstrates that AFB is lower and reproductive rates are higher in deprived neighborhoods, in line with reduced expected durations of reproductive lifespan. This study demonstrates that there exists substantial phenotypic plasticity for reproductive parameters in contemporary humans and that much of this plasticity is still expressed in relation to mortality profiles.

5.2 Cognitive Buffering

Returning to equation [13] and the associated discussion, it may also be the case that larger brains themselves aid survival and are in fact a crucial facilitator of greater longevity (e.g., Sol 2009a, 2009b). Allman, McLaughlin, and Hakeem (1993, 121) suggest that, in any given environmental conditions, "the longer the life-span of the animal, the more likely it is to encounter severe crises during its lifetime." These researchers therefore argue that "species with longer life-spans would have larger brains in order to sustain individuals through the more severe crises likely to occur in a longer life" (121). Building on studies such as those of Reader and Laland (2002), reported above, Sol (2009a, 2009b) has marshalled a substantial body of data to suggest that larger-brain animals are better at devising flexible behavioral solutions to novel challenges, and that these solutions enhance rates of survival in the wild. For example, due to higher innovation rates in their regions of origin, larger-brain avian species show greater survivorship and rates of establishment when introduced to novel habitats during conservation initiatives (Sol and Lefebvre 2000; Sol et al. 2002; Sol, Duncan, et al. 2005). A similar pattern is found during mammalian reintroductions (Sol et al. 2008), leading Sol, Duncan, et al. (2005, 5464) to conclude that enlarged brains "function, and hence may have evolved, to deal with changes in the environment."

The focus on the challenges created by fluctuating environments accords with recent research in evolutionary anthropology (e.g., Potts 1996, 1998, 2013; Potts and Faith 2015; Kingston 2007; Grove 2011a, 2011b, 2017), much of which suggests that increasing paleoclimatic variability over the past 5–6 million years has placed a premium on behavioral versatility. In establishing the variability selection hypothesis, for example, Potts (1998, 85) notes the effects of climatic variability in promoting the evolution of "complex structures or behaviors that are designed to respond to novel and unpredictable adaptive settings." Though numerous routes toward behavioral versatility may arise under variability selection, many may be directly supported by encephalization, and Potts stresses the benefits that arise from "a large brain . . . that is effective in processing external data and generating complex cognitive responses" (85). Analyses by numerous researchers (e.g., Ash and Gallup 2007; Bailey and Geary 2009; Grove 2012) have found positive relationships between increasing climatic variability and encephalization during human evolution. Combined with the arguments of researchers such as Sol (2009a), based on analyses of contemporary mammalian and avian taxa, this suggests that a particular stimulus to encephalization may occur when animals repeatedly encounter environmental novelty. Since hominins have evolved slower life histories against the backdrop of increasing climatic variability, the survival benefits of encephalization must have been considerable (Grove 2017).

6. Conclusions for Human Success

The argument outlined in this chapter is that human success or, to paraphrase Foley (1987), the particular constellation of traits that led to human uniqueness evolved in concert with our increasing reliance on cultural relative to biological means in dealing with environmental challenges. To the extent that our species can be regarded as successful, this success stems from the amplification of widespread relationships between life history, encephalization, and culture that form part of our primate heritage. There exists a cognitive continuum, corresponding to a considerable degree to the fast/slow life-history continuum, between animals that reproduce rapidly, have small brains, and rely primarily on the genetic system to track environmental change, and those animals that reproduce slowly, have large brains, and rely primarily on the cultural system to track environmental change. Of course highly beneficial genetic mutations, when they arise, would still be expected to spread as rapidly as generation time allows through populations of the latter type, but such populations would be shifted toward a preponderance of cultural solutions to environmental challenges. The pressing issue, however, concerns identification of the factors leading a species to occupy a particular position on this continuum. Human success stems from an evolutionary trajectory that has positioned our species at what might be termed the "cultural" extreme of the continuum—we demonstrate slower reproductive rates, larger brains, and a greater reliance on culture than any of our primate relatives—but what caused our ancestors to evolve toward this extreme?

The key to answering this question may lie in the difference between an empirical record of paleoenvironmental change and the way in which a given species actually experiences and responds to that change. Essentially, a species can respond to environmental change either by shifting its geographic range so as to maintain equilibrium conditions or by adapting to that change in situ so as to maintain sufficient fitness under the new regime. These responses are not mutually exclusive, though the former is by far the more common over evolutionary timescales. When adaptation forms part of the solution, the form it takes depends on the rate of environmental change: when the rate of change is slow relative to the generation time of the animal, directional selection may be possible; when the rate of change is fast relative to the generation time of the animal, selection for greater tolerance (via increasing niche breadth, generalism, or plasticity) is the likely outcome (e.g., Grove 2014). Critically, change that occurs *within the lifetime* of an individual can be mitigated only via learning or other forms of phenotypic or behavioral plasticity (e.g., Stephens 1991).

When climatic change is minimal, rapidly and slowly reproducing animals will experience—and adapt to—the environment in very much the same way: both will track environmental change via their mean phenotype, with relatively narrow environmental tolerance. But as change increases, animals of the two types will adapt in quite different ways: rapidly reproducing animals will continue to track environmental change via the mean phenotype, whereas slowly reproducing animals will be forced to increase their environmental tolerance considerably (Grove 2014). Results of a theoretical model designed to examine the responses of these two strategies are presented in Figure 7.7. The basic difference between the responses of rapidly and slowly reproducing animals explains the variety of strategies found within the extant species of an order such as the primates but does not explain why a particular species came to evolve a particular strategy in the first place.

The answer to this second problem lies at the interface between the geographic or habitat shifts adopted by a species in response to climate change and the adaptive changes (if any) that those shifts require. Chimpanzees, for example, are largely restricted to forest-based frugivory (e.g., Potts 2004), and their populations are likely to have expanded, contracted, and fragmented in line with the dynamics of this habitat type. Even "savannah chimpanzees" derive the majority of their calories from the forested areas of their habitats (Schoeninger, Moore, and Sept 1999), and the australopithecine dietary niche may have been similarly narrow. By the time of early *Homo*, however, hominin diet had diversified, with *Homo rudolfensis* and *Homo habilis* showing distinct signs of omnivory and the occupation of a broader array of habitat types (e.g., Ungar, Grine, and Teaford 2006). There is some evidence that the paranthropines also had somewhat

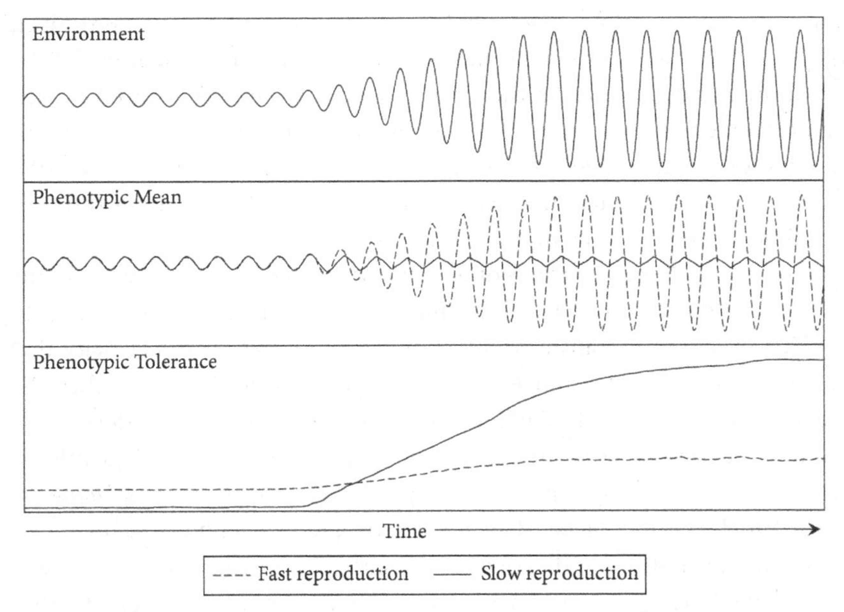

Figure 7.7 Output of an evolutionary algorithm (simplified from Grove 2014) for species with fast and slow reproductive rates. The upper panel shows the environment to which both species are adapting. When the amplitude of environmental change is small, both species can track the environment via their phenotypic means (middle panel). When the amplitude of environmental change increases, the rapidly reproducing species can persist with this strategy; the slowly reproducing species, however, shifts the weight of adaptation toward increasing phenotypic tolerance (lower panel), as the phenotypic mean is no longer capable of tracking environmental change.

broader diets than the australopithecines (e.g., Wood and Strait 2004). Such dietary generality, and the associated expansion of niche breadth, would have reduced an energetic constraint on brain growth (particularly via increased meat and marrow consumption in *Homo*), led to larger or less fragmented populations, and accelerated the nascent reliance on technology.

An increased reliance on technology—or on other forms of cultural behavior—brings many advantages, but it also creates problems for a species. Potts (2004) argues that chimpanzees and other great apes exploited their considerable intelligence in extracting as much energy as possible from a particular habitat type, and that their specialization on the relatively narrow forest-frugivory niche placed them in a "cognitive trap," linking them inextricably to this habitat and, more recently, to its largely anthropogenic decline. Thus chimpanzees, for example, minimized the extent of environmental change that they actually encountered by expanding and contracting their populations as the area covered by this habitat waxed and waned. As demonstrated elegantly by Sol, Lefebvre, and Rodriguez-Teijeiro (2005) for avian taxa, the cognitive requirements of a single, relatively unchanging habitat are considerably less than those of multiple or frequently changing habitat types.

While the cultural capacities of the other great apes have evolved to refine habitat and dietary specializations, those of the genus *Homo* are adapted to provide broader generalism or, more accurately, a degree of plasticity that supports a considerable range of interchangeable specializations. Thus the *Homo* cognitive trap is of a different nature but is no less severe. In developing cultural specializations for multiple habitat types, our genus consequently established an underlying generalist biology. We were, by the time of *Homo sapiens* if not before, hopelessly dependent upon a series of cultural technologies to obtain our multiple and varied food sources. This evolutionary route was crucial in allowing us, as a genus, to expand beyond our native habitat rather than following the generic ape strategy of expanding only as and when that native habitat itself expanded (Foley and Lahr 1997; Lahr and Foley 1998).

Adopting the cultural route has enabled geographic and habitat expansion, but it has also created an exponential series of new challenges. Proponents of the behavioral drive hypothesis (e.g., Wyles, Kunkel, and Wilson 1983; Wilson 1985; West-Eberhard 1989) argue that behavioral innovation can accelerate biological evolution by exposing animals to novel selective pressures.

Though limited in number, there are convincing empirical demonstrations of this effect (e.g., Huey, Hertz, and Sinervo 2003; Sol, Stirling, and Lefebvre 2005; see also Duckworth 2009). An important alternative to the behavioral drive hypothesis is the "behavioral inhibition" hypothesis, which states that if a population can adapt to an environmental change behaviorally, it effectively negates the selection pressures that would have been exerted by that change on the genetic system, and should therefore inhibit evolutionary change (Bogert 1949; Sol, Stirling, and Lefebvre 2005). Intriguingly, the logical structures of the behavioral drive and behavioral inhibition hypotheses are almost identical and can be presented as differing only in the primacy of the behavioral or the environmental change (Duckworth 2009). If the behavioral change occurs in response to an environmental change and acts primarily to maintain the selective status quo, then behavioral inhibition may occur. If the behavioral change occurs in the absence of environmental change but allows the population to expand its niche along some axis, then behavioral drive may occur. This distinction is directly mirrored by the distinction between counteractive niche construction, during which organisms counteract a prior change in their selective environment, and inceptive niche construction, during which organisms initiate a change in their selective environment (Odling-Smee, Laland, and Feldman 2003).

Both of these hypotheses are plausible, and in the niche construction format they are simply alternatives that could have been adopted simultaneously by different species, or even sequentially by the same species. Yet, whether counteractive or inceptive, one must consider the speed of the response required in relation to the life history of the organism. If a rapid solution to a novel selective pressure is required in a slowly reproducing species, it would be more beneficial, and more likely, for the solution itself to arise via a combination of innovation and social learning rather than via the biological system. This alternative response to novel selective pressures effectively establishes a feedback loop: the further a species falls into the trap of responding via cultural means, the more likely that species is to have to do so in the future, leading to an exponential increase in the breadth and sophistication of the cultural repertoire. The initial differences in the cultural capabilities of ancestral *Pan* and the lineage leading to *Homo* may have been slight, but they may have been sufficient to tip *Homo* alone into this feedback loop. The amplification of these differences over the course of hominin evolution was what led ultimately to the success of our species.

Acknowledgments

I would like to thank Hugh Desmond and Grant Ramsey for organizing and inviting me to participate in the Human Success: Evolutionary Origins and Ethical Implications conference at KU Leuven in June 2019. The comments of Hugh Desmond and two anonymous reviewers led to considerable clarification of some key issues and to a more robust series of analyses.

References

Aiello, L. C. 1992. "Allometry and the Analysis of Size and Shape in Human Evolution." *Journal of Human Evolution* 22 (2): 127–147. doi:10.1016/0047-2484(92)90034-7.

Aiello, L. C., and R. I. M. Dunbar. 1993. "Neocortex Size, Group Size, and the Evolution of Language." *Current Anthropology* 34 (2): 184–193. doi:10.1086/204160.

Aiello, L. C., and P. Wheeler. 1995. "The Expensive Tissue Hypothesis: The Brain and the Digestive System in Human and Primate Evolution." *Current Anthropology* 36 (2): 199–221. doi:10.1086/204350.

Allio, R., S. Donega, N. Galtier, and B. Nabholz. 2017. "Large Variation in the Ratio of Mitochondrial to Nuclear Mutation Rate across Animals: Implications for Genetic Diversity and the Use of Mitochondrial DNA as a Molecular Marker." *Molecular Biology and Evolution* 34 (11): 2762–2772. doi:10.1093/molbev/msx197.

Allman, J., T. McLaughlin, and A. Hakeem. 1993. "Brain Weight and Life Span in Primate Species." *Proceedings of the National Academy of Sciences of the United States of America* 90 (1): 118–122. doi:10.1073/pnas.90.1.118.

Arnold, C., L. J. Matthews, and C. L. Nunn. 2010. "The 10k Trees Website: A New Online Resource for Primate Phylogeny." *Evolutionary Anthropology* 19 (3): 114–118. doi:10.1002/evan.20251.

Ash, J., and G. G. Gallup. 2007. "Paleoclimatic Variation and Brain Expansion during Human Evolution." *Human Nature: An Interdisciplinary Biosocial Perspective* 18 (2): 109–124. doi:10.1007/s12110-007-9015-z.

Bailey, W. J., D. H. A. Fitch, D. A. Tagle, J. Czelusniak, J. L. Slightom, and M. Goodman. 1991. "Molecular Evolution of the Psi-eta-globin Gene Locus: Gibbon Phylogeny and the Hominoid Slowdown." *Molecular Biology and Evolution* 8 (2): 155–184.

Bailey, D. H., and D. C. Geary. 2009. "Hominid Brain Evolution." *Human Nature: An Interdisciplinary Biosocial Perspective* 20 (1): 67–79. doi:10.1007/s12110-008-9054-0.

Barrickman, N. L., M. L. Bastian, K. Isler, and C. P. van Schaik. 2008. "Life history Costs and Benefits of Encephalization: A Comparative Test Using Data from Long-Term Studies of Primates in the Wild." *Journal of Human Evolution* 54 (5): 568–590. doi:10.1016/j.jhevol.2007.08.012.

Barton, R. A., and I. Capellini. 2011. "Maternal Investment, Life Histories, and the Costs of Brain Growth in Mammals." *Proceedings of the National Academy of Sciences of the United States of America* 108 (15): 6169–6174. doi:10.1073/pnas.1019140108.

Bentley-Condit, V. K., and E. O. Smith. 2010. "Animal Tool Use: Current Definitions and an Updated Comprehensive Catalog." *Behaviour* 147 (2): 185–A32. doi:10.1163/000579509x12512865686555.

Berger, L. R., J. Hawks, D. J. de Ruiter, S. E. Churchill, P. Schmid, L. K. Delezene, T. L. Kivell, et al. 2015. "Homo naledi, a New Species of the Genus Homo from the Dinaledi Chamber, South Africa." *Elife* 4: 286–299. doi:10.7554/eLife.09560.

Besenbacher, S., C. Hvilsom, T. Marques-Bonet, T. Mailund, and M. H. Schierup. 2019. "Direct Estimation of Mutations in Great Apes Reconciles Phylogenetic Dating." *Nature Ecology & Evolution* 3 (2): 286–292. doi:10.1038/s41559-018-0778-x.

Bielby, J., G. M. Mace, O. R. P. Bininda-Emonds, M. Cardillo, J. L. Gittleman, K. E. Jones, C. D. L. Orme, and A. Purvis. 2007. "The Fast-Slow Continuum in Mammalian Life History: An Empirical Reevaluation." *American Naturalist* 169 (6): 748–757. doi:10.1086/516847.

Bleiweiss, R. 1998. "Relative-Rate Tests Aid Biological Causes of Molecular Evolution in Hummingbirds." *Molecular Biology and Evolution* 15 (5): 481–491. doi:10.1093/oxfordjournals.molbev.a025947.

Bogert, C. M. 1949. "Thermoregulation in Reptiles, a Factor in Evolution." *Evolution* 3 (3): 195–211. doi:10.2307/2405558.

Bromham, L. 2002. "Molecular Clocks in Reptiles: Life History Influences Rate of Molecular Evolution." *Molecular Biology and Evolution* 19 (3): 302–309. doi:10.1093/oxfordjournals.molbev.a004083.

Bromham, L. 2009. "Why Do Species Vary in Their Rate of Molecular Evolution?" *Biology Letters* 5 (3): 401–404. doi:10.1098/rsbl.2009.0136.

Bromham, L. 2011. "The Genome as a Life-History Character: Why Rate of Molecular Evolution Varies between Mammal Species." *Philosophical Transactions of the Royal Society B-Biological Sciences* 366 (1577): 2503–2513. doi:10.1098/rstb.2011.0014.

Bromham, L., A. Rambaut, and P. H. Harvey. 1996. "Determinants of Rate Variation in Mammalian DNA Sequence Evolution." *Journal of Molecular Evolution* 43 (6): 610–621. doi:10.1007/bf02202109.

Coqueugniot, H., and J. J. Hublin. 2007. "Endocranial Volume and Brain Growth in Immature Neandertals." *Periodicum Biologorum* 109 (4): 379–385.

Coqueugniot, H., J. J. Hublin, F. Veillon, F. Houet, and T. Jacob. 2004. "Early Brain Growth in Homo erectus and Implications for Cognitive Ability." *Nature* 431 (7006): 299–302. doi:10.1038/nature02852.

Damuth, J. 1981a. "Home Range, Home Range Overlap, and Species Energy Use among Herbivorous Mammals." *Biological Journal of the Linnean Society* 15 (3): 185–193. doi:10.1111/j.1095-8312.1981.tb00758.x.

Damuth, J. 1981b. "Population Density and Body Size in Mammals." *Nature* 290 (5808): 699–700. doi:10.1038/290699a0.

Damuth, J. 1987. "Interspecific Allometry of Population Density in Mammals and Other Animals: The Independence of Body Mass and Population Energy Use." *Biological Journal of the Linnean Society* 31 (3): 193–246. doi:10.1111/j.1095-8312.1987.tb01990.x.

Deaner, R. O., R. A. Barton, and C. P. van Schaik. 2003. "Primate Brains and Life Histories: Renewing the Connection." In *Primate Life Histories and Socioecology*, edited by P. M. Kappeler and M. E. Pereira, 233–265. Chicago: University of Chicago Press.

DeCasien, A. R., N. A. Thompson, S. A. Williams, and M. R. Shattuck. 2018. "Encephalization and Longevity Evolved in a Correlated Fashion in Euarchontoglires but Not in Other Mammals." *Evolution* 72 (12): 2617–2631. doi:10.1111/evo.13633.

Drummond, A. J., S. Y. W. Ho, M. J. Phillips, and A. Rambaut. 2006. "Relaxed Phylogenetics and Dating with Confidence." *Plos Biology* 4 (5): 699–710. doi:10.1371/journal.pbio.0040088.

Du, A., A. M. Zipkin, K. G. Hatala, E. Renner, J. L. Baker, S. Bianchi, K. H. Bernal, and B. A. Wood. 2018. "Pattern and Process in Hominin Brain Size Evolution Are Scale-Dependent." *Proceedings of the Royal Society B-Biological Sciences* 285 (1873): n.p. doi:10.1098/rspb.2017.2738.

Duckworth, R. A. 2009. "The Role of Behavior in Evolution: A Search for Mechanism." *Evolutionary Ecology* 23 (4): 513–531. doi:10.1007/s10682-008-9252-6.

Duda, P., and J. Zrzavy. 2013. "Evolution of Life History and Behavior in Hominidae: Towards Phylogenetic Reconstruction of the Chimpanzee-Human Last Common Ancestor." *Journal of Human Evolution* 65 (4): 424–446. doi:10.1016/j.jhevol.2013.07.009.

Elango, N., J. Lee, Z. G. Peng, Y. H. E. Loh, and S. V. Yi. 2009. "Evolutionary Rate Variation in Old World Monkeys." *Biology Letters* 5 (3): 405–408. doi:10.1098/rsbl.2008.0712.

Elango, N., J. W. Thomas, S. V. Yi, and NISC Comparative Sequencing Program. 2006. "Variable Molecular Clocks in Hominoids." *Proceedings of the National Academy of Sciences of the United States of America* 103 (5): 1370–1375. doi:10.1073/pnas.0510716103.

Emery, N. J. 2006. "Cognitive Ornithology: The Evolution of Avian Intelligence." *Philosophical Transactions of the Royal Society B-Biological Sciences* 361 (1465): 23–43. doi:10.1098/rstb.2005.1736.

Emery, N. J., and N. S. Clayton. 2004. "The Mentality of Crows: Convergent Evolution of Intelligence in Corvids and Apes." *Science* 306 (5703): 1903–1907. doi:10.1126/science.1098410.

Engen, S., and B. E. Saether. 2017. "R- and K-Selection in Fluctuating Populations Is Determined by the Evolutionary Trade-off between Two Fitness Measures: Growth Rate and Lifetime Reproductive Success." *Evolution* 71 (1): 167–173. doi:10.1111/evo.13104.

Estrada, A., P. A. Garber, A. B. Rylands, C. Roos, E. Fernandez-Duque, A. Di Fiore, K. A. I. Nekaris, et al. 2017. "Impending Extinction Crisis of the World's Primates: Why Primates Matter." *Science Advances* 3 (1): n.p. doi:10.1126/sciadv.1600946.

Falk, D., C. Hildebolt, K. Smith, M. J. Morwood, T. Sutikna, P. Brown, Jatmiko, E. W. Saptomo, B. Brunsden, and F. Prior. 2005. "The Brain of LB1, Homo floresiensis." *Science* 308 (5719): 242–245. doi:10.1126/science.1109727.

Foley, R. A. 1987. *Another Unique Species: Patterns in Human Evolutionary Ecology*. London: Longman.

Foley, R., and M. M. Lahr. 1997. "Mode 3 Technologies and the Evolution of Modern Humans." *Cambridge Archaeological Journal* 7 (1): 3–36. doi:10.1017/s0959774300001451.

Fu, Q. M., A. Mittnik, P. L. F. Johnson, K. Bos, M. Lari, R. Bollongino, C. K. Sun, et al. 2013. "A Revised Timescale for Human Evolution Based on Ancient Mitochondrial Genomes." *Current Biology* 23 (7): 553–559. doi:10.1016/j.cub.2013.02.044.

Gillooly, J. F., A. P. Allen, G. B. West, and J. H. Brown. 2005. "The Rate of DNA Evolution: Effects of Body Size and Temperature on the Molecular Clock." *Proceedings of the National Academy of Sciences of the United States of America* 102 (11): 140–145. doi:10.1073/pnas.0407735101.

Goodman, M. 1961. "Role of Immunochemical Differences in Phyletic Development of Human Behavior." *Human Biology* 33 (2): 131–162.

Goodman, M. 1962. "Evolution of Immunologic Species Specificity of Human Serum Proteins." *Human Biology* 34 (2): 104–150.

Goodman, M. 1963. "Man's Place in the Phylogeny of the Primates as Reflected in Serum Proteins." In *Classification and Human Evolution*, edited by S. L. Washburn, 204–234. Chicago: Aldine.

Goodman, M. 1981. "Decoding the Pattern of Protein Evolution." *Progress in Biophysics & Molecular Biology* 38 (2): 105–164. doi:10.1016/0079-6107(81)90012-2.

Goodman, M. 1985. "Rates of Molecular Evolution: The Hominoid Slowdown." *Bioessays* 3 (1): 9–14. doi:10.1002/bies.950030104.

Grove, M. 2011a. "Change and Variability in Plio-Pleistocene Climates: Modelling the Hominin Response." *Journal of Archaeological Science* 38 (11): 3038–3047. doi:10.1016/j.jas.2011.07.002.

Grove, M. 2011b. "Speciation, Diversity, and Mode 1 Technologies: The Impact of Variability Selection." *Journal of Human Evolution* 61 (3): 306–319. doi:10.1016/j.jhevol.2011.04.005.

Grove, M. 2012. "Orbital Dynamics, Environmental Heterogeneity, and the Evolution of the Human Brain." *Intelligence* 40 (5): 404–418. doi:10.1016/j.intell.2012.06.003.

Grove, M. 2014. "Evolution and Dispersal under Climatic Instability: A Simple Evolutionary Algorithm." *Adaptive Behavior* 22 (4): 235–254. doi:10.1177/1059712314533573.

Grove, M. 2017. "Environmental Complexity, Life History, and Encephalisation in Human Evolution." *Biology & Philosophy* 32 (3): 395–420. doi:10.1007/s10539-017-9564-4.

Grove, M., E. Pearce, and R. I. M. Dunbar. 2012. "Fission-Fusion and the Evolution of Hominin Social Systems." *Journal of Human Evolution* 62 (2): 191–200. doi:10.1016/j.jhevol.2011.10.012.

Groves, C. P. 2005. "Order Primates." In *Mammal Species of the World*, 3rd ed., edited by D. E. Wilson and D. M. Reeder, 111–184. Baltimore, MD: Johns Hopkins University Press.

Guatelli-Steinberg, D. 2009. "Recent Studies of Dental Development in Neandertals: Implications for Neandertal Life Histories." *Evolutionary Anthropology* 18 (1): 9–20. doi:10.1002/evan.20190.

Harvey, P. H., and M. D. Pagel. 1991. *The Comparative Method in Evolutionary Biology*. Oxford: Oxford University Press.

Harvey, P. H., D. E. L. Promislow, and A. F. Read. 1989. "Causes and Correlates of Life History Differences among Mammals." In *Comparative Socioecology*, edited by R. A. Foley and V. Standen, 305–318. Oxford: Blackwell.

Herbert, G., and S. Easteal. 1996. "Relative Rates of Nuclear DNA Evolution in Human and Old World Monkey Lineages." *Molecular Biology and Evolution* 13 (7): 1054–1057. doi:10.1093/oxfordjournals.molbev.a025656.

Herculano-Houzel, S., K. Catania, P. R. Manger, and J. H. Kaas. 2015. "Mammalian Brains Are Made of These: A Dataset of the Numbers and Densities of Neuronal and Nonneuronal Cells in the Brain of Glires, Primates, Scandentia, Eulipotyphlans, Afrotherians and Artiodactyls, and Their Relationship with Body Mass." *Brain Behavior and Evolution* 86 (3–4): 145–163. doi:10.1159/000437413.

Huelsenbeck, J. P., B. Larget, and D. Swofford. 2000. "A Compound Poisson Process for Relaxing the Molecular Clock." *Genetics* 154 (4): 1879–1892.

Huey, R. B., P. E. Hertz, and B. Sinervo. 2003. "Behavioral Drive versus Behavioral Inertia in Evolution: A Null Model Approach." *American Naturalist* 161 (3): 357–366. doi:10.1086/346135.

Isler, K., E. C. Kirk, J. M. A. Miller, G. A. Albrecht, B. R. Gelvin, and R. D. Martin. 2008. "Endocranial Volumes of Primate Species: Scaling Analyses Using a Comprehensive and Reliable Data Set." *Journal of Human Evolution* 55 (6): 967–978. doi:10.1016/j.jhevol.2008.08.004.

Isler, K., and C. P. van Schaik. 2009a. "The Expensive Brain: A Framework for Explaining Evolutionary Changes in Brain Size." *Journal of Human Evolution* 57 (4): 392–400. doi:10.1016/j.jhevol.2009.04.009.

Isler, K., and C. P. Van Schaik. 2009b. "Why Are There So Few Smart Mammals (but So Many Smart Birds)?" *Biology Letters* 5 (1): 125–129. doi:10.1098/rsbl.2008.0469.

Isler, K., and C. P. van Schaik. 2012a. "Allomaternal Care, Life History and Brain Size Evolution in Mammals." *Journal of Human Evolution* 63 (1): 52–63. doi:10.1016/j.jhevol.2012.03.009.

Isler, K., and C. P. van Schaik. 2012b. "How Our Ancestors Broke through the Gray Ceiling Comparative Evidence for Cooperative Breeding in Early Homo." *Current Anthropology* 53: S453–S465. doi:10.1086/667623.

Janik, V. M. 2009. "Acoustic Communication in Delphinids." In *Advances in the Study of Behavior*, vol. 40, edited by M. Naguib, K. Zuberbuhler, N. S. Clayton and V. M. Janik, 123–157. Amsterdam: Elsevier.

Jerison, H. J. 1973. *Evolution of the Brain and Intelligence*. New York: Academic Press.

Jones, K. E., J. Bielby, M. Cardillo, S. A. Fritz, J. O'Dell, C. D. L. Orme, . . . A. Purvis. 2009. "PanTHERIA: A Species-Level Database of Life History, Ecology, and Geography of Extant and Recently Extinct Mammals." *Ecology* 90 (9): 2648–2648. https://doi.org/10.1890/08-1494.1.

Keightley, P. D., and A. Eyre-Walker. 2007. "Joint Inference of the Distribution of Fitness Effects of Deleterious Mutations and Population Demography Based on Nucleotide Polymorphism Frequencies. *Genetics* 177 (4): 2251–2261. doi:10.1534/genetics.107.080663.

Keightley, P. D., G. V. Kryukov, S. Sunyaev, D. L. Halligan, and D. J. Gaffney. 2005. "Evolutionary Constraints in Conserved Nongenic Sequences of Mammals." *Genome Research* 15 (10): 1373–1378. doi:10.1101/gr.3942005.

Kelley, J., and G. T. Schwartz. 2012. "Life-History Inference in the Early Hominins Australopithecus and Paranthropus." *International Journal of Primatology* 33 (6): 1332–1363. doi:10.1007/s10764-012-9607-2.

Kim, S. H., N. Elango, C. Warden, E. Vigoda, and S. V. Yi. 2006. "Heterogeneous Genomic Molecular Clocks in Primates." *Plos Genetics* 2 (10): 1527–1534. doi:10.1371/journal.pgen.0020163.

Kimura, M. 1968. "Evolutionary Rate at the Molecular Level." *Nature* 217 (5129): 624–626. doi:10.1038/217624a0.

Kimura, M. 1983. *The Neutral Theory of Molecular Evolution*. Cambridge: Cambridge University Press.

Kimura, M. 1991. "The Neutral Theory of Molecular Evolution—A Review of Recent Evidence. *Japanese Journal of Genetics* 66 (4): 367–386. doi:10.1266/jjg.66.367.

Kingston, J. D. 2007. "Shifting Adaptive Landscapes: Progress and Challenges in Reconstructing Early Hominid Environments." In *Yearbook of Physical Anthropology*, vol. 50, edited by S. Stinson, 20–58. Hoboken, NJ: Wiley.

Kohne, C. 1970. "Evolution of Higher-Organism DNA." *Quarterly Review of Biophysics* 3: 327–375.

Koop, B. F., M. Goodman, P. Xu, K. Chan, and J. L. Slightom. 1986. "Primate Eta-globin DNA Sequences and Man's Place among the Great Apes." *Nature* 319 (6050): 234–238. doi:10.1038/319234a0.

Kosiol, C., T. Vinar, R. R. da Fonseca, M. J. Hubisz, C. D. Bustamante, R. Nielsen, and A. Siepel. 2008. "Patterns of Positive Selection in Six Mammalian Genomes." *Plos Genetics* 4 (8): 234–237. doi:10.1371/journal.pgen.1000144.

Kuzawa, C. W., and J. M. Bragg. 2012. "Plasticity in Human Life History Strategy Implications for Contemporary Human Variation and the Evolution of Genus Homo." *Current Anthropology* 53: S369–S382. doi:10.1086/667410.

Lack, D. 1954. *The Natural Regulation of Animal Numbers*. Oxford: Clarendon Press.

Lahr, M. M., and R. A. Foley. 1998. "Towards a Theory of Modern Human Origins: Geography, Demography, and Diversity in Recent Human Evolution." In *Yearbook of Physical Anthropology*, vol. 41, edited by C. Ruff, 137–176. Hoboken, NJ: Wiley.

Laird, C. D., B. L. McConaughy, and B. J. McCarthy. 1969. "Rate of Fixation of Nucleotide Substitutions in Evolution. *Nature* 224 (5215): 149–154. doi:10.1038/224149a0.

Lande, R., S. Engen, and B. E. Saether. 2017. "Evolution of Stochastic Demography with Life History Tradeoffs in Density-Dependent Age-Structured Populations." *Proceedings of the National Academy of Sciences of the United States of America* 114 (44): 11582–11590. doi:10.1073/pnas.1710679114.

Lanfear, R., J. A. Thomas, J. J. Welch, T. Brey, and L. Bromham. 2007. "Metabolic Rate Does Not Calibrate the Molecular Clock." *Proceedings of the National Academy of Sciences of the United States of America* 104 (39): 15388–15393. doi:10.1073/pnas.0703359104.

Lanfear, R., J. J. Welch, and L. Bromham. 2010. "Watching the Clock: Studying Variation in Rates of Molecular Evolution between Species." *Trends in Ecology & Evolution* 25 (9): 495–503. doi:10.1016/j.tree.2010.06.007.

Lee, P. C. 2012. "Growth and Investment in Hominin Life History Evolution: Patterns, Processes, and Outcomes." *International Journal of Primatology* 33 (6): 1309–1331. doi:10.1007/s10764-011-9536-5.

Lehtonen, J., and R. Lanfear. 2014. "Generation Time, Life History and the Substitution Rate of Neutral Mutations." *Biology Letters* 10 (11): 1–4. doi:10.1098/rsbl.2014.0801.

Leigh, S. R. 2006. "Brain Ontogeny and Life History in Homo erectus." *Journal of Human Evolution* 50 (1): 104–108. doi:10.1016/j.jhevol.2005.02.008.

Leigh, S. R., and G. E. Blomquist. 2007. "Life History." In *Primates in Perspective*, edited by C. Campbell, A. Fuentes, K. MacKinnon, S. Bearder and R. Stumpf, 396–407. Oxford: Oxford University Press.

Lepage, T., D. Bryant, H. Philippe, and N. Lartillot. 2007. "A General Comparison of Relaxed Molecular Clock Models." *Molecular Biology and Evolution* 24 (12): 2669–2680. doi:10.1093/molbev/msm193.

Li, W. H., D. L. Ellsworth, J. Krushkal, B. H. J. Chang, and D. Hewett-Emmett. 1996. "Rates of Nucleotide Substitution in Primates and Rodents and the Generation Time Effect Hypothesis." *Molecular Phylogenetics and Evolution* 5 (1): 182–187. doi:10.1006/mpev.1996.0012.

Li, W. H., and M. Tanimura. 1987. "The Molecular Clock Runs More Slowly in Man Than in Apes and Monkeys." *Nature* 326 (6108): 93–96. doi:10.1038/326093a0.

Li, W. H., M. Tanimura, and P. M. Sharp. 1987. "An Evaluation of the Molecular Clock Hypothesis Using Mammalian DNA Sequences." *Journal of Molecular Evolution* 25 (4): 330–342. doi:10.1007/bf02603118.

Loewe, L., and W. G. Hill. 2010. "The Population Genetics of Mutations: Good, Bad and Indifferent." *Philosophical Transactions of the Royal Society B-Biological Sciences* 365 (1544): 1153–1167. doi:10.1098/rstb.2009.0317.

Lovejoy, C. O., A. H. Burstein, and K. G. Heiple. 1972. "Primate Phylogeny and Immunological Distance." *Science* 176 (4036): 803–805. doi:10.1126/science.176.4036.803.

MacArthur, R. H., and E. O. Wilson. 1967. *The Theory of Island Biogeography*. Princeton, NJ: Princeton University Press.

Martin, A. P., G. J. P. Naylor, and S. R. Palumbi. 1992. "Rates of Mitochondrial DNA Evolution in Sharks are Slow Compared with Mammals." *Nature* 357 (6374): 153–155. doi:10.1038/357153a0.

Martin, A. P., and S. R. Palumbi. 1993. "Body Size, Metabolic Rate, Generation Time, and the Molecular Clock." *Proceedings of the National Academy of Sciences of the United States of America* 90 (9): 4087–4091. doi:10.1073/pnas.90.9.4087.

Martin, R. D. 1996. "Scaling of the Mammalian Brain: The Maternal Energy Hypothesis." *News in Physiological Sciences* 11: 149–156. doi:10.1152/physiologyonline.1996.11.4.149.

McComb, K., and S. Semple. 2005. "Coevolution of Vocal Communication and Sociality in Primates." *Biology Letters* 1 (4): 381–385. doi:10.1098/rsbl.2005.0366.

Moorjani, P., C. E. G. Amorim, P. F. Arndt, and M. Przeworski. 2016. "Variation in the Molecular Clock of Primates." *Proceedings of the National Academy of Sciences of the United States of America* 113 (38): 10607–10612. doi:10.1073/pnas.1600374113.

Mulcahy, N. J., and J. Call. 2006. "Apes Save Tools for Future Use." *Science* 312 (5776): 1038–1040. doi:10.1126/science.1125456.

Muthukrishna, M., M. Doebeli, M. Chudek, and J. Henrich. 2018. "The Cultural Brain Hypothesis: How Culture Drives Brain Expansion, Sociality, and Life History." *Plos Computational Biology* 14 (11): n.p. doi:10.1371/journal.pcbi.1006504.

Nabholz, B., S. Glemin, and N. Galtier. 2008. "Strong Variations of Mitochondrial Mutation Rate across Mammals—The Longevity Hypothesis." *Molecular Biology and Evolution* 25 (1): 120–130. doi:10.1093/molbev/msm248.

Navarrete, A. F., S. M. Reader, S. E. Street, A. Whalen, and K. N. Laland. 2016. "The Coevolution of Innovation and Technical Intelligence in Primates." *Philosophical Transactions of the Royal Society B-Biological Sciences* 371 (1690): n.p. doi:10.1098/rstb.2015.0186.

Nettle, D. 2010. "Dying Young and Living Fast: Variation in Life History across English Neighborhoods." *Behavioral Ecology* 21 (2): 387–395. doi:10.1093/beheco/arp202.

Nielsen, R. 2005. "Molecular Signatures of Natural Selection." *Annual Review of Genetics* 39: 197–218. doi:10.1146/annurev.genet.39.073003.112420.

Nikolaev, S. I., J. I. Montoya-Burgos, K. Popadin, L. Parand, E. H. Margulies, S. E. Antonarakis, and NIHISCCS Program. 2007. "Life-History Traits Drive the Evolutionary Rates of Mammalian Coding and Noncoding Genomic Elements." *Proceedings of the National Academy of Sciences of the United States of America* 104 (51): 20443–20448. doi:10.1073/pnas.0705658104.

Odling-Smee, F. J., K. N. Laland, and M. W. Feldman. 2003. *Niche Construction: The Neglected Process in Evolution*. Princeton, NJ: Princeton University Press.

Ohta, T. 1992. "The Nearly Neutral Theory of Molecular Evolution." *Annual Review of Ecology and Systematics* 23: 263–286. doi:10.1146/annurev.es.23.110192.001403.

Ohta, T. 1993. "An Examination of the Generation-Time Effect on Molecular Evolution." *Proceedings of the National Academy of Sciences of the United States of America* 90 (22): 10676–10680. doi:10.1073/pnas.90.22.10676.

Orme, D., R. Freckleton, G. Thomas, T. Petzoldt, S. Fritz, N. Isaac, and W. Pearse. 2018. "The Caper Package: Comparative Analysis of Phylogenetics and Evolution in R." *R Packages* 5: 1–36.

Overington, S. E., J. Morand-Ferron, N. J. Boogert, and L. Lefebvre. 2009. "Technical Innovations Drive the Relationship between Innovativeness and Residual Brain Size in Birds." *Animal Behaviour* 78 (4): 1001–1010. doi:10.1016/j.anbehav.2009.06.033.

Pagel, M. 1999a. "Inferring the Historical Patterns of Biological Evolution." *Nature* 401 (6756): 877–884. doi:10.1038/44766.

Pagel, M. 1999b. "The Maximum Likelihood Approach to Reconstructing Ancestral Character States of Discrete Characters on Phylogenies." *Systematic Biology* 48 (3): 612–622. doi:10.1080/106351599260184.

Pianka, E. R. 1970. "R-Selection and K-Selection." *American Naturalist* 104 (940): 592–599. doi:10.1086/282697.

Potts, R. 1998. "Variability Selection in Hominid Evolution." *Evolutionary Anthropology* 7 (3): 81–96. doi:10.1002/(sici)1520-6505(1998)7:3<81::aid-evan3>3.0.co;2-a.

Potts, R. 2004. "Paleoenvironmental Basis of Cognitive Evolution in Great Apes." *American Journal of Primatology* 62 (3): 209–228. doi:10.1002/ajp.20016.

Potts, R. 2013. "Hominin Evolution in Settings of Strong Environmental Variability." *Quaternary Science Reviews* 73: 1–13. doi:10.1016/j.quascirev.2013.04.003.

Potts, R., and J. T. Faith. 2015. "Alternating High and Low Climate Variability: The Context of Natural Selection and Speciation in Plio-Pleistocene Hominin Evolution." *Journal of Human Evolution* 87: 5–20. doi:10.1016/j.jhevol.2015.06.014.

R Core Team. 2020. "R: A Language and Environment for Statistical Computing." *R Foundation for Statistical Computing, Vienna, Austria.* https://www.R-project.org/.

Reader, S. M. 2003. "Innovation and Social Learning: Individual Variation and Brain Evolution." *Animal Biology* 53 (2): 147–158. doi:10.1163/157075603769700340.

Reader, S. M., Y. Hager, and K. N. Laland. 2011. "The Evolution of Primate General and Cultural Intelligence." *Philosophical Transactions of the Royal Society B-Biological Sciences* 366 (1567): 1017–1027. doi:10.1098/rstb.2010.0342.

Reader, S. M., and K. N. Laland. 2002. "Social Intelligence, Innovation, and Enhanced Brain Size in Primates." *Proceedings of the National Academy of Sciences of the United States of America* 99 (7): 4436–4441. doi:10.1073/pnas.062041299.

Rendell, L., and H. Whitehead. 2001. "Culture in Whales and Dolphins." *Behavioral and Brain Sciences* 24 (2): 309–382. doi:10.1017/s0140525x0100396x.

Revell, L. J. 2012. "Phytools: An R Package for Phylogenetic Comparative Biology (and Other Things)." *Methods in Ecology and Evolution* 3 (2): 217–223. doi:https://doi.org/10.1111/j.2041-210X.2011.00169.x.

Reznick, D., M. J. Bryant, and F. Bashey. 2002. "R- and K-Selection Revisited: The Role of Population Regulation in Life-History Evolution." *Ecology* 83 (6): 1509–1520. doi:10.2307/3071970.

Robson, S. L., and B. Wood. 2008. "Hominin Life History: Reconstruction and Evolution." *Journal of Anatomy* 212 (4): 394–425. doi:10.1111/j.1469-7580.2008.00867.x.

Rowe, D. L., and R. L. Honeycutt. 2002. "Phylogenetic Relationships, Ecological Correlates, and Molecular Evolution within the Cavioidea (Mammalia, Rodentia)." *Molecular Biology and Evolution* 19 (3): 263–277. doi:10.1093/oxfordjournals.molbev.a004080.

Saclier, N., C. M. Francois, L. Konecny-Dupre, N. Lartillot, L. Gueguen, L. Duret, F. Malard, C. J. Douady, and T. Lefebure. 2018. "Life History Traits Impact the Nuclear Rate of Substitution but Not the Mitochondrial Rate in Isopods." *Molecular Biology and Evolution* 35 (12): 2900–2912. doi:10.1093/molbev/msy184.

Sarich, V. M., and A. C. Wilson. 1967a. "Immunological Time Scale for Hominid Evolution." *Science* 158 (3805): 1200–1203. doi:10.1126/science.158.3805.1200.

Sarich, V. M., and A. C. Wilson. 1967b. "Rates of Albumin Evolution in Primates." *Proceedings of the National Academy of Sciences of the United States of America* 58 (1): 142–147. doi:10.1073/pnas.58.1.142.

Sarich, V. M., and A. C. Wilson. 1973. "Generation Time and Genomic Evolution in Primates." *Science* 179 (4078): 1144–1147. doi:10.1126/science.179.4078.1144.

Scally, A., and R. Durbin. 2012. "Revising the Human Mutation Rate: Implications for Understanding Human Evolution." *Nature Reviews Genetics* 13 (10): 745–753. doi:10.1038/nrg3295.

Scerri, E. M. L., M. G. Thomas, A. Manica, P. Gunz, J. T. Stock, C. Stringer, M. Grove, et al. 2018. "Did Our Species Evolve in Subdivided Populations across Africa, and Why Does It Matter?" *Trends in Ecology & Evolution* 33 (8): 582–594. doi:10.1016/j.tree.2018.05.005.

Schoeninger, M. J., J. Moore, and J. M. Sept. 1999. "Subsistence Strategies of Two Savanna Chimpanzee Populations: The Stable Isotope Evidence." *American Journal of Primatology* 49 (4): 297–314. doi:10.1002/(sici)1098-2345(199912)49:4<297::aid-ajp2>3.0.co;2-n.

Schuppli, C., S. M. Graber, K. Isler, and C. P. van Schaik. 2016. "Life History, Cognition and the Evolution of Complex Foraging Niches." *Journal of Human Evolution* 92: 91–100. doi:10.1016/j.jhevol.2015.11.007.

Smaers, J. B., C. S. Mongle, and A. Kandler. 2016. "A Multiple Variance Brownian Motion Framework for Estimating Variable Rates and Inferring Ancestral States." *Biological Journal of the Linnean Society* 118 (1): 78–94. doi:10.1111/bij.12765.

Smith, B. H., and R. L. Tompkins. 1995. "Towards a Life-History of the Hominidae." *Annual Review of Anthropology* 24: 257–279. doi:10.1146/annurev.an.24.100195.001353.

Smith, R. J. 1996. "Biology and Body Size in Human Evolution—Statistical Inference Misapplied." *Current Anthropology* 37 (3): 451–481. doi:10.1086/204505.

Smith, T. M., P. Tafforeauc, D. J. Reid, J. Pouech, V. Lazzari, J. P. Zermeno, D. Guatelli-Steinberg, et al. 2010. "Dental Evidence for Ontogenetic Differences between Modern Humans and Neanderthals." *Proceedings of the National Academy of Sciences of the United States of America* 107 (49): 20923–20928. doi:10.1073/pnas.1010906107.

Smith, T. M., M. Toussain, D. J. Reid, A. J. Olejniczak, and J. J. Hublin. 2007. "Rapid Dental Development in a Middle Paleolithic Belgian Neanderthal." *Proceedings of the National Academy of Sciences of the United States of America* 104 (51): 20220–20225. doi:10.1073/pnas.0707051104.

Snyder, W. S., M. J. Cook, E. S. Nasset, L. R. Karhausen, G. P. Howells, and I. H. Tipton. 1975. *Report of the Task Group on Reference Man*. Oxford: Pergamon.

Sokal, R. R., and F. J. Rohlf. 1995. *Biometry*. 3rd ed. New York: W. H. Freeman.

Sol, D. 2009a. "The Cognitive-Buffer Hypothesis for the Evolution of Large Brains." In *Cognitive Ecology II*, edited by R. Dukas and R. M. Ratcliffe, 111–134. Chicago: University of Chicago Press.

Sol, D. 2009b. "Revisiting the Cognitive Buffer Hypothesis for the Evolution of Large Brains." *Biology Letters* 5 (1): 130–133. doi:10.1098/rsbl.2008.0621.

Sol, D., S. Bacher, S. M. Reader, and L. Lefebvre. 2008. "Brain Size Predicts the Success of Mammal Species Introduced into Novel Environments." *American Naturalist* 172: S63–S71. doi:10.1086/588304.

Sol, D., R. P. Duncan, T. M. Blackburn, P. Cassey, and L. Lefebvre. 2005. "Big Brains, Enhanced Cognition, and Response of Birds to Novel Environments." *Proceedings of the National Academy of Sciences of the United States of America* 102 (15): 5460–5465. doi:10.1073/pnas.0408145102.

Sol, D., and L. Lefebvre. 2000. "Behavioural Flexibility Predicts Invasion Success in Birds Introduced to New Zealand." *Oikos* 90 (3): 599–605. doi:10.1034/j.1600-0706.2000.900317.x.

Sol, D., L. Lefebvre, and J. D. Rodriguez-Teijeiro. 2005. "Brain Size, Innovative Propensity and Migratory Behaviour in Temperate Palaearctic Birds." *Proceedings of the Royal Society B-Biological Sciences* 272 (1571): 1433–1441. doi:10.1098/rspb.2005.3099.

Sol, D., D. G. Stirling, and L. Lefebvre. 2005. "Behavioral Drive or Behavioral Inhibition in Evolution: Subspecific Diversification in Holarctic Passerines." *Evolution* 59 (12): 2669–2677. doi:10.1111/j.0014-3820.2005.tb00978.x.

Sol, D., S. Timmermans, and L. Lefebvre. 2002. "Behavioural Flexibility and Invasion Success in Birds." *Animal Behaviour* 63: 495–502. doi:10.1006/anbe.2001.1953.

Steiper, M. E., and E. R. Seiffert. 2012. "Evidence for a Convergent Slowdown in Primate Molecular Rates and Its Implications for the Timing of Early Primate Evolution." *Proceedings of the National Academy of Sciences of the United States of America* 109 (16): 6006–6011. doi:10.1073/pnas.1119506109.

Steiper, M. E., N. M. Young, and T. Y. Sukarna. 2004. "Genomic Data Support the Hominoid Slowdown and an Early Oligocene Estimate for the Hominoid-Cercopithecoid Divergence." *Proceedings of the National Academy of Sciences of the United States of America* 101 (49): 17021–17026. doi:10.1073/pnas.0407270101.

Stephens, D. W. 1991. "Change, Regularity, and Value in the Evolution of Animal Learning." *Behavioral Ecology* 2 (1): 77–89. doi:10.1093/beheco/2.1.77.

Street, S. E., A. F. Navarrete, S. M. Reader, and K. N. Laland. 2017. "Coevolution of Cultural Intelligence, Extended Life History, Sociality, and Brain Size in Primates." *Proceedings of the National Academy of Sciences of the United States of America* 114 (30): 7908–7914. doi:10.1073/pnas.1620734114.

Thierry, B. 2007. "Unity in Diversity: Lessons from Macaque Societies." *Evolutionary Anthropology* 16 (6): 224–238. doi:10.1002/evan.20147.

Thomas, J. A., J. J. Welch, R. Lanfear, and L. Bromham. 2010. "A Generation Time Effect on the Rate of Molecular Evolution in Invertebrates." *Molecular Biology and Evolution* 27 (5): 1173–1180. doi:10.1093/molbev/msq009.

Thornton, A., and K. McAuliffe. 2015. "Cognitive Consequences of Cooperative Breeding? A Critical Appraisal." *Journal of Zoology* 295 (1): 12–22. doi:10.1111/jzo.12198.

Thornton, A., K. McAuliffe, S. R. X. Dall, E. Fernandez-Duque, P. A. Garber, and A. J. Young. 2016. "Fundamental Problems with the Cooperative Breeding Hypothesis: A Reply to Burkart & van Schaik." *Journal of Zoology* 299 (2): 84–88. doi:10.1111/jzo.12351.

Tsuboi, M., W. van der Bijl, B. T. Kopperud, J. Erritzoe, K. L. Voje, A. Kotrschal, K. E. Yopak, S. P. Collin, A. N. Iwaniuk, and N. Kolm. 2018. "Breakdown of Brain-Body Allometry and the Encephalization of Birds and Mammals." *Nature Ecology & Evolution* 2 (9): 1492–1500. doi:10.1038/s41559-018-0632-1.

Ungar, P. S., F. E. Grine, and M. F. Teaford. 2006. "Diet in Early Homo: A Review of the Evidence and a New Model of Adaptive Versatility." *Annual Review of Anthropology* 35: 209–228. doi:10.1146/annurev.anthro.35.081705.123153.

Vitousek, P. M., H. A. Mooney, J. Lubchenco, and J. M. Melillo. 1997. "Human Domination of Earth's Ecosystems." *Science* 277 (5325): 494–499. doi:10.1126/science.277.5325.494.

Walker, R., M. Gurven, K. Hill, H. Migliano, N. Chagnon, R. De Souza, G. Djurovic, et al. 2006. "Growth Rates and Life Histories in Twenty-Two Small-Scale Societies." *American Journal of Human Biology* 18 (3): 295–311. doi:10.1002/ajhb.20510.

Welch, J. J., O. R. P. Bininda-Emonds, and L. Bromham. 2008. "Correlates of Substitution Rate Variation in Mammalian Protein-Coding Sequences." *BMC Evolutionary Biology* 8: 53. doi:10.1186/1471-2148-8-53.

West-Eberhard, M. J. 1989. "Phenotypic Plasticity and the Origins of Diversity." *Annual Review of Ecology and Systematics* 20: 249–278. doi:10.1146/annurev.es.20.110189.001341.

Wheeler, B. C. 2009. "Monkeys Crying Wolf? Tufted Capuchin Monkeys Use Anti-predator Calls to Usurp Resources from Conspecifics." *Proceedings of the Royal Society B-Biological Sciences* 276 (1669): 3013–3018. doi:10.1098/rspb.2009.0544.

Whiten, A., and E. van de Waal. 2017. "Social Learning, Culture and the 'Socio-cultural Brain' of Human and Non-human Primates." *Neuroscience and Biobehavioral Reviews* 82: 58–75. doi:10.1016/j.neubiorev.2016.12.018.

Wilson, A. C. 1985. "The Molecular Basis of Evolution." *Scientific American* 253 (4): 164–173. doi.org/10.1038/scientificamerican1085-164.

Wilson, A. C., and V. M. Sarich. 1969. "A Molecular Time Scale for Human Evolution." *Proceedings of the National Academy of Sciences of the United States of America* 63 (4): 1088–1093. doi:10.1073/pnas.63.4.1088.

Wood, B., and E. K. Boyle. 2016. "Hominin Taxic Diversity: Fact or Fantasy?" *American Journal of Physical Anthropology* 159: 37–78. doi:10.1002/ajpa.22902.

Wood, B., and D. Strait. 2004. "Patterns of Resource Use in Early Homo and Paranthropus." *Journal of Human Evolution* 46 (2): 119–162. doi:10.1016/j.jhevol.2003.11.004.

Woolfit, M. 2009. "Effective Population Size and the Rate and Pattern of Nucleotide Substitutions." *Biology Letters* 5 (3): 417–420. doi:10.1098/rsbl.2009.0155.

Wright, J., G. H. Bolstad, Y. G. Araya-Ajoy, and N. J. Dingemanse. 2019. "Life-History Evolution under Fluctuating Density-Dependent Selection and the Adaptive Alignment of Pace-of-Life Syndromes." *Biological Reviews* 94 (1): 230–247. doi:10.1111/brv.12451.

Wu, C. I., and W. H. Li. 1985. "Evidence for Higher Rates of Nucleotide Substitution in Rodents Than in Man." *Proceedings of the National Academy of Sciences of the United States of America* 82 (6): 1741–1745. doi:10.1073/pnas.82.6.1741.

Wyles, J. S., J. G. Kunkel, and A. C. Wilson. 1983. "Birds, Behavior, and Anatomical Evolution." *Proceedings of the National Academy of Sciences of the United States of America–Biological Sciences* 80 (14): 4394–4397. doi:10.1073/pnas.80.14.4394.

Yi, S. J., D. L. Ellsworth, and W. H. Li. 2002. "Slow Molecular Clocks in Old World Monkeys, Apes, and Humans." *Molecular Biology and Evolution* 19 (12): 2191–2198. doi:10.1093/oxfordjournals.molbev.a004043.

Yi, S. V. 2013. "Morris Goodman's Hominoid Rate Slowdown: The Importance of Being Neutral." *Molecular Phylogenetics and Evolution* 66 (2): 569–574. doi:10.1016/j.ympev.2012.07.031.

Yoder, A. D., and Z. H. Yang. 2000. "Estimation of Primate Speciation Dates Using Local Molecular Clocks." *Molecular Biology and Evolution* 17 (7): 1081–1090. doi:10.1093/oxfordjournals.molbev.a026389.

Zuckerkandl, E., and L. Pauling. 1962. "Molecular Disease, Evolution and Genic Heterogeneity." In *Horizons in Biochemistry*, edited by M. Kasha and B. Pullman, 189–225. New York: Academic Press.

Zuckerkandl, E., and L. Pauling. 1965. "Evolutionary Divergence and Convergence in Proteins." In *Evolving Genes and Proteins*, edited by V. Bryon and H. J. Vogel, 97–166. New York: Academic Press.

8

A Gene-Culture Coevolutionary Perspective on Human Success

Kathryn Demps and Peter J. Richerson

1. Introduction

As we have seen in previous chapters, several lines of ancestral apes had the same shot at evolutionary success as the one that culminated in *Homo sapiens*. In the intervening millennia, our own ancestors replaced other intelligent, social, slow-growing, tool-using apes. It is likely that we will soon be the only ape species on the planet without a continued and conscious conservation effort for the remnant populations of the other extant apes. Our geographical expansions across the globe and population increases have led us into a level of success that no other species has achieved, as defined in terms of our total biomass, or our ecological dominance of most of the Earth's terrestrial and amphibious biomes (although not yet our species' longevity; see Antón, chapter 6, this volume; also see Desmond and Ramsey, chapter 1 for a discussion of other cultural measures of success). Our impact on the Earth's biogeochemistry is so large that some have proposed to erect a new geological era—the Anthropocene—to recognize this "achievement" (Crutzen 2006; Crutzen and Stoermer 2000; and see part III, this volume). We will present evidence that the development of the Anthropocene has been recent and explosive. Humans seem to have been an uncommon—even endangered—species, with human populations estimated to be as low as 10,000 to 20,000 individuals at times over the past two million years (Huff et al. 2010; Ambrose 1998). The first glimmerings of the explosion were the beginning of a long period of slow population growth and an increase in technological sophistication that began after roughly 50 Ka. Our genus *Homo* first appeared about 2 Ma and our species *H. sapiens* about 200 Ka, so we are a late-blooming success even by the standards of the historical depth of our genus and species. The questions we want to address in this paper

Kathryn Demps and Peter J. Richerson, *A Gene-Culture Coevolutionary Perspective on Human Success* In: *Human Success*. Edited by: Hugh Desmond and Grant Ramsey, Oxford University Press.
DOI: 10.1093/oso/9780190096168.003.0008

are: Why are we such an outstanding success? What explains why our success comes late in our evolutionary history? Why is our success so dangerously explosive?

2. Big Brains Plus Culture

Commentators on our success often begin with our unusually large brains for our body size (as discussed by Grove, chapter 7, this volume). There is less consensus on why large brains were adaptive for our ancestors. Many commentators speak of our "intelligence." Some have argued that our big brain endows individual humans with a powerful domain—general "improvisational" intelligence that allows us to solve novel complex problems individually (Cosmides and Tooby 2001; Pinker 2010). This would have developed along with specific cognitive models that evolved in the Pleistocene (Tooby and Cosmides 1992). Others hypothesize that big brains allow us to be a sociocultural species, spreading the cognitive effort of innovation among many, so that cultural adaptations can arise relatively quickly (compared to genetic evolution) by accumulating the innovations of others, past and present (Boyd, Richerson, and Henrich 2011; Mesoudi 2011). If so, individuals do not need to have an implausibly powerful individual intelligence to account for the complexity of our technologies, languages, and societies (Dean et al. 2014). Instead, we are shrewd plagiarists, the costs of big brains becoming bearable in a feedback loop between cumulative cultural learning, dependence on high-quality extracted foods, and extensions of the life cycle to slowly fund the capacity of our species to dominate Earth's ecosystems.

Two kinds of evidence suggest that the cultural hypothesis is the more plausible. First, our brains have been very large for about 200,000 years, but our success is rather more recent (Richerson and Boyd 2013). Second, the microevolution of artifacts like dinner forks and paper clips shows that they are the product of a rather limited inventive intelligence on the part of individuals (Petroski 1992)—our past is littered with more evidence of fine-tuning than of great leaps forward. An additional complicating factor is group size, which may have coevolved with brain size and the increasing demands of social relationships. However, this would also coincide with the ability to draw on larger populations to maintain cultural adaptations (Henrich 2004). If modern hunter-gatherer societies could ever be taken as a model for prehistoric societies, it seems as if we would have had functional societies much in excess of the predicted 150 members, based on comparative brain size with other social primates (Bird et al. 2019; also see Vermeij, chapter 5, this volume). Increasing brain size may also have coevolved with brain structure reorganization, but we are unable to evaluate this hypothesis with current data. At this point, we will set aside these unanswered questions and focus on what we can observe in the archaeological record and among modern populations: the evolution of a cumulative, complex culture.

Compared to other primates, our large brains make it possible to combine flexible individual learning with knowledge, beliefs, and skills that we can learn socially and cooperatively. This process allows individual humans to learn from the costly trial-and-error experiences of others and for populations of humans to quickly and competently respond to novel environments without waiting around for genetically predisposed behavior to catch up with new and varying circumstances. Cultures show long-term patterns of descent with modification that are the signature of an evolutionary process (Mace and Jordan 2011). Adaptations beyond the capabilities of what even the smartest of us can achieve can accumulate in populations so as to weave hard-learned lessons together with future success. So, although big brains may have been a precursor to our demographic success, we have to look to other adaptations to generate the demographic success of human expansion in recent prehistory.

3. Culture as an Adaptation

Culture is a Lamarckian system of inheritance that ties capacities for individual learning and selective social learning to an inheritance system. When the environments of parents and offspring resemble each other, when others are willing to teach, and when offspring are good selective imitators and individual learners, it can increase fitness to learn from others rather than just learning for oneself (Boyd and Richerson 1985; Rogers 1988; McElreath and Strimling 2008; Demps et al. 2012).

Because culture can evolve more rapidly than genes, we can lean on social learning to track spatial and temporal environmental variation more efficiently than by relying on genes for behavioral and psychological adaptations (Perreault 2012). Of course, if environments change slowly or if spatial variation is seldom encountered, selection on genes adapts organisms well enough without needing to rely on a costly system of cultural evolution. On the other hand, if environments vary quickly enough, acquiring ideas from others risks already being outmoded and not fit to the current environment, so individuals should rely on learning for themselves. Culture is expected to thrive in the middle ground, where the ideas of others are still useful enough to help track a changeable environment (Boyd and Richerson 2005). In general, natural selection will favor individuals adjusting their strategy of information acquisition to fit the statistics of environmental variation and the costs of various strategies for acquiring useful information (Richerson 2019).

Actual patterns of environmental variation predict that social learning should be common; this, indeed, is the case (Laland 2017). However, humans are unusual in the number of traits that are transmitted culturally and in the evolution of complex multipart cultural items that evolve cumulatively over substantial spans of time (Tennie, Call, and Tomasello 2009). Other mammalian lineages also evolved larger brains in the Pleistocene, but humans hold down the tail of the distribution of brain size relative to body size (Jerison 1973) as well as in the sophistication of our culture and our ability to meet novel environmental challenges (even when we create them ourselves!).

Human cultural adaptations fall into two large classes: technology and social organization. Technology includes both tools and the related natural history knowledge needed to make and deploy tools (Boyd, Richerson, and Henrich 2013). An example is cutting implements, such as stone tools—durable, efficient, and with a well-known evolution. Stone tools were also used to make other tools out of wood, hide, plant fibers, and the like, and this class of technology also includes material items like boats and clothing. Since these items are much less durable than stone, it is much more difficult to suss out their significance in the archaeological record. We can use tools to support an incredible variety of subsistence systems. Without needing to resort to specialized biological adaptations, we can hunt, gather, herd, domesticate, and fish resources in environments ranging from tropical forests to deserts, high plateaus to ocean floors. Humans are genetically a single species, but culturally we acclimate to different environments as if we were thousands of different locally specialized species. Even rather generalized hunting and gathering adaptations combined with the Acheulean hand axe allowed *Homo erectus* to rapidly colonize the Old World's temperate regions (Rightmire, Lordkipanidze, and Vekua 2006).

Tool-based foraging often takes advantage of resources that are difficult and risky to extract. Social organization fosters the interpersonal exchanges and risk mitigation that level dependence on these resources within social groups. Meat, one of the least regularly dependable foraged resources, also tends to be the most widely shared. Strategic kindness regarding extracted foods, supported by culturally bound social organization, may especially increase fitness for individuals in challenging ecological contexts.

Humans are likewise quite diverse as regards our social organization (Flannery and Marcus 2012; Jordan et al. 2013). The processes that give rise to complex social organizations are controversial, but one suggestion is that human cultural variation uniquely predisposes people to selection at the level of groups (Richerson et al. 2016). Darwin (1874, 178–179) alluded to this process in *The Descent of Man* as a potential type of natural selection unique to humans in order to explain, in principle, our prosocial emotions: "A tribe including many members who, from possessing in a high degree the spirit of patriotism, fidelity, obedience, courage, and sympathy, were always ready to aid one another, and to sacrifice themselves for the common good, would be victorious over most other tribes; and this would be natural selection." Even among hunters and gatherers, tool use and social organization are quite diverse, and at least some of that diversity is adaptive (Steward 1955). Strategic kindness to conspecifics may become especially relevant when culture makes possible inheritance, variation, and competition at the group level.

Why are we the only species to evolve cumulative culture? Part of the answer to our cultural heritage is our ancestry in the ape lineage. Apes have relatively large brains, are social, and have hands adapted for dexterity. Our australopithecine ancestors using bipedal locomotion allowed our hands to become specialized for intricate dexterity combined with considerable power (Kivell et al. 2016). Humans have capitalized on these inheritances to develop complex, locally adapted technologies, first, to extract resources requiring specialized tools and, second, to develop culturally transmitted cooperative social adaptations suited to different environments (Hill, Barton, and Hurtado 2009). But several species of australopiths lived for around two million years before one of them evolved into our genus, *Homo*. An early proposal argued that the evolution of free hands in the australopith lineage set off a positive evolutionary feedback between bipedality, increasing tool use, and bigger brains (Washburn 1960). This may have been linked to a move from forested to savannah landscapes (Potts 1998). However, the survival of only one of these australopith and *Homo* lineages suggests that we must look beyond the australopithecine preadaptations for a complete account of human evolution.

Language is often nominated as the critical human adaptation and is often described as a basically innate adaptation (Pinker 1994). Other authors argue that it is largely a cultural adaptation (Deutscher 2005; Everett 2012; Hurford 2011; Tomasello 2005). It plays essential roles in the transmission of technology (Dean et al. 2012; Liebenberg 1990) and social organization (Richerson and Boyd 2010). Language leaves no direct traces in the archaeological record until the invention of writing in the Holocene, so its evolutionary history is speculative. Our ancestors might have come quite far on the basis of imitation and mimetic communication. No doubt language—even written language—is essential to operate the complex social institutions of modern societies, but conceivably it could be a development of the past 50,000 years, like so much technology and language-related symbolic activity. But if culture is so adaptive, why do we see big brains long before the explosion of cumulative culture and population growth in our lineage in the past 50,000 years?

4. The Adaptive Costs of Big Brains

Brain tissue is metabolically expensive, accounting for 50% of the energy expenditure in young children and 16% in adults (Aiello and Wheeler 1995). The large heads of babies and small children require substantial resource transfers long before immatures can support themselves. Brain glucose demand peaks in childhood, which corresponds with a slowing in body-weight growth to divert resources to brain development (Kuzawa et al. 2014). Brains also take a lot of time to grow and learn in order to be useful. In contemporary foraging-horticultural societies, the age at which juveniles begin to produce more calories than they consume is about 18 years (Kaplan et al. 2000). In our species' evolutionary history, diet quality for adults and juveniles alike had to improve to provide the protein and fats required to build and maintain larger and larger human brains. Instead of investing in larger jaws and teeth to masticate low-quality abundant foods, as some ancient hominins did, our ancestors benefited from their capabilities to create knowledge, tools, and cooking to increase the type, quality, and digestibility of ancestral diets. Socially transmitted extraction techniques, coupled with intragroup food transfers, can provide the calories and macronutrients necessary for developing large brains, which can in turn support imitation and learning to extract high-quality, hard-to-get foods. Aiello and Wheeler (1995) argue that the metabolic costs of brains were traded off against similarly expensive gut tissue, exacerbating our need to exploit nutrient-dense foods. These foods require not only great skill, specialized technology, and hard work to exploit, but also a social, cooperative temperament to pool high-quality resources in order to create a dependable diet. Early on, much of the advantage of culture was dissipated in acquiring the nutrients needed to support the neurological basis of the capacity for culture.

Thus, as so often with adaptations, there are trade-offs. Any parent of small children will recognize the demands of successfully rearing a slow-maturing, large-brained, and culturally dependent social primate. Sometime in the past two million years, cooperative breeding and care by alloparents would have become imperative to successfully raising highly dependent, expensive, and slow-growing children (Burkart, Hrdy, and Van Schaik 2009; Hrdy 2009). Human children are born closer together than other apes' offspring and thus also overlap during periods of dependence and, at the same time, are less likely to die during young childhood. Humans thus have a demographic advantage relative to other apes. Under ideal conditions we can double our population size each generation, whereas the maximum rates of increase of other apes are far lower. Parents would not be able to achieve this feat without the support of partners, relatives, and friends (Hrdy 2009).

Humans also seem to have included large amounts of scavenged and hunted meat in their diets—even in the Lower Pleistocene—to support their still relatively modest brains. This would have brought our ancestors into competition with a diverse Pleistocene predator/scavenger guild focused on medium-size and large herbivores (Caro and Stoner 2003; Rodríguez-Gómez et al. 2017), another ecological cost of our large brain. Late Pleistocene *Homo neanderthalensis* seems to have been highly carnivorous and ecologically subordinate to lions and therefore rather uncommon as an apex predator that was also preyed upon (Churchill 2014). Similarly, anatomically modern humans seem to have been quite rare until about 50 Ka (Atkinson, Gray, and Drummond 2008). The high overhead costs of the big brain–with-culture adaptation seem to have led us into a fugitive niche where we were anything but an outstanding success for almost all of the Pleistocene history of our genus. Big brains may support improved powers of innovation and learning, but they have high overhead costs. Still, big brains for our body size seem to be doing a lot of work for us, but why only us? And why only so late in time?

5. A Macroevolutionary Approach

Over the past two million years, we and our hominin relatives increasingly came to depend on cultural adaptations. During this time, humans and our direct ancestors adapted through a gene-culture coevolutionary process that selected for the cognitive and behavioral abilities to cooperate with conspecifics as parts of cultural groups. Plausibly, the increasing importance of cultural adaptation in our lineage, and the increasing evolution of cognitive skills to support cultural adaptations, were driven by increases in the intermediate millennial- and submillennial-scale climate variation. Variation at these timescales would favor specifically cultural mechanisms over purely individual processes like improvisational intelligence, on the one hand, or purely genetic adaptations on the other (Richerson and Boyd 2013). The Pleistocene also led to the continuation of the long-term trend of the cooling of the poles relative to the equator, and hence more spatial environmental variability (see Antón, chapter 6, this volume). The Pleistocene exhibits a pattern of increasing millennial- and submillennial-scale variation at least over the past 400,000–800,000 years (Loulergue et al. 2008; Martrat et al. 2007). This led our successful ancestors to build upon our ancient foundations in the past couple of hundred thousand years with further increases in cultural complexity, a dependency on high-quality diets and cooperation, and brain size (McBrearty and Brooks 2000; Will, Conard, and Tyron 2019). The evolution of large brains and sophisticated culture may merely have adapted to the increase in millennial- and submillennial-scale environmental variation that favors the cultural mode of adaptation.

Humans' potential for rapid cultural evolution can make us into superpredators that drive prey to extinction (Genet et al. 2021: Model 3). For all but the recent few millennia, humans were hunter-gatherers. Hunter-gatherers are essentially predators on the plants and animals from which they derive their subsistence. Models of predator-prey dynamics tend to generate complex patterns, such as cycles, chaotic variation, and extinction (Lampert and Hastings 2016). The megafaunal extinctions that occurred whenever the technologically sophisticated hunters of the past 50,000 years Ka expanded from Africa into regions occupied by nonhuman apex predators testify to the human capacity for superpredation (Barnosky et al. 2004; Prescott et al. 2012). Early in the last glacial period in southern Africa two short-lived innovative cultures, the Still Bay and Howison's Poort, might represent episodes of superpredation followed by collapse (Jacobs et al. 2008), but the Terminal

Pleistocene extinctions were least severe in Africa and Eurasia and most severe in the Americas and Australia, which had no long history of coevolution with increasingly sophisticated hunters.

Early on, members of the genus *Homo* may have used simple social learning to achieve a wider geographical distribution than other apes, but not until complex culture appears do we achieve ecological dominance. The Dimanisi *Homo erectus* people reached the Caucasus by 1.8 Ma, not long after this species first appeared in Africa (Gabunia et al. 2000). Recent stone tool data suggest that Oldowan toolmakers reached China even earlier (Zhu et al. 2018). Their simple stone tools represent the earliest tradition of toolmaking, lacking the complexity and beauty of later tool kits. Members of the genus *Homo* did not become common until relatively recently. In western Eurasia, Neanderthals seem to have been quite rare compared to the anatomically modern humans who replaced them after about 40 Ka. How rare is rare? Estimates based on genetic variation range from the order of a thousand to the order of tens of thousands, albeit divided into small isolated subgroups (Rogers, Bohlender, and Huff 2017). Churchill (2014) suggests that Neanderthals faced heavy competition from other top carnivores, such as lions, wolves, and hyenas, and were not the dominant member of the carnivore guild in western Eurasia.

Tomasello, Kruger, and Ratner (1993) proposed that a cultural ratchet produced cumulative cultural complexity in humans. Our ability to transmit complex cultural items accurately by imitation and teaching meant that successive generations could add innovations to an existing tool or social rule, gradually building up cultural traits that were far more sophisticated than single individuals could invent on their own. Theory suggests that cultural complexity should also be a function of population size (Henrich 2004). The larger a population, the more likely complex cultural traits are to survive. And the more innovators there are, the more likely it is that improvements to existing designs will be introduced. Field data (Kline and Boyd 2010) and experimental evidence (Derex et al. 2013) support the theoretical modeling. Maintaining specialized knowledge like ethnobotany and open-ocean sailing without the benefit of the internet or written word requires a diversity of specialists. Small, isolated populations, such as the archaeological remains on the island of Tasmania indicate, cannot maintain or improve specialized cultural knowledge (Henrich 2004).

One hypothesis to consider is that until the very late Pleistocene, humans were uncommon to rare due to a combination of the high overhead of big brains and fierce competition from other carnivores. Some direct evidence suggests that humans were uncommon to rare in the Pleistocene (Bocquet-Appel et al. 2005; Li and Durbin 2011; Sánchez-Quinto and Lalueza-Fox 2015). We know of no evidence that members of the genus *Homo* were ever very common until the Holocene (the past 10,700 years). Culture allowed an increase in "niche breadth" (leading to increased geographical range), but possibly without actively moving into niches of other predators (thereby avoiding competitive exclusion dynamics). The fitness overhead imposed by our large brain, and our inability to build up populations large enough to take full advantage of culture, left a succession of *Homo* species in small evolutionary culs-de-sac.

To judge from mitochondrial DNA, anatomically modern humans began to break out of the cul-de-sac around 50 Ka as we spread around the world (Atkinson, Gray, and Drummond 2008). It is probably not coincidental that complex, cumulative cultural art and artifacts start to become prevalent around this time—possibly in a positive feedback loop with population size. It is also probably not coincidental that long, high-resolution ice and ocean cores record the most intense millennial-scale variation in climates from 50 to 11 Ka (Loulergue et al. 2008; Martrat et al. 2007). Humans at that time perhaps had a decisive advantage over other top carnivores in the form of cultural adaptations that could better track such intense high-frequency climate change (Richerson, Boyd, and Bettinger 2009). Populations began to increase, making more complex tools and more complex societies possible. Improved projectile weapons like spear throwers allowed us, for the first time, to dominate encounters with other apex predators, such as lions, from which we formerly had to flee.

6. A Holocene Cultural and Demographic Explosion

As a result of increasing population densities coupled with an intense dependence on complex culture, the end of the Pleistocene and the beginning of the Holocene marked a dramatic shift in human cultural and genetic evolution (Bar-Yosef 1998; Richerson, Boyd, and Bettinger 2001). Human subsistence systems shifted emphasis from the hunting of game to more plant-based diets. The reason is probably that plant-rich diets generally require a careful balancing of different plant species because any one species typically leads to deficiency diseases if eaten exclusively. Plants tend to be too low in protein, lacking in essential amino acids, short on vitamins and minerals, and often protected by toxins. The Mesoamerican maize-beans-squash trio is an example of a plant dietary complex that is reasonably balanced. The trouble in the Pleistocene was that rapid, high-amplitude climate variation on the timescale of centuries to millennia prevented a sufficiently stable plant assemblage from persisting long enough for balanced plant-rich diets to evolve (Richerson, Boyd, and Bettinger 2001). As the climate stabilized 11,500 years ago, experimentation with plants began, leading to plant (and animal) domestication. Still, it took a few thousand years for the first fully agricultural systems to evolve. Hunting and gathering systems without agriculture also intensified their use of plants and other tough resources, such as Arctic marine mammals. Most California Native Americans did not practice agriculture, but their use of gathered plant resources increased substantially over the Holocene (Wohlgemuth 2010). Similarly, the high Arctic adaptive sequence that today's Inuit inherited has a long Holocene trajectory of increasing sophistication (Prentiss, Walsh, and Foor 2017).

Once we acquired domesticated plants and animals beginning about 11 Ka (Richerson, Boyd, and Bettinger 2001), the stage was set for the evolution of a runaway mutualism in which improved domesticates, larger population sizes, improved farming technology, improved transport, and larger economies and societies led to huge increases in human populations. The fundamental dynamics of the agricultural system is that of a mutualism in which the exploiters invest in the populations they exploit (Efferson 2008; Genet, Davidson, Richerson, and Efferson 2021: Model 5). Growing and increasingly connected populations increased the rate of technological and social innovation (Henrich 2004; Powell, Shennan, and Thomas 2009; Richerson, Boyd, and Bettinger 2009; Shennan 2001). Population growth, presumably a Malthusian response to increasingly efficient production systems, was excruciatingly slow in the first half of the Holocene. It may have been as fast in intensifying hunter-gatherers as among agriculturalists (Zahid, Robinson, and Kelly 2016). Nevertheless, the positive feedback between population size, technical sophistication, and agricultural productivity gradually built up. Finally, in the past few centuries, the growth of human populations, and their social and technological sophistication, became explosive (Cohen 1995). Mathematical models suggest that this dramatic hockey stick–shaped trajectory just resulted from the nonlinear positive feedback from population size to the rate of innovation (Korotayev, Malkov, and Khaltourina 2006). There is no need to posit fundamental revolutions in the Holocene, like an industrial revolution, in the sense of a change in the structure or parameters of the basic dynamic model change. At least in models, the nonlinear feedback generated by population increase accelerating the rate of technological evolution and technological change accelerating population increase is all that is required.

Over the past two centuries, the most recent demographic transition has decoupled population growth from economic growth (Newson and Richerson 2009). Although birth rates have dropped even in the face of increasing resources, they still remain higher than death rates, resulting in global population increase. On top of this, growth in affluence per capita has continued to increase, impacting the Earth's resources and environments. Thomas Malthus is quaking in his boots; a transition to sustainability is a fraught project, stalled because of the complexities of managing a global common pool of resources in a world politically dominated by unruly, competitive nation-states (Brooks et al. 2018; Neumayer 2013; Pezzey and Burke 2014).

If on average the growth of human populations can be modeled as a smooth hyperexponential curve, local trajectories seem to be characterized by boom-and-bust dynamics. Such dynamics can be explained by internal sociodemographic processes (Turchin and Nefedov 2009), internal ecological setbacks (Diamond 2005), external environmental shocks like drought, or external political shocks such as the catastrophic conquest of the Americas by Europeans. Interestingly, marked boom-and-bust dynamics characterize the earliest agricultural societies of western Europe and likely other early Holocene societies (Shennan et al. 2013). The Anthropocene has generated an environmental crisis that threatens a rather dramatic biosphere-wide bust. The climatically calm Holocene did not lead to a quiet life for humans!

7. Human Success?

Zhou Enlai is reputed to have answered the question "Was the French Revolution a success?" with the quip "Too early to say." It is tempting to answer the question of human success in the same way. In the Pleistocene, humans seem to have been an uncommon, if not endangered, series of species. All the other human species are extinct, and our own species might have been reduced to a few thousand people in Africa before we began our late Paleolithic increase. Holocene humans became common, but as members of societies prone to wild booms and spectacular collapses. In the past few centuries humans have become the largest population of large mammals on Earth. We have spread to every terrestrial habitat, and we exploit the oceans for food and as the superhighways of our globalized economies. Our

Anthropocene impacts on climate and biodiversity are alarming. The harm we could do to ourselves and the rest of the biosphere in an exchange of nuclear weapons is terrifying to contemplate. Will the Anthropocene turn out to be a thin layer of marine sediments rich in anthropogenic detritus followed by a return to more normal detritus after humans again become rare or absent? Or will it mark the beginning of the human discovery of something that might actually warrant the term "civilization"? Too early to say!

References

Aiello, L. C., and P. Wheeler. 1995. "The Expensive-Tissue Hypothesis: The Brain and the Digestive System in Human and Primate Evolution." *Current Anthropology* 36 (2): 199–221.

Ambrose, S. H. 1998. "Late Pleistocene Human Population Bottlenecks, Volcanic Winter, and Differentiation of Modern Humans." *Journal of Human Evolution* 34 (6): 623–651.

Atkinson, Q. D., R. D. Gray, and A. J. Drummond. 2008. "mtDNA Variation Predicts Population Size in Humans and Reveals a Major Southern Asian Chapter in Human Prehistory." *Molecular Biology and Evolution* 25 (2): 468–474.

Barnosky, A. D., P. L. Koch, R. S. Feranec, S. L. Wing, and A. B. Shabel. 2004. "Assessing the Causes of Late Pleistocene Extinctions on the Continents." *Science* 306 (5693): 70–75.

Bar-Yosef, O. 1998. "On the Nature of Transitions: The Middle to Upper Paleolithic and the Neolithic Revolution." *Cambridge Archaeological Journal* 8 (2): 141–163.

Bird, D. W., R. B. Bird, B. F. Codding, and D. W. Zeanah. 2019. "Variability in the Organization and Size of Hunter-Gatherer Groups: Foragers Do Not Live in Small-Scale Societies." *Journal of Human Evolution* 131: 96–108.

Bocquet-Appel, J.-P., P.-Y. Demars, L. Noiret, and D. Dobrowsky. 2005. "Estimates of Upper Paleolithic Meta-population Size in Europe from Archaeological Data." *Journal of Archaeological Science* 32: 1656–1668.

Boyd, R., and P. J. Richerson. 1985. *Culture and the Evolutionary Process*. Chicago: University of Chicago Press.

Boyd, R., and P. J. Richerson. 2005. *The Origin and Evolution of Cultures*. New York: University of Oxford Press.

Boyd, R., P. J. Richerson, and J. Henrich. 2011. "The Cultural Niche: Why Social Learning Is Essential for Human Adaptation." *Proceedings of the National Academy of Sciences* 108 (Supplt 2): 10918–10925.

Boyd, R., P. J. Richerson, and J. Henrich. 2013. "The Cultural Evolution of Technology: Facts and Theories." In *Cultural Evolution: Society, Technology, Language, and Religion*, edited by P. J. Richerson and M. H. Christiansen, 119–142. Strüngmann Forum Reports, vol. 12, Julia Lupp, general editor. Cambridge, MA: MIT Press.

Brooks, J. S., T. M. Waring, M. Borgerhoff Mulder, and P. J. Richerson. 2018. "Applying Cultural Evolution to Sustainability Challenges: An Introduction to the Special Issue." *Journal of Sustainability Science* 13 (1): 1–8.

Burkart, J. M., S. B. Hrdy, and C. P. Van Schaik. 2009. "Cooperative Breeding and Human Cognitive Evolution." *Evolutionary Anthropology* 18: 175–186.

Caro, T. M., and C. J. Stoner. 2003. "The Potential for Interspecific Competition among African Carnivores." *Biological Conservation* 110 (1): 67–75.

Churchill, S. E. 2014. *Thin on the Ground: Neandertal Biology, Archeology and Ecology*. Hoboken, NJ: John Wiley and Sons.

Cohen, J. E. 1995. *How Many People Can the Earth Support?* New York: Norton.

Cosmides, L., and J. Tooby. 2001. "Unravelling the Enigma of Human Intelligence: Evolutionary Psychology and the Multimodular Mind." In *The Evolution of Intelligence*, edited by R. J. Sternberg and J. C. Kaufman, 145–199. Hillsdale, NJ: Erlbaum.

Crutzen, P. J. 2006. "The Anthropocene." In *Earth System Science in the Anthropocene*, edited by E. Ehlers and T. Krafft, 13–18. Berlin: Springer.

Crutzen, P. J., and E. F. Stoermer. 2000. "The Anthropocene." *Royal Swedish Academy of Sciences IGBP Newsletter* 41: 17–18.

Darwin, C. 1874. *The Descent of Man and Selection in Relation to Sex*. 2 vols. New York: American Home Library.

Dean, L. G., R. L. Kendal, S. J. Schapiro, B. Thierry, and K. N. Laland. 2012. "Identification of the Social and Cognitive Processes Underlying Human Cumulative Culture." *Science* 335 (6072): 1114–1118.

Dean, L. G., G. L. Vale, K. N. Laland, E. Flynn, and R. L. Kendal. 2014. "Human Cumulative Culture: A Comparative Perspective." *Journal of Biological Reviews* 89 (2): 284–301.

Demps, K., F. Zorondo-Rodríguez, C. García, and V. Reyes-García. 2012. "Social Learning across the Life Cycle: Cultural Knowledge Acquisition for Honey Collection among the Jenu Kuruba, India." *Evolution and Human Behavior* 33 (5): 460–470. doi:http://dx.doi.org/10.1016/j.evolhumbehav.2011.12.008.

Derex, M., M.-P. Beugin, B. Godelle, and M. Raymond. 2013. "Experimental Evidence for the Influence of Group Size on Cultural Complexity." *Nature* 503 (7476): 389–391.

Deutscher, G. 2005. *The Unfolding of Language: An Evolutionary Tour of Mankind's Greatest Invention*. New York: Henry Holt.

Diamond, J. 2005. *Collapse: How Societies Choose to Fail or Succeed*. New York: Viking.

Donald, M. 1991. *Origins of the Modern Mind: Three Stages in the Evolution of Culture and Cognition*. Cambridge, MA: Harvard University Press.

Efferson, C. 2008. "Prey-Producing Predators: The Ecology of Human Intensification." *Nonlinear Dynamics, Psychology, and Life Sciences* 12: 55–74.

Everett, D. L. 2012. *Language: The Cultural Tool*. New York: Pantheon.

Flannery, K., and J. Marcus. 2012. *The Creation of Inequality: How Our Prehistoric Ancestors Set the Stage for Monarchy, Slavery, and Empire*. Cambridge, MA: Harvard University Press.

Gabunia, L., A. Vekua, D. Lordkipanidze, C. C. Swisher III, R. Ferring, A. Justus, M. Nioradze, et al. 2000. "Earliest Pleistocene Hominid Cranial Remains from Dmanisi, Republic of Georgia: Taxonomy, Geological Setting, and Age. *Science* 288: 1019–1025.

Genet, R., C. Davidson, P. J. Richerson, and C. Efferson. 2021. "Dynamic Models of Human Systems." *Cultural Evolution Society*. https://learn.culturalevolutionsociety.org/.

Henrich, J. 2004. "Demography and Cultural Evolution: Why Adaptive Cultural Processes Produced Maladaptive Losses in Tasmania." *American Antiquity* 69 (2): 197–214.

Hill, K., M. Barton, and A. M. Hurtado. 2009. "The Emergence of Human Uniqueness: Characters Underlying Behavioral Modernity." *Evolutionary Anthropology* 18: 174–187.

Hrdy, S. B. 2009. *Mothers and Others: The Evolutionary Origins of Mutual Understanding*. Cambridge, MA: Harvard University Press.

Huff, C. D., J. Xing, A. R. Rogers, D. Witherspoon, and L. B. Jorde. 2010. "Mobile Elements Reveal Small Population Size in the Ancient Ancestors of Homo sapiens." *Proceedings of the National Academy of Sciences* 107 (5): 2147–2152.

Hurford, J. R. 2011. *The Origins of Grammar: Language in the Light of Evolution II*. Oxford: Oxford University Press.

Jacobs, Z., R. G. Roberts, R. F. Galbraith, H. J. Deacon, R. Grün, A. Mackay, P. Mitchell, R. Vogelsang, and L. Wadley. 2008. "Ages for the Middle Stone Age of Southern Africa: Implications for Human Behavior and Dispersal." *Science* 322: 733–735.

Jerison, H. J. 1973. *Evolution of the Brain and Intelligence*. New York: Academic Press.

Jordan, F. M., C. P. van Schaik, P. Francois, H. Gintis, D. Haun, D. J. Hruschka, M. A. Janssen, et al. 2013. "Cultural Evolution of the Structure of Human Groups." In *Cultural Evolution: Society, Technology, Language, and Religion*, edited by P. J. Richerson and M. H. Christiansen, 87–116. Strüngmann Forum Report, vol. 12, J. Lupp, general editor. Cambridge MA: MIT Press.

Kaplan, H., K. Hill, J. Lancaster, and A. M. Hurtado. 2000. "A Theory of Human Life History Evolution: Diet, Intelligence, and Longevity." *Evolutionary Anthropology* 9 (4): 156–185.

Kivell, T. L., P. Lemelin, B. G. Richmond, and D. Schmitt. 2016. *The Evolution of the Primate Hand*. Berlin: Springer.

Kline, M. A., and R. Boyd. 2010. "Population Size Predicts Technological Complexity in Oceania." *Proceedings of the Royal Society B* 277: 2559–2564.

Korotayev, A. V., A. S. Malkov, and D. A. Khaltourina. 2006. *Introduction to Social Macrodynamics: Secular Cycles and Millennial Trends*. Moscow: Russian State University for the Humanities and Russian Academy of Sciences.

Kuzawa, C. W., H. T. Chugani, L. I. Grossman, L. Lipovich, O. Muzik, P. R. Hof, D. E. Wildman, C. C. Sherwood, W. R. Leonard, and N. Lange. 2014. "Metabolic Costs and Evolutionary Implications of Human Brain Development." *Proceedings of the National Academy of Sciences* 111 (36): 13010–13015. doi:10.1073/pnas.1323099111.

Laland, K. N. 2017. *Darwin's Unfinished Symphony: How Culture Made the Human Mind*. Princeton, NJ: Princeton University Press.

Lampert, A., and A. Hastings. 2016. "Stability and Distribution of Predator-Prey Systems: Local and Regional Mechanisms and Patterns." *Ecology Letters* 19 (3): 279–288.

Li, H., and R. Durbin. 2011. "Inference of Human Population History from Individual Whole-Genome Sequences." *Nature* 475 (7357): 493–496.

Liebenberg, L. 1990. *The Art of Tracking: The Origin of Science*. Cape Town: David Phillip.

Loulergue, L., A. Schilt, R. Spahni, V. Masson-Delmotte, T. Blunier, B. Lemieux, J.-M. Barnola, D. Raynaud, T. F. Stocker, and J. Chappellaz. 2008. "Orbital and Millennial-Scale Features of Atmospheric CH4 over the Past 800,000 Years." *Nature* 453 (7193): 383–386.

Mace, R., and F. M. Jordan. 2011. "Macro-evolutionary Studies of Cultural Diversity: A Review of Empirical Studies of Cultural Transmission and Cultural Adaptation." *Philosophical Transactions of the Royal Society B* 366 (1563): 402–411.

Martrat, B., J. O. Grimalt, N. J. Shackleton, L. de Abreu, M. A. Hutterli, and T. F. Stocker. 2007. "Four Climate Cycles of Recurring Deep and Surface Water Destabilizations on the Iberian Margin." *Science* 317 (5837): 502–507.

McBrearty, S., and A. S. Brooks. 2000. "The Revolution That Wasn't: A New Interpretation of the Origin of Modern Human Behavior." *Journal of Human Evolution* 39 (5): 453–563.
McElreath, R., and P. Strimling. 2008. "When Natural Selection Favors Imitation of Parents." *Current Anthropology* 49 (2): 307–316.
Mesoudi, A. 2011. *Cultural Evolution: How Darwinian Theory Can Explain Human Culture and Synthesize the Social Sciences*. Chicago: University of Chicago Press.
Neumayer, E. 2013. *Weak versus Strong Sustainability: Exploring the Limits of Two Opposing Paradigms*. 4th ed. Cheltenham: Edward Elgar.
Newson, L., and P. J. Richerson. 2009. "Why Do People Become Modern: A Darwinian Mechanism." *Population and Development Review* 35 (1): 117–158.
Perreault, C. 2012. "The Pace of Cultural Evolution." *PLOS ONE* 7 (9): e45150.
Petroski, H. 1992. *The Evolution of Useful Things*. New York: Vintage Books.
Pezzey, J. C. V., and P. J. Burke. 2014. "Towards a More Inclusive and Precautionary Indicator of Global Sustainability." *Ecological Economics* 106 (Suppl C): 141–154.
Pinker, S. 1994. *The Language Instinct: How the Mind Creates Language*. New York: William Morrow.
Pinker, S. 2010. "The Cognitive Niche: Coevolution of Intelligence, Sociality, and Language. *Proceedings of the National Academy of Sciences* 107 (Suppl 2): 8993–8999.
Potts, R. 1998. "Environmental Hypotheses of Hominin Evolution." *American Journal of Physical Anthropology* 107 (S27): 93–136.
Powell, A., S. Shennan, and M. G. Thomas. 2009. "Late Pleistocene Demography and the Appearance of Modern Human Behavior." *Science* 324: 1298–1301.
Prentiss, A. M., M. J. Walsh, and T. A. Foor. 2017. "Evolution of Early Thule Material Culture: Cultural Transmission and Terrestrial Ecology." *Human Ecology* 46: 633–650.
Prescott, G. W., D. R. Williams, A. Balmford, R. E. Green, and A. Manica. 2012. "Quantitative Global Analysis of the Role of Climate and People in Explaining Late Quaternary Megafaunal Extinctions." *Proceedings of the National Academy of Sciences USA* 109 (12): 4527–4531.
Richerson, P. J. 2019. "An Integrated Bayesian Theory of Phenotypic Flexibility." *Behavioural Processes* 161: 54–64.
Richerson, P., R. Baldini, A. Bell, K. Demps, K. Frost, V. Hillis, S. Mathew, et al. 2016. "Cultural Group Selection Plays an Essential Role in Explaining Human Cooperation: A Sketch of the Evidence, Together with Commentaries and Authors' Response." *Behavioral and Brain Sciences* 39 (e 30): 1–68.
Richerson, P. J., and R. Boyd. 2010. "Why Possibly Language Evolved." *Biolinguistics* 4 (2–3): 289–306.
Richerson, P. J., and R. Boyd. 2013. "Rethinking Paleoanthropology: A World Queerer Than We Supposed." In *Evolution of Mind, Brain, and Culture*, edited by G. Hatfield and H. Pittman, 263–302. Philadelphia: University of Pennsylvania Museum of Archaeology and Anthropology.
Richerson, P. J., R. Boyd, and R. L. Bettinger. 2001. "Was Agriculture Impossible during the Pleistocene but Mandatory during the Holocene? A Climate Change Hypothesis." *American Antiquity* 66 (3): 387–411.
Richerson, P. J., R. Boyd, and R. L. Bettinger. 2009. "Cultural Innovations and Demographic Change." *Human Biology* 81 (2–3): 211–235.
Rightmire, G. P., D. Lordkipanidze, and A. Vekua. 2006. "Anatomical Descriptions, Comparative Studies and Evolutionary Significance of the Hominin Skulls from Dmanisi, Republic of Georgia." *Journal of Human Evolution* 50 (2): 115–141.

Rodríguez-Gómez, G., J. Rodríguez, J. A. Martín-González, and A. Mateos. 2017. "Evaluating the Impact of Homo-Carnivore Competition in European Human Settlements during the Early to Middle Pleistocene." *Quaternary Research* 88 (1): 129–151.

Rogers, A. R. 1988. "Does Biology Constrain Culture?" *American Anthropologist* 90 (4): 819–831.

Rogers, A. R., R. J. Bohlender, and C. D. Huff. 2017. "Early History of Neanderthals and Denisovans." *Proceedings of the National Academy of Sciences USA* 114 (37): 9859–9863.

Sánchez-Quinto, F., and C. Lalueza-Fox. 2015. "Almost 20 Years of Neanderthal Palaeogenetics: Adaptation, Admixture, Diversity, Demography and Extinction." *Philosophical Transactions of the Royal Society B* 370 (1660): 20130374.

Shennan, S. 2001. "Demography and Cultural Innovation: A Model and Its Implications for the Emergence of Human Culture." *Cambridge Archaeological Journal* 11 (1): 5–16.

Shennan, S., S. S. Downey, A. Timpson, K. Edinborough, S. Colledge, T. Kerig, K. Manning, and M. G. Thomas. 2013. "Regional Population Collapse Followed Initial Agriculture Booms in Mid-Holocene Europe." *Nature Communications* 4: 2486.

Steward, J. H. 1955. *Theory of Culture Change: The Methodology of Multilinear Evolution*. Urbana: University of Illinois Press.

Tennie, C., J. Call, and M. Tomasello. 2009. "Ratcheting Up the Ratchet: On the Evolution of Cumulative Culture." *Philosophical Transactions of the Royal Society B: Biological Sciences* 364 (1528): 2405–2415.

Tomasello, M. 2005. *Constructing a Language: A Usage-Based Theory of Language Acquisition*. Cambridge, MA: Harvard University Press.

Tomasello, M., A. C. Kruger, and H. H. Ratner. 1993. "Cultural Learning." *Behavioral and Brain Sciences* 16: 495–552.

Tooby, J., and L. Cosmides. 1992. "The Psychological Foundations of Culture." In *The Adapted Mind: Evolutionary Psychology and the Generation of Culture*, edited by J. Barkow, L. Cosmides, and J. Tooby, 19–136. New York: Oxford University Press.

Turchin, P., and S. A. Nefedov. 2009. *Secular Cycles*. Princeton, NJ: Princeton University Press.

Washburn, S. L. 1960. "Tools and Human Evolution." *Scientific American* 203 (3): 3–15.

Will, M., N. J. Conard, and C. A. Tyron. 2019. "Timing and Trajectory of Cultural Evolution on the African Continent 200,000–30,000 Years Ago." In *Modern Human Origins and Dispersal*, edited by Y. Sahle, H. Reyes-Centeno, and C. Bentz, 25–72. Tübingen: Kerns Verlag.

Wohlgemuth, E. 2010. "Plant Resource Structure and the Prehistory of Plant Use in Central Alta California." *California Archaeology* 2 (1): 57–76.

Zahid, H. J., E. Robinson, and R. L. Kelly. 2016. "Agriculture, Population Growth, and Statistical Analysis of the Radiocarbon Record." *Proceedings of the National Academy of Sciences* 113 (4): 931–935.

Zhu, Z., R. Dennell, W. Huang, Y. Wu, S. Qiu, S. Yang, Z. Rao, et al. 2018. "Hominin Occupation of the Chinese Loess Plateau since about 2.1 Million Years Ago." *Nature* 559: 608–612.

PART III

HUMAN SUCCESS IN THE ANTHROPOCENE

9

Anthropocene Patterns in Stratigraphy as a Perspective on Human Success

Jan Zalasiewicz, Mark Williams, and Colin Waters

1. A Biological Measure of Success

There can be little doubt about the success of *Homo sapiens*, viewed simply in terms of the numbers of our species and from the standpoint of the early 21st century by our widespread modification of the planet. Human population now stands at 7.98 billion (as we write in 2022), and for the past few years has been growing at some 80 million per year,[1] which can be thought of as adding to the world, each month, a major city—and its people—such as Rio de Janeiro or Hyderabad. Our primate cousins have been left far behind in this advance. Our closest relative, the common chimpanzee, numbers perhaps 300,000 individuals (though estimates range down to 170,000), or enough to fill Barcelona's famous Nou Camp stadium for three football matches. Of the ~500 species of primates on Earth (Estrada et al. 2017), humans make up more than 99.9% of all the individuals.

This success can be measured in other ways. Humans are approximately 10-fold the mass of wild terrestrial mammals (Smil 2011, Table 2). Most of the mass of land mammals is made up of the small number of species that our species has redesigned by selective breeding, in order that we may eat them or their products (e.g., milk) or use their labor. Standing populations of cattle, pigs, and sheep in 2019 exceeded four billion (FAOSTAT 2021). When birds are included, the numbers of individuals are quite astonishing: broiler chickens number ~23 billion (Bennett et al. 2018). Simply in terms of numbers of individuals, these chickens might therefore be thought to be more successful than *Homo sapiens*, though their brief and sorry existence challenges this assessment. (The lifespan of a broiler chicken, for instance—a form completely unable to live in the wild—averages six weeks; Bennett et al. 2018.)

Jan Zalasiewicz, Mark Williams, and Colin Waters, *Anthropocene Patterns in Stratigraphy as a Perspective on Human Success* In: *Human Success*. Edited by: Hugh Desmond and Grant Ramsey, Oxford University Press.
 DOI: 10.1093/oso/9780190096168.003.0009

These numbers are not simply the replacement of wild animals by a combination of domesticated animals and humans, which combined now have a mass more than 20-fold that of wild land mammals (Bar-on, Philips, and Milo 2018; see also Smil 2011). The zoologist and paleontologist Anthony Barnosky (2008) estimated that there has been an increase in terrestrial vertebrate biomass by something like an order of magnitude over geological background levels, the result of human landscape transformation into croplands—feedstock for those domesticated animals—amplified by a rough doubling of the surface phosphorus and nitrogen cycles, the former by extraction from rocks and the latter by fixation of atmospheric nitrogen using the Haber-Bosch process (Canfield, Glazer, and Falkowski 2010; Filippelli 2002). Humans have also, uniquely for a terrestrially evolved species, effectively become top predators in the oceanic realm. In 2018 some 84.4 million tons of fish were captured from the seas. Of these, those overfished species have steadily risen since the 1970s to one-third of assessed fisheries (FAO 2020). The proportion of the world's primary biological production that is now appropriated, directly or indirectly, to support our species' present numbers is something on the order of 25% to 40%, depending on how estimates are made (Williams et al. 2015, 2016).

This geologically unique pattern of species dominance might therefore be considered a form of success, though, as we shall see, it is by no means an unmitigated one, particularly as we look into the future. It may be analyzed in many ways. One of these is to examine the geological effects, and the geological history, of this rise to dominance, through the prism of the Anthropocene concept, which suggests that human influence on Earth has become profound enough to usher in a new geological epoch (Crutzen and Stoermer 2000; Crutzen 2002). This concept places the current situation and its recent development within the much longer context of planetary history and allows the word "success" to be explored by comparing fluxes of matter and energy between "natural" and "human" systems via the imprints they leave on the Earth's geological record.

2. The Geological Marks of Human Success

The notion that human success could have long-term consequences to our planet has been suggested more or less since geology emerged as a distinct science in the 18th century—though, until recently, such suggestions were dismissed as far-fetched by most geologists. In arguably the first science-based Earth history, the Comte de Buffon's (2018) *Les Époques de la Nature*, written in 1778, the seventh and last of the epochs that de Buffon used to describe Earth history represents the time when human actions combined with natural forces to change conditions on our planet's surface; among the phenomena that Buffon noted or intuited (with optimism) were regional climate warming through deforestation, urbanization, and a proliferation of new varieties of animals and plants through selective breeding. A century later, George Perkins Marsh, known as "America's first conservationist," and the Italian geologist/cleric Antonio Stoppani (who proposed the term "Anthropozoic") described forms of large-scale human impact, themes also variously addressed in the early 20th century by such scientists as Vladimir Vernadsky in his explorations of the biosphere and his effective founding of the discipline of biogeochemistry, and Robert Sherlock in his thorough summations of the volumes of materials moved by mining and quarrying (Steffen et al. 2011).

Though they agreed about the facts of human impact, geologists mostly considered these phenomena—however striking they may appear—as brief and transient features of an Earth that was not only known by then to be many millions of years old but that was also clearly dynamic, with (for example) large-scale changes to patterns of oceans and continents. By such a measure, human impact, it was thought, could never be more than trivial (e.g., Berry 1925).

The transience of human impact began to be more seriously questioned (though not, for the most part, by geologists) in the later part of the 20th century, when there developed a growing appreciation of the potential scale and long-term effects of anthropogenic climate change, and also of the accelerating scale of human-driven change to the biosphere. In 1992, the environmental science writer Andrew Revkin suggested the term "Anthrocene" in a book on global warming. This term did not take root, and nor did "Homogenocene," proposed at the end of the century by the biologist Michael Samways (1999) to describe a biosphere being transformed by the translocation of nonnative species. However, when the hugely influential atmospheric chemist Paul Crutzen (a Nobel laureate for his work on unraveling the mechanisms that lay behind the destruction of the ozone layer) improvised the term "Anthropocene" at a scientific meeting in Mexico in 2000, the idea fell on more fertile ground. Crutzen's intervention was a reflex reaction to his growing irritation with the use of the word "Holocene" (formally, the geological epoch that we live in) to describe a world that he saw as increasingly moving away from the conditions of a typical interglacial phase of the Quaternary (Crutzen and Stoermer 2000; Crutzen 2002), citing changes such as marked perturbations to the carbon, sulfur, and nitrogen cycles and widespread replacement of natural habitat and biotas by cropland and domesticated animals. Crutzen was a key part of the large Earth System science community, then involved in such major initiatives as the International Geosphere-Biosphere Programme (IGBP), and the term was quickly adopted by colleagues of his such as Will Steffen, then IGBP director, and spread throughout that community, becoming increasingly used in their publications (e.g., Steffen, Crutzen, and McNeill 2007).

Crutzen, in his development of the term, referred to it as a geological "epoch" and explicitly proposed that the Holocene had terminated (around the beginning of the Industrial Revolution in the late 18th century, he suggested). However, neither he nor his colleagues were scientists of the kind—stratigraphers—who specialize in the geological analysis of Earth history and who maintain the Geological Time Scale (GTS: more formally, the International Chronostratigraphic Chart, https://stratigraphy.org/timescale/). The formal process of creating, or amending, units is lengthy and necessarily bureaucratic in order to maintain nomenclatural stability, and each change must be agreed upon by a succession of committees related to the International Commission on Stratigraphy (ICS). As the term "Anthropocene" began to be used more extensively, largely by members of the Earth System science community, it had gone through none of this formal process.

When geologists began to notice the term, a key question was whether it could make *geological* sense as well as sense to scientists working on contemporary global change, and sufficiently so to pass through all of the hurdles needed for acceptance on to the GTS. What was needed here was a demonstration not only of the geohistorical significance of the term but also of the distinctiveness of its expression in geological strata—albeit very recent strata. For virtually all of our planet's history, our direct evidence comes only from the rocks that have formed over that enormous span of time. For (eventual) acceptance as a formal Anthropocene Epoch by the ICS, the demonstration of a distinct and clearly definable Anthropocene Series (the formal term for a time-restricted unit of strata, parallel with the epoch) was a necessary—but perhaps not sufficient—condition.

Initial geological consideration of the Anthropocene by a team under the auspices of the Geological Society of London suggested that the

Anthropocene possessed merit as a potential formal geological unit (Zalasiewicz et al. 2008) and led to the setting up of a working group of the Subcommission on Quaternary Stratigraphy (a component of the ICS) to examine the case further. Over several years, this group collated further evidence to demonstrate the geological reality and distinctiveness of the Anthropocene (Waters et al. 2016; Zalasiewicz, Waters, Williams, et al. 2019) in the form of what one might term sedimentary residues of human success, in expanding their numbers, in modifying the Earth to suit their perceived needs, and the resultant wastes generated as a consequence of such expansion. These residues within recent strata—termed "proxy stratigraphic signals" by geologists—include such things as the following:

- The greenhouse gases carbon dioxide and methane trapped in bubbles in layers of the ice caps of Greenland and Antarctica show sharp increases after industrialization, reflecting enormous energy output from the burning of hydrocarbons (Waters et al. 2016 and references therein).
- Changes in ratios of carbon isotopes and upturns in abundance of particles of black carbon and fly ash can be observed during the mid-20th century in other sedimentary strata, which also signal this extraordinary human energy output (Elsig et al. 2009; Rose 2015).
- Reductions in the flux of river sediments reaching the oceans reflect the construction of >58,000 large dams, more than 95% of them since the mid-20th century, for irrigation, hydroelectric generation, water supply, flood control, and recreation (Syvitski and Kettner 2011; Syvitski et al. 2020).
- Changed proportions of nitrogen isotopes in lake sediments and Arctic icesheets (the latter also showing a doubling of nitrate concentrations) are a proxy for the growth in fertilizer use (Holtgrieve et al. 2011; Wolfe et al. 2013).
- Widely distributed residues of pesticides and other industrial persistent organic pollutants (Bigus, Tobiszewski, and Namieśnik 2014).
- Artificial radionuclides mainly from fallout related to nuclear device testing (Waters et al. 2015).
- Widely, if irregularly, distributed fragments of concrete (Waters and Zalasiewicz 2018) and other artificial rocks associated with human infrastructure and novel mineral-like compounds (Hazen et al. 2017).

- Widely distributed "technofossils," many made of durable new "minerals" such as plastics (Zalasiewicz et al. 2014, 2016; Zalasiewicz, Gabbott, and Waters 2019).
- Abundant introduced nonnative species (Seebens et al. 2017), those moved both deliberately or accidentally by human activities, and leaving fossilized remains (e.g., Williams et al. 2019).

Such tangible geological markers provide a sufficient basis for a formal proposal to be made over the next few years to add the Anthropocene to the GTS. This is not to say that the Anthropocene is universally accepted as a concept or is uniquely interpreted. Critiques include the suggestions that (1) it is simply too brief—currently approximating to a single lifetime—and insubstantial materially to amount to epoch status; (2) it reflects human rather than planetary history; (3) it is based on future rather than past events; and (4) it is a politically driven rather than scientific term (e.g., Autin and Holbrook 2012; Gibbard and Walker 2014; Finney and Edwards 2016; Santana 2018). These critiques can be answered by noting that the Anthropocene may be brief but has already acquired a rich, extensive, and distinctive material (stratal) record (see above) reflecting substantial and in part irreversible change to the Earth System, as great or greater than that associated with previous epoch-scale transformations. This implies that the term may be used in a wide variety of contexts, including sociopolitical ones, but that it has a solid geological basis (Zalasiewicz et al. 2017, 2021).

Nevertheless, there is more than one "Anthropocene" used in discourse today (Zalasiewicz et al. 2021). For instance, within human-centered disciplines such as archaeology and anthropology, focus is placed upon the origin of demonstrable human impact where it appears, giving rise to an "early anthropogenic concept" that can locally extend through the entire Holocene (and even into the preceding Pleistocene; e.g., Ruddiman et al. 2015; Ellis et al. 2016; Scott 2017; Ruddiman 2018) wherever signs of significant human activity (e.g., hunting, farming) appear in the record. This is a defensible concept, akin to archaeological time terms such as "Neolithic" and "Bronze Age," which vary in duration from place to place depending on local cultural history—but it is wholly distinct from Anthropocene as considered in geological and Earth System science terms (Zalasiewicz, Waters, Head, et al. 2019) and described in detail in Zalasiewicz, Waters, Williams, and colleagues (2019) that we employ here.

Furthermore, the Anthropocene, as considered in geological/Earth System science terms, has two expressions. First, it represents a distinctive history of human activities, including unprecedented energy use, driving a wide array of physical, chemical, and biological transformations of the Earth's surface. Geologists would call this an Anthropocene Epoch, an "abstract" and intangible unit of time and history, which would have a single, carefully selected starting point. The precise starting point is still being deliberated but will be somewhere within the mid-20th-century interval and likely marked by the near-simultaneous spread of radionuclides from atmospheric H-bomb tests in the early 1950s (Waters et al. 2015). Second, the Anthropocene is a physical unit of recent sediment layers (geologists would term this unit the Anthropocene Series) which include the material, geological evidence of these processes, including traces of plastics, radionuclides, preserved remains of introduced species, and so on. It is the exact parallel of the Anthropocene Epoch and is its physical expression.

This physical expression is hugely important, as for almost all of Earth history (i.e., for all the time before written records began), it provides the *only* guide to the course and nature of that history. In the terms of this volume, it is the only way to ground-truth the scale of human "success," by allowing comparison of human-driven changes with changes that took place in prehistoric times. For instance, the rate of increase in the Anthropocene of atmospheric carbon dioxide levels since industrialization can be compared with natural increases that took place over the past 800,000 years of the Ice Ages, as preserved within the ice layers of Antarctica. The Anthropocene rise is already greater than that between typical glacial and interglacial intervals, and the rate of rise is more than 100 times more rapid (Waters et al. 2016), demonstrating modern human capacity to effect extraordinary levels of change to the atmosphere.

Notwithstanding the discussions taking place around these differing perspectives, these geological signs that we associate here with the Anthropocene might be seen as reflecting one measure of human success. As regards the mechanisms that drove them, though, it is instructive to look at the pattern of the development of these signals over deep time. These provide a strikingly clear pattern, which must lie at the heart of any attempts to explain how human success arose.

3. The Timing of Human Success

One way of looking at human success is via tangible measures of planetary impact, and perhaps the clearest measure comprises anthropogenic energy use, proxied by perturbations to atmospheric carbon dioxide levels, in relation to the known history of the human species.

Homo sapiens, as a fossil morphospecies distinct from the fossils of other hominin species (all now extinct), we now know, extends back some 300,000 years (Richter et al. 2017; Hublin et al. 2017). For some five-sixths of that time *H. sapiens*, although tool-using, showed little sign of being anything other than one of a few coexisting primate species with limited wider impact on the world (see Demps and Richerson, chapter 8, this volume).

Some 70,000–50,000 years ago there is evidence of culturally modern humans, though this development seems not to be reflected in any detectable change in fossil anatomy at that time. Evidence of more sophisticated behavioral patterns include symbolic artifacts (e.g., Henshilwood et al. 2002) and an evolving stone tool kit of the Aurignacian cultural type, but a more widely detectable geological expression of emerging human success unfolded from that point where humans evolved more effective hunting methods—reflected in their stone tools—over the next tens of millennia, which led to waves of extinctions of terrestrial megafaunal species (see Sandom et al. 2014), extinctions that continue today.

This change to the terrestrial food web might have had wider repercussions on the course of the transition from the last glacial into the present interglacial phase. This climate transition in general forms part of a pattern of metronomically regular glacial-interglacial transitions that extend back millions of years (Lisiecki and Raymo 2005). They have their basis in astronomical variations in the Earth's spin and orbit, and hence in sunlight received (Hays, Imbrie, and Shackleton 1976). The exact pattern of transition might have been influenced by factors such as human-induced extinction of mammoths contributing to a rapid increase in dwarf tree cover, in turn leading to a high-latitude warming effect (Doughty, Wolf, and Field 2010), though this remains highly speculative.

The transition from the Pleistocene into the Holocene showed a typical pattern for a glacial-interglacial transition of a rapid, if stepped, rise in atmospheric CO_2 levels from ~180 to ~260 ppm over some six millennia, a major amplifier in the warming that then took place. This was followed by a gentle, almost imperceptible decline over the next 4,000 years. This pattern is typical of interglacial phases, where the gentle decline is often followed by a steeper decline that leads into the next glacial phase. In the case of the Holocene, though, this decline slowly changed, over a millennium or so (~7000 years ago), into an equally slow and gently overall upward trend that carried CO_2 levels to ~280 ppm immediately prior to the onset of the Industrial Age.

This initial gentle rise in CO_2 levels has been attributed to the development of sufficient levels of farming to drip-feed this greenhouse gas into the air, mainly as a consequence of deforestation, with a later similarly gentle rise in methane levels starting about 5,000 years ago that has been attributed to rice cultivation in East Asia and the spread of livestock farming across Africa and Europe (Ruddiman 2003, 2013). It does not mark the start of farming (a development that began in the Late Pleistocene and became widespread from the Americas to East Asia shortly after the start of the Holocene), but it was argued that farming then attained sufficient development to affect levels of this greenhouse gas—a critical parameter of the Earth System—at a time when human population numbers had risen to a few million people. This assertion is controversial, as there is geochemical evidence that most, at least, of this slight extra input of atmospheric CO_2 came from outgassing from the oceans, while the mid-Holocene methane rise is more plausibly, if not certainly, anthropogenic (see discussion in Zalasiewicz, Waters, Head, et al. 2019 and references therein). Whether or not these slow and modest greenhouse gas changes were human-driven, the resultant slight but steady warming effect may, serendipitously, have maintained the equable climatic conditions of the Holocene by preventing the return of glaciation (Ganopolski, Winkelmann, and Schellnhuber 2016).

Similar arguments have been applied to small "wobbles" in CO_2 superimposed upon this overall Late Holocene rising trend. For instance, a roughly half-century-scale dip of ~10 ppm at ~1610 CE has been called the Orbis spike (Lewis and Maslin 2015) and has been ascribed to European colonization of the Americas, genocide of the Indigenous Americans, and consequent forest regrowth to sequester the CO_2. (It has been suggested, too, as an appropriate beginning for the Anthropocene.) More recent geochemical evidence, though, casts doubt on this suggested link between CO_2 and colonization of North America (Rubino et al. 2016).

Hence, over most of the ~11,700 years of the Holocene, the growing development of human civilizations across the world, with the rise and fall of major dynasties and sociocultural phases and developments (such as the spread of writing and then printing), led to growing and spreading environmental changes linked to agriculture and urbanization. These developments might have engendered (but, on current evidence, did not certainly engender) a slight but steady influence on atmospheric CO_2 and CH_4 levels and, perhaps through this, on climate. This overall pattern may be viewed as representing steady human success, and perhaps even sustainable human success—had it been sustained, of course.

The next developments show a clear break from that pattern. Sometime around the end of the 18th century, atmospheric CO_2 levels began to rise from the ~280 ppm that they had slowly attained. The rise accelerated from the mid-19th century, and then again from the mid-20th century, with the rise from that time on (still ongoing, and indeed accelerating) accounting for the bulk of the ~125 ppm rise in CO_2 levels and being over 100 times faster than the (geologically rapid) rise in levels around the Pleistocene-Holocene boundary. A comparable—if not even more marked—change occurred with a greater than doubling of CH_4 levels over the same time interval. The rise in CO_2 levels can certainly be attributed to the burning of fossil fuels, as this rise correlates with the enormous increase in their exploitation, and their characteristic "light" carbon isotopic signature is now pervasively present in the atmosphere and in recent geological materials, such as wood and calcium carbonate shells as well as within human skeletons and teeth (so we personally preserve the signal we created), made from this component (Waters et al. 2016). The methane has more complex origins, including from escaped natural gas and land use changes.

This is a clear break from Holocene trends, as Crutzen noted (and, indeed, from the trends of these gases throughout the last million years, at least, of the Quaternary Period). He suggested the end of the 18th century, coinciding with the beginning of the Industrial Revolution (and incidentally the approximate time the human population exceeded the one billion mark), as the beginning of the Anthropocene. In "historical" terms, such a boundary would make good sense, but for the stratigraphical boundary of a geological time unit, the key criterion is to be able to trace it, synchronously and in a practical geological sense, within strata using the best evidence possible. The mid-18th-century changes are not as clearly or sharply seen, or as globally widespread, in recent sedimentary layers as are those associated with the mid-20th century, where the spread of fly ash, rapidly changing carbon isotope patterns, microplastics, pesticide residues, and (especially sharply) artificial radionuclides from the subaerial nuclear bomb tests allow the demarcation of a clear geological boundary (Waters et al. 2016). These geological patterns can be mapped closely onto the global changes identified as representing the Great Acceleration of population, industrialization, and globalization (Steffen, Crutzen, and McNeill 2007; Steffen et al. 2015), are associated with accelerating patterns of species translocations (Seebens et al. 2017), and are the optimal candidate to represent a pragmatic geological interpretation of the Anthropocene.

It is this "acceleration of human success" (as one might term it if using the criteria of population growth and industrial/economic growth) that needs to be explained—how it happened and also why it happened at this particular time (and no other) in human history. This is not really a problem to be tackled geologically, but some geological analogs may be brought to bear on the question. There is, of course, the wider question of what this unprecedented upswing, with all its many and deep impacts on the Earth System, holds in store for the chances of "success" for future human generations.

4. Discussion

The pattern of evolution of collective human activity, when viewed through the signatures of a few key Earth System parameters and placed in a geological context, does not suggest a pattern of more or less gradual human progress, interrupted by setbacks, as in the impression that might be gained from simplified historical accounts. Rather, for all of the complexity and reality of the interplay and evolution of different political, cultural, social, and military developments preserved within recorded history, this perspective suggests a pattern of long periods of stasis, interrupted by intervals of major innovation, often arising at multiple sites but within particular timeframes, such as the beginnings of domestication of the landscape during the Early and Middle Holocene.

There is the long period of negligible planetary impact prior to cultural modernity. For more than 200,000 years *Homo sapiens* seems to have played a small role in the ecosystems it inhabited. Then came changes (cognitive and cultural) that led to more effective hunting, that set in train the pattern of rolling extinctions from some 50 millennia ago. Then, some 10 millennia ago, began the slow and complex transition to more settled agrarian communities (Scott 2017). For thousands of years after farming had begun, a baseline of activity was sustained that maintained human population at a few 10s to 100 million, saw complex and shifting patterns of social organization that included nations and empires, brought environmental changes to (mainly) the terrestrial realm, but in which a general stability of the Earth System, including climate and sea level, was maintained (Syvitski et al. 2020).

Then came an explosive rise in activity after the multiple impacts of industrialization, the Great Acceleration, and the agricultural Green Revolution. We now live in an Anthropocene world that is still evolving rapidly from decade to decade and which might be held to be the apotheosis (or perhaps apogee) of human success.

This kind of pattern might be thought to resemble a form of punctuated equilibrium at one level, and some of the large-scale trends and step changes in Earth history at another. Earth history has been characterized by such changes in the past. For example, there are the different patterns of evolution before and after the Cambrian explosion of animal life at the beginning of the Phanerozoic Eon (~539 million years ago). A hundred million years later, the land surface was slowly transformed from being largely barren to being clothed by complex terrestrial ecosystems.

Some of these step changes that restructured the biosphere show considerable delay from the origin of some evolutionary novelty to this novelty's assuming a state of importance or dominance. For instance, vertebrates have for so long held dominant positions in both marine and terrestrial food webs that it seems counterintuitive to recall that they originated early in the Cambrian Period (over 500 million years ago), and then remained rare and obscure for over 100 million years, until their spectacular rise in the late Silurian and Devonian periods (Zalasiewicz and Williams 2018).

The step-like pattern of human progress, although highly temporally compressed by comparison, nevertheless has seen greatly increased levels of planetary consequence with each step, and so this perspective may be useful.

The step change that brought in the Anthropocene after a long period of human cultural evolution, and of a near-stasis of the Earth System, is clearly the one that needs focusing on, not least because—if sustained—it may represent a planetary change. A change, that is, that reverberates throughout the entire Earth System, on a par with those major changes noted above. The change is signaled most clearly by the rise in atmospheric CO_2 levels—now at levels probably not experienced for three million years—that indicate the increasing massive energy consumption of humans and of the by-products of that consumption, greenhouse gases. This is occurring alongside a nitrogen perturbation that may be the greatest of its kind for over two billion years (Canfield, Glazer, and Falkowski 2010).

The Anthropocene step change comprised a transition from overall slow population growth to rapid population growth, an enormous increase in energy expenditure—indeed, more energy has been expended by humans in the 70 years since 1950 than throughout the entire preceding 11 millennia of the Holocene (Syvitski et al. 2020)—and hyperrapid technological and industrial evolution associated with globalization and greater social interconnectedness. This all took place apparently with no significant inherent increase in individual human capabilities, mentally or physically. Hence, other factors must have come into play—factors that had been excluded for many centuries beforehand.

Some of these factors were and are clearly technological, interlinked with behavioral changes. Thus, one well-known key element was the acquisition of the ability to increase food production through the use of the Haber-Bosch process in nitrogen fixation—a process developed in the early 20th century that, in itself, by some estimates, keeps about half the current human population alive (Erisman et al. 2008). The Haber-Bosch process has been called the "detonator" of the human "population explosion" (Smil 1999). However, many other technological developments have made this population explosion possible, one of the most prominent being the invention of antibiotics, which has prevented many deaths.

Our energy expenditure has greatly increased through the ability to locate hydrocarbon resources deep in the ground and to extract them. Access to this fossil energy (Syvitski et al. 2020) lies behind many other human developments, such as the building and maintenance of a complex urban infrastructure. In the past few decades, computers and the internet have become critical to maintaining this infrastructure—and are impacting social and organizational factors that have likewise played a key role—for instance, the kind of financial architectures (e.g., the Bretton Woods system) developed after World War II to allow economic expansion.

In reality, the technological and organizational/social factors are tightly intertwined and operate via a form of coevolution, where advances in one factor lead to cascades of changes elsewhere in the system, a phenomenon well studied by sociologists and ethicists of technology (e.g., Sandler 2014). However, as an effective way of considering these coevolutionary processes at a planetary scale, we would like to introduce the technosphere concept as proposed by Peter Haff. In brief, this concept translates the almost infinite complexity of human-driven processes into simpler physical and dynamical terms. From a geological perspective this is useful as it allows the human processes to be compared with other forms of planetary behavior.

The technosphere, in Haff's (2014, 2019) sense, may be thought of as a new "sphere" of the Earth, joining the lithosphere, hydrosphere, cryosphere, atmosphere, and biosphere. It is linked with (and is dependent on) all of these, as all the other spheres of this planet are mutually interlinked. The technosphere in Haff's sense comprises all of the industrial, technical, and transport systems manufactured by humans, including modern agricultural systems. Humans are an integral component, together with farm animals and crops (many of which have been technologically modified, and some, like the broiler chicken, to the point of not being able to exist without continuous technological support), and also the human social systems—social, educational, financial, political, military—that combine to keep this new sphere functioning. Its dynamics include certain new features not possessed by the other spheres. For instance, bulk solids transport on Earth has always been directed by gravity, as in rockfalls or river flows, or by heat-driven currents, on long timescales in continental drift, and on short timescales in wind-blown desert dunes and coastal sediments. Now bulk solids transport (of coal, minerals, bulk aggregates) takes place on Earth at fluxes that far exceed those of natural surface sedimentary processes (Cooper et al. 2018), via road, rail, ship, and airplane, along paths determined by what Haff (e.g., 2016) terms "human purpose." It is one symptom of novel planetary behavior.

Despite the operation of human purpose within it, humans here do not so much create and direct the technosphere as a whole (not least as the technosphere is emerging despite the division of humans into large numbers of competing, and frequently warring, groups) but are bound within it. The technosphere, for now at least, needs humans to keep it functioning—while, conversely, it is the continuous functioning of the technosphere that now allows very large numbers of humans to be kept alive. The beginning of the Anthropocene may be thought to approximate to the transition from a patchwork of proto-technospheres of a more or less regional pattern, into a unified and tightly integrated (despite the tribal nature of human societies) planetary system.

In this perspective, the "success" of humans has come at the price of being bound up in a very new planetary system. This system is evolving rapidly—far more rapidly than are humans as a biological species—partly by its own internal dynamics and partly through external pressures and exigencies, such as the increasing numbers of humans to be accommodated within it, the need for greater and newer material and energy resources to keep it functional, and the need to cope with emerging problems. The latter problems include large amounts of waste products (the technosphere, unlike the biosphere, is very poor at recycling), changes in climate that are a direct consequence of its functioning and that impact both its human components and the biosphere more generally, as well as other unforeseen consequences of novel subsystems (such as the impact of the internet upon sociopolitical activity).

The continuing success of humans, therefore, might be thought to depend on how well the technosphere can be steered, or nudged, into a form and function where it can maintain a long-term coexistence with the other spheres, particularly the biosphere, which is important to sustaining it. If such steering does not take place, or proves unsuccessful, then the current parasitizing of the biosphere will lead to its fatal degradation, a development that would undermine both the technosphere and its dependent human components. This is an entirely novel challenge within an entirely novel situation both for the planet and for its human population. Success in this challenge seems by no means guaranteed.

Note

1. Worldometer, http://www.worldometers.info/world-population/. Accessed October 29, 2022.

References

Autin, W. J., and J. M. Holbrook. 2012. "Is the Anthropocene an Issue of Stratigraphy or Pop Culture?" *GSA Today* 22 (7): 60–61. doi:10.1130/G153GW.1.

Barnosky, A. D. 2008. "Megafauna Biomass Tradeoff as a Driver of Quaternary and Future Extinctions." *Proceedings of the National Academy of Sciences (USA)* 105: 11543–11548.

Bar-On, Y. M., R. Philips, and R. Milo. 2018. "The Biomass Distribution on Earth." *Proceedings of the National Academy of Sciences of the United States of America* 115: 6506–6511.

Bennett, C. E., R. Thomas, M. Williams, J. Zalasiewicz, M. Edgeworth, H. Miller, B. Coles, A. Foster, E. J. Burton, and U. Marume. 2018. "The Broiler Chicken as a Signal of a Human Reconfigured Biosphere." *Royal Society Open Science* 5: 180325.

Berry, E. W. 1925. "The Term Psychozoic." *Science* 44: 16.

Bigus, P., M. Tobiszewski, and J. Namieśnik. 2014. "Historical Records of Organic Pollutants in Sediment Cores." *Marine Pollution Bulletin* 78 (1): 26–42.

Canfield, D. E., A. N. Glazer, and P. G. Falkowski. 2010. "The Evolution and Future of Earth's Nitrogen Cycle." *Science* 330: 192–196.

Cooper, A. H., T. J. Brown, S. J. Price, J. R. Ford, and C. N. Waters. 2018. "Humans Are the Most Significant Global Geomorphological Driving Force of the 21st Century." *Anthropocene Review* 5 (3): 222–229.

Crutzen, P. J. 2002. "Geology of Mankind." *Nature* 415 (January 3): 23.

Crutzen, P. J., and E. F. Stoermer. 2000. "The 'Anthropocene.'" *Global Change IGBP Newsletter* 41: 17–18.

de Buffon, G.-L. L. 2018. *The Epochs of Nature*. Edited and translated by J. Zalasiewicz, A.-S. Milon, and M. Zalasiewicz. Chicago: University of Chicago Press.

Doughty, C. E., A. Wolf, and C. B. Field. 2010. "Biophysical Feedbacks between the Pleistocene Megafauna Extinction and Climate: The First Human-Induced Global Warming?" *Geophysical Research Letters* 37 (15): L15703. doi:10.1029/2010GL043985.

Ellis, E., M. Maslin, N. Boivin, and A. Bauer. 2016. "Involve Social Scientists in Defining the Anthropocene." *Nature* 540: 192–193.

Elsig, J., J. Schmitt, D. Leuenberger, R. Schneider, M. Eyer, M. Leuenberger, F. Joos, H. Fischer, and T. F. Stocker. 2009. "Stable Isotope Constraints on Holocene Carbon Cycle Changes from an Antarctic Ice Core." *Nature* 461 (7263): 507–510.

Erisman, J. W., M. A. Sutton, J. Galloway, Z. Klimont, and W. Winiwarter. 2008. "How a Century of Ammonia Synthesis Changed the World." *Nature Geoscience* 1: 636–639.

Estrada, A., P. A. Garber, A. B. Rylands, C. Roos, E. Fernandez-Duque, A. Di Fiore, et al. 2017. "Impending Extinction Crisis of the World's Primates: Why Primates Matter." *Science Advances* 3 (1): e1600946.

FAO (Food and Agriculture Organization of the United Nations). 2018. *The State of World Fisheries and Aquaculture 2018—Meeting the Sustainable Development Goals*. FAO, Rome.

FAO (Food and Agriculture Organization of the United Nations). 2020. "The State of World Fisheries and Aquaculture 2020." http://www.fao.org/state-of-fisheries-aquaculture.

FAOSTAT (Food and Agriculture Organization of the United Nations—Statistics Division). 2021. "Data: Food and Agriculture Organization of the United Nations." http://www.fao.org/faostat/en/#data.

Filippelli, G. 2002. "The Global Phosphorus Cycle." *Reviews in Mineralogy and Geochemistry* 48: 391–425.

Finney, S. C., and L. E. Edwards. 2016. "The 'Anthropocene' Epoch: Scientific Decision or Political Statement?" *GSA Today* 26 (2–3): 4–10.

Ganopolski, A., R. Winkelmann, and H. J. Schellnhuber. 2016. "Critical Insolation–CO_2 Relation for Diagnosing Past and Future Glacial Inception." *Nature* 529: 200–203. doi:10.1038/nature16494.

Gibbard, P. L., and M. J. C. Walker. 2014. "The Term 'Anthropocene' in the Context of Formal Geological Classification." In *A Stratigraphical Basis for the Anthropocene*, edited by C. N. Waters, J. A. Zalasiewicz, M. Williams, M. A. Ellis, and A. M. Snelling, 29–37. Special Publications 395. London: Geological Society.

Haff, P. K. 2014. "Technology as a Geological Phenomenon: Implications for Human Well-Being." In *A Stratigraphical Basis for the Anthropocene*, edited by C. N. Waters, J. A. Zalasiewicz, M. Williams, M. A. Ellis, and A. M. Snelling, 301–309. Special Publications 395. London: Geological Society of London.

Haff, P. K. 2016. "Purpose in the Anthropocene: Dynamical Role and Physical Basis." *Anthropocene* 16: 54–60.

Haff, P. K. 2019. "The Technosphere and Its Physical Stratigraphic Record." In *The Anthropocene as a Geological Time Unit: A Guide to the Scientific Evidence and Current Debate*, edited by J. Zalasiewicz, C. N. Waters, M. Williams, and C. P. Summerhayes, 137–155. Cambridge: Cambridge University Press.

Hays, J. D., J. Imbrie, and N. J. Shackleton. 1976. "Variations in the Earth's Orbit: Pacemaker of the Ice Ages." *Science* 194: 1121.

Hazen, R. M., E. S. Grew, M. J. Origlieri, and R. T. Downs. 2017. "On the Mineralogy of the 'Anthropocene Epoch.'" *American Mineralogist* 102: 595–611.

Henshilwood, C. S., F. d'Errico, R. Yates, Z. Jacobs, C. Tribolo, G. A. T. Duller, N. Mercier, et al. 2002. "Emergence of Modern Human Behaviour: Middle Stone Age Engravings from South Africa." *Science* 295: 1278–1280.

Holtgrieve, G. W., D. E. Schindler, W. O. Hobbs, P. R. Leavitt, E. J. Ward, L. Buting, C. Guangjie, et al. 2011. "A Coherent Signature of Anthropogenic Nitrogen Deposition to Remote Watersheds of the Northern Hemisphere." *Science* 334 (6062): 1545–1548.

Hublin, J.-J., A. Ben-Ncer, S. E. Bailey, S. E. Freidline, S. Neubauer, M. M. Skinner, I. Bergmann, et al. 2017. "New Fossils from Jebel Irhoud, Morocco and the Pan-African Origin of *Homo sapiens*." *Nature* 546: 289–292.

Lewis, S. L., and M. A. Maslin. 2015. "Defining the Anthropocene." *Nature* 519: 171–180.

Lisecki, L., and M. E. Raymo. 2005. "A Pliocene-Pleistocene Stack of 57 Globally Distributed Benthic $\delta^{18}O$ Records." *Paleoceanography* 20: PA1003. doi:10.1029/2004PA001071.

Revkin, A. C. 1992. *Global Warming: Understanding the Forecast (American Museum of Natural History, Environmental Defense Fund)*. New York: Abbeville Press.

Richter, D., R. Grün, R. Joannes-Boyau, T. E. Steele, F. Amani, M. Rué, P. Fernandes, et al. 2017. "The Age of the Hominin Fossils from Jbel Irhoud, Morocco, and the Origins of the Middle Stone Age." *Nature* 546: 293–296.

Rose, N. L. 2015. "Spheroidal Carbonaceous Fly-Ash Particles Provide a Globally Synchronous Stratigraphic Marker for the Anthropocene." *Environmental Science & Technology* 49: 4155–4162.

Rubino, M., D. M. Etheridge, C. M. Trudinger, C. E. Allison, P. J. Rayner, I. Enting, R. Mulvaney, et al. 2016. "Low Atmospheric CO_2 Levels during the Little Ice Age Due to Cooling-Induced Terrestrial Uptake." *Nature Geoscience* 9: 691–694.

Ruddiman, W. F. 2003. "The Atmospheric Greenhouse Era Began Thousands of Years Ago." *Climatic Change* 61: 261–293.

Ruddiman, W. F. 2013. "The Anthropocene." *Annual Review of Earth and Planetary Science Letters* 41: 45–68.

Ruddiman, W. F. 2018. "Three Flaws in Defining a Formal 'Anthropocene.'" *Progress in Physical Geography: Earth and Environment* 42 (4): 451–461.

Ruddiman, W. F., E. C. Ellis, J. O. Kaplan, and D. Q. Fuller. 2015. "Defining the Epoch We Live In. *Science* 348: 38–39.

Samways, M. 1999. "Translocating Fauna to Foreign Lands: Here Comes the Homogenocene." *Journal of Insect Conservation* 3: 65–66.

Sandler, R. L., ed. 2014. *Ethics and Emerging Technologies*. London: Palgrave Macmillan UK. https://doi.org/10.1057/9781137349088.

Sandom, C., S. Faurby, B. Sandel, B., and J.-C. Svenning. 2014. "Global Late Quaternary Megafauna Extinctions Linked to Humans, Not Climate Change." *Proceedings of the Royal Society B* 281 (20133254).

Santana, C. 2018. "Waiting for the Anthropocene." *British Journal for the Philosophy of Science* 70 (4): 1073–1096. doi:10.1093/bjps/axy022.

Scott, J. C. 2017. *Against the Grain: A Deep History of the Earliest States*. New Haven, CT: Yale University Press.

Seebens, H., T. M. Blackburn, E. E. Dyer, P. Genovesi, P. E. Hulme, J. M. Jeschke, S. Pagad, et al. 2017. "No Saturation in the Accumulation of Alien Species Worldwide." *Nature Communications* 8: 14435.

Smil, V. 1999. "Detonator of the Population Explosion." *Nature* 400 (6743): 415. https://doi.org/10.1038/22672.

Smil, V. 2011. "Harvesting the Biosphere: The Human Impact." *Population and Development Review* 37 (4): 613–636.

Steffen, W., W. Broadgate, L. Deutsch, O. Gaffney, and C. Ludwig. 2015. "The Trajectory of the Anthropocene: The Great Acceleration." *Anthropocene Review* 2 (1): 81–98.

Steffen, W., P. J. Crutzen, and J. R. McNeill. 2007. "The Anthropocene: Are Humans Now Overwhelming the Great Forces of Nature?" *Ambio* 36: 614–621.

Steffen, W., J. Grinevald, P. Crutzen, and J. McNeill. 2011. "The Anthropocene: Conceptual and Historical Perspectives." *Philosophical Transactions of the Royal Society A* 369: 842–867.

Steffen, W., A. Sanderson, P. D. Tyson, J. Jäger, P. A. Matson, B. Moore III, and F. Oldfield. 2004. *Global Change and the Earth System: A Planet under Pressure*. IGBP Book Series. Berlin: Springer-Verlag.

Syvitski, J. P. M., and A. Kettner. 2011. "Sediment Flux and the Anthropocene." *Philosophical Transactions of the Royal Society A* 369 (1938): 957–975.

Syvitski, J., C. N. Waters, J. Day, J. D. Milliman, C. Summerhayes, W. Steffen, J. Zalasiewicz, et al. 2020. "Extraordinary Human Energy Consumption and Resultant Geological Impacts Beginning around 1950 CE Initiated the Proposed Anthropocene Epoch." *Communications Earth & Environment* 1: 32. doi.org/10.1038/s43247-020-00029-y.

Waters, C. N., J. P. M. Syvitski, A. Gałuszka, G. J. Hancock, J. Zalasiewicz, A. Cearreta, J. Grinevald, et al. 2015. "Can Nuclear Weapons Fallout Mark the Beginning of the Anthropocene Epoch?" *Bulletin of the Atomic Scientists* 71 (3): 46–57.

Waters, C. N., and J. Zalasiewicz. 2018. "Concrete: The Most Abundant Novel Rock Type of the Anthropocene." In *The Encyclopedia of the Anthropocene*, edited by D. A. DellaSala and M. I. Goldstein, vol. 1, 75–85. Oxford: Elsevier. https://doi.org/10.1016/B978-0-12-809665-9.09775-5.

Waters, C. N., J. Zalasiewicz, C. Summerhayes, A. D. Barnosky, C. Poirier, A. Galuszka, A. Cearreta, et al. 2016. "The Anthropocene Is Functionally and Stratigraphically Distinct from the Holocene." *Science* 351 (6269): 137.

Williams, M., J. Zalasiewicz, D. Aldridge, V. Bault, M. Head, and A. D. Barnosky. 2019. "The Biostratigraphic Signal of the Neobiota." In *The Anthropocene as a Geological Time Unit*, edited by J. Zalasiewicz, C. Waters, M. Williams, and C. Summerhayes, 119–127. Cambridge: Cambridge University Press.

Williams, M., J. Zalasiewicz, P. K. Haff, C. Schwägerl, A. D. Barnosky, and E. C. Ellis. 2015. "The Anthropocene Biosphere." *Anthropocene Review* 2: 196–219.

Williams, M., J. Zalasiewicz, C. Waters, M. Edgeworth, C. Bennett, A. D. Barnosky, E. C. Ellis, et al. 2016. "The Anthropocene: A Conspicuous Stratigraphical Signal of Anthropogenic Changes in Production and Consumption across the Biosphere." *Earth's Future* 4: 34–53.

Wolfe, A. P., W. O. Hobbs, H. H. Birks, J. P. Briner, S. Holmgren, Ó. A. Ingólfsson, S. S. Kaushal, et al. 2013. "Stratigraphic Expressions of the Holocene-Anthropocene Transition Revealed in Sediments from Remote Lakes." *Earth-Science Reviews* 116: 17–34.

Zalasiewicz, J., S. E. Gabbott, and C. N. Waters. 2019. "Plastic Waste: How Plastic Has Become Part of the Earth's Geological Cycle." In *Waste: A Handbook for Management*, edited by T. M. Letcher and D. A. Vallero, 443–452. 2nd ed. New York: Elsevier.

Zalasiewicz, J., C. N. Waters, E. C. Ellis, M. J. Head, D. Vidas, W. Steffen, J. A. Thomas, et al. 2021. "The Anthropocene: Comparing Its Meaning in Geology (Chronostratigraphy) with Conceptual Approaches Arising in Other Disciplines." *Earth's Future* 9 (3): e2020EF001782. https://doi.org/10.1029/2020EF001896.

Zalasiewicz, J., C. N. Waters, M. J. Head, C. Poirier, C. P Sumerhayes, R. Leinfelder, J. Grinevald, et al. 2019. "A Formal Anthropocene Is Compatible with but Distinct from Its Diachronous Anthropogenic Counterparts: A Response to W. F. Ruddiman's 'Three Flaws in Defining a Formal Anthropocene.'" *Progress in Physical Geography* 43: 319–333.

Zalasiewicz, J., C. N. Waters, J. A. Ivar do Sul, P. L. Corcoran, A. D. Barnosky, A. Cearreta, M. Edgeworth, et al. 2016. "The Geological Cycle of Plastics and Their Use as a Stratigraphic Indicator of the Anthropocene." *Anthropocene* 13: 4–17.

Zalasiewicz, J., C. N. Waters, M. Williams, and C. P. Summerhayes, eds. 2019. *The Anthropocene as a Geological Time Unit: A Guide to the Scientific Evidence and Current Debate*. Cambridge: Cambridge University Press.

Zalasiewicz, J., C. N. Waters, A. P. Wolfe, A. D. Barnosky, A. Cearreta, M. Edgeworth, E. C. Ellis, et al. 2017. "Making the Case for a Formal Anthropocene: An Analysis of Ongoing Critiques." *Newsletters on Stratigraphy* 50: 205–226.

Zalasiewicz, J., and M. Williams. 2018. *Skeletons: The Frame of Life*. Oxford: Oxford University Press.

Zalasiewicz, J., M. Williams, A. Smith, T. L. Barry, A. L. Coe, P. R. Bown, P. Brenchley, et al. 2008. "Are We Now Living in the Anthropocene?" *GSA Today* 18 (2): 4–8.

Zalasiewicz, J., M. Williams, C. N. Waters, A. D. Barnosky, and P. Haff. 2014. "The Technofossil Record of Humans." *Anthropocene Review* 1: 34–43.

10

Utter Success and Extensive Inequity

Assessing Processes, Patterns, and Outcomes of the Human Niche in the Anthropocene

Agustín Fuentes

1. Introduction

When thinking about the evolution of human success, and how we measure it in a contemporary context, there are a number of areas to clarify. The first is understanding what we mean by evolutionary success in general. The second is what we understand by human evolutionary success, which is an assessment with specific regard to the human lineage, that is, the genus *Homo*, particularly over the Pleistocene, when our genus originated and radiated. The third is what we understand as evolutionary success in contemporary *Homo sapiens*—humans today. There is a consensus about patterns of evolutionary processes by which we might assess the evolutionary success of a lineage or species. However, evolutionary processes generate continuities and discontinuities in the genomes, morphologies, behaviors, and overall trajectories of organisms. Thus, one might expect that in any given lineage the specific series of discontinuities that the lineage experiences relative to other related lineages (which is, in fact, what defines that lineage) might create contexts and patterns under which the specifics of how one might assess evolutionary success in that lineage differs from that of other lineages.

For example, in the Caniforma (a suborder of the order Carnivora in Mammalia) the Canidae lineage (dogs writ large) and the Pinnipedia lineage (seals writ large) are closely related groups sharing many evolutionary continuities. However, they are also characterized by significant discontinuities, such as aquatic versus terrestrial adaptations that create very different ecological parameters for the measuring of "success" in regard to morphological, physiological, and behavioral processes and characteristics. This is not to say that the actual evolutionary processes and basic pressures of

Agustín Fuentes, *Utter Success and Extensive Inequity* In: *Human Success*. Edited by: Hugh Desmond and Grant Ramsey, Oxford University Press. © Oxford University Press 2023. DOI: 10.1093/oso/9780190096168.003.0010 reproduction, energetic balance, and avoiding predation/fatality are unique to a given lineage—they are not. Such core processes and challenges are shared by all organisms. However, the patterns of responses to pressures and how the evolutionary processes play out in a given lineage can be distinctive because of the particularities of the trajectory, ecology, and specific phylogenetic history of that lineage. Thus, one might have to modify assessments or develop lineage-specific determinants or descriptions of aspects of evolutionary success in a given lineage relative to more general assumptions/measures of such processes.

2. Evolutionary Success across Theoretical Frameworks

It is clear that the standard notion of evolutionary success refers to some form of demographic and genomic continuity across time of a given lineage, species, or population. This is seen when the lineage, species, or population is (a) not extinct, (b) has sufficient reproductive population size to maintain some level of genomic diversity and/or physiological or behavioral plasticity sufficient to offer variation that can potentially respond to existing pressures, and (c) is not dramatically decreasing in range or overall demographic size. Since we tend to think of evolutionary processes as driving cross-generational change or constancy in a given environment, then a general measure of success would be a population's abilities or capacities to thrive over time—that is, to maintain numeric consistency and/or expand its numeric representation and other demographic variables, relative to a given environment or environments.

In standard evolutionary theory (SET) (also termed "neo-Darwinism"; e.g., Wilson 1975; Futuyma 2017; Uller and Laland 2019) something that we could term "population success" is measured as either maintenance at certain numeric population levels or expansion and diversification into more types of environments by a population (or a number of populations of the same species). However, it is important to note that in the SET this populational success is seen as emerging from the success of specific individuals with the better "fit" variants (an individual's success at passing along its genomic content within the population). With selection at the individual level within a population structure, the better fit variants held by the more successful individuals become pervasive in that population (and eventually across the species of interest). The lineage or the species that is capable of effectively maintaining or expanding population levels in given environments due to the aggregate success of better-fit individuals in its population(s) can be described as evolutionarily successful. At greater than 7.8 billion *Homo sapiens* currently living across the entire planet, one can, in the simplest of SET terms, describe humans as evolutionarily successful.

Such a traditional assessment is tied to fitness. But how "fitness" is defined and measured is not always clear or consistent across different modes of assessment (see Uller and Laland 2019). There is a range of uses of this term, but in most cases, when speaking about lineages, "fitness" is used as a shorthand for the best suite of characteristics (traits) to survive and reproductively succeed in a given set of environmental pressures. Fitness, then, is either that constellation of traits within a population or a specific trait in that population that leads those individuals with that constellation or trait, and its underlying, heritable genetic component, to contribute more offspring to subsequent generations, on average, than other individuals with similar but not identical traits.[1] This emerges from the most basic conceptualization of natural selection: there is variation in a given trait, and certain variants do better by providing individuals with those traits a higher likelihood, on average, of successful reproduction in a given environmental context (or across multiple such contexts).

This is the standard measure for success: traits (and the individuals who have them) that do better in a given environment offer on average a higher likelihood of successfully navigating environmental challenges in that given environment. Fitness in this context must have an underlying genetic basis, and thus be biologically heritable. Traits that are more fit will then facilitate the passing of a particular genotype from generation to generation, such that over time that genotype will be overrepresented in subsequent populations. This is success in SET: adaptation via natural selection shaping populations across time. Those lineages that are successful have acquired the adaptions necessary for success, while those that do not eventually go extinct.

In contrast to the SET, there is an expanded framework, the extended evolutionary synthesis (EES; Laland et al. 2015), in which evolutionary processes related to potential success also include multiple modes of inheritance, not just genetic inheritance. These are epigenetic inheritance, behavioral inheritance, and (particularly in humans) symbolic or cultural inheritance. In the EES, the ways in which success in the traditional fitness sense may be measured or expanded is increased/diversified because of these multiple modes of inheritance (Laland et al. 2015; Uller and Laland 2019). Classic genotype fitness due to the action of selection is not the only way to assess a population's capacity or trajectory in regard to maintaining or expanding its numbers across time. Thus, in the EES, it is not just the maintenance or increase in the frequency of a specific genotype that is necessarily the main or even primary factor influencing a population's demographic and cross-generational success. We might see the passing on of epigenomic patterns, of particular behavioral patterns, or of particular cultural or symbolic patterns, without specific genetic underpinnings, that also facilitate a population's ability to maintain itself in a given environment or to even expand in that environment or into additional environments (see also chapters by Rosslenbroich, Grove, and Demps and Richerson in this volume).

The critical difference between SET and the EES is not the ultimate measure of success, that is, the ability of a population to maintain itself or to expand in a given environment or across environments. Rather, a core difference between SET and the EES lies in the ways in which the critical evolutionary units that facilitate a population doing well, maintaining itself, or expanding across time in a given environment are transmitted (Laland et al. 2015; Uller and Laland 2019; Baedke, Fábregas-Tejeda, and Vergara-Silva 2020). Basically, the difference is in what those units facilitating success *are*. This is aptly framed by Kevin Laland and colleagues (2014, 162) when they state that "organisms are constructed in development, not simply 'programmed' to develop by genes. Living things do not evolve to fit into pre-existing environments, but co-construct and coevolve with their environments, in the process changing the structure of ecosystems."

In SET, the critical units are ultimately genetic (the genetic basis underlying more fit variants). While the interface with pressures (selection) is primarily occurring with the variants themselves (phenotypes), it's the underlying genetic representation (genotype) that is the key to adaptive fitness, and thus evolutionary success. In the EES, the phenotypes that lead to success in a given population may be influenced via the inheritance of particular genetic components (DNA sequences) but may also include the inheritance of epigenetic elements (non-DNA cellular components or processes that alter DNA sequence expression/function). Additionally, evolutionarily relevant traits may pass from generation to generation via nongenetically triggered means, such as behavioral transmission (learning, teaching, ratcheting), via inheritance of modified ecologies, or via cultural or symbolic inheritance (e.g., Whitehead et al. 2019; Demps and Richerson, this volume)—or, most likely, some combination of all of these. Thus, the levels and structures of the processes of relevance to evolutionary success are more varied in the EES than in the SET.

3. Success in the Genus *Homo*

Over the past ~7 million years there have been at least seven genera of hominins, but for the past million years or so the only extant representatives of the hominin lineage are members of the genus *Homo*. In the past at least 30,000–40,000 years, only one variety of *Homo*, *Homo sapiens*—contemporary humans—constitutes the entirety of living representatives of the genus. Across most of the Pleistocene, there were multiple populations of genus *Homo*, but whether they were different subspecies, different species, or just different populations is unclear at this point. Given the lack of clarity or resolution, I will not engage the debate about species distinctions in later Pleistocene *Homo* (see Ackermann, Mackay, and Arnold 2015; Antón 2013 and this volume; Kissel and Fuentes 2018, 2021; Scerri et al. 2018; Scerri, Chikhi, and Thomas 2019; Wood and Boyle 2016). Rather, I will offer an overview of the *Homo* lineage in the Pleistocene and refer to contemporary humans only at the species level (e.g., *Homo sapiens*).

We have a relatively robust understanding of the key patterns and processes of human evolution across the Pleistocene (see Figure 10.1). Starting two to three million years ago the human niche developed via heightened abilities for cooperation and collaboration, resulting in avoiding predators, making and sharing stone tools, widening dietary options and enhancing nutritional return, cooperatively caring for and raising young, and the initial stages of reshaping local ecologies (Antón, Potts, and Aiello 2014; Foley 2016; Fuentes 2017a, 2018; van Schaik 2016; see also Antón, Grove, and Potts chapters in this volume). Well before one million years ago, populations of our genus, *Homo*, were spread across much of the Afro-Eurasian land masses, fanning out and exploring new landscapes, developing new approaches to deal with the challenges of the diverse environments encountered. Throughout this process they became increasingly capable of drawing on a range of cognitive and social resources, on their histories and experiences, and combining them with an emerging imagination to think beyond the "here and now" and to develop mental representations in order to see and feel and know about the world and their lives (Fuentes 2017a, 2019; Kissel and Fuentes 2018; Grove, this volume).

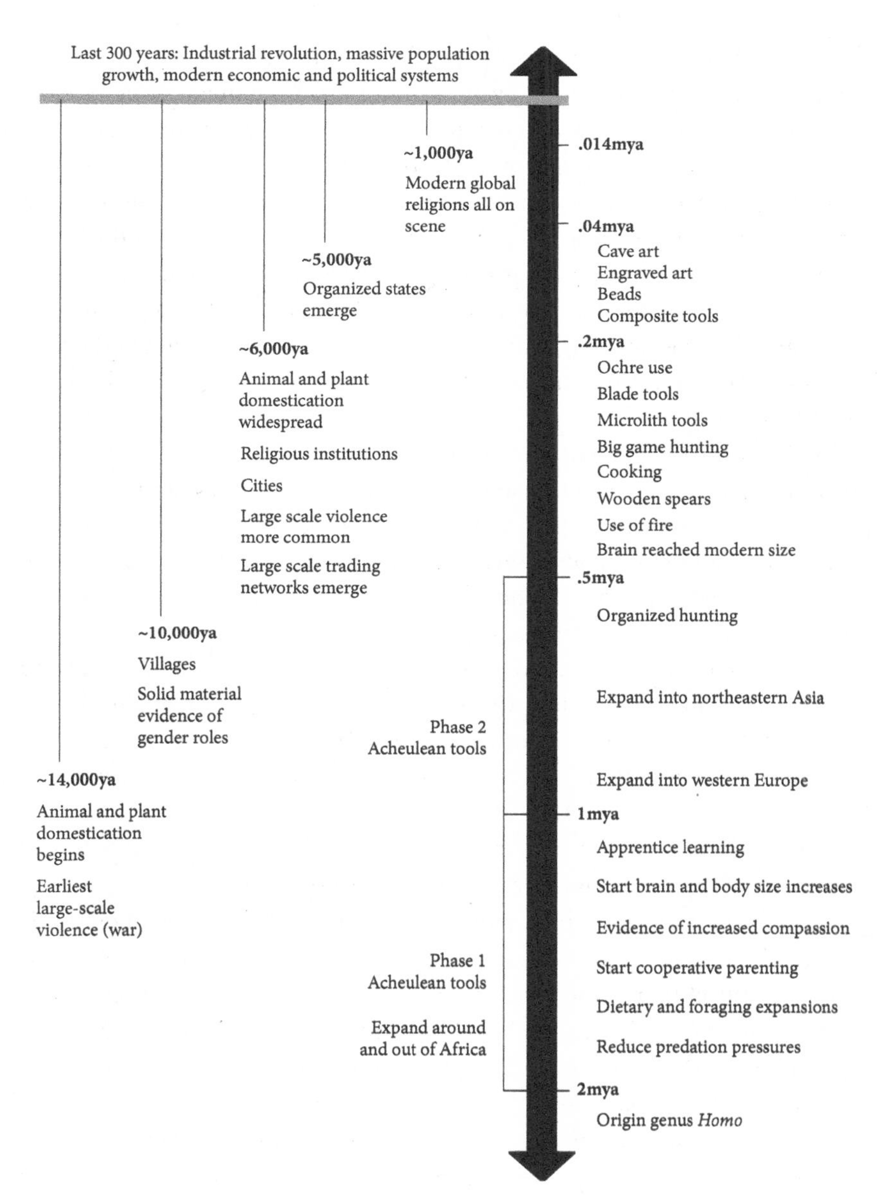

Figure 10.1 These are key events in Pleistocene human evolution likely related to aspects of human success. "Ya" refers to "years ago" and "mya" refers to "million of years ago."

Source: Reprinted from Fuentes 2018 with permission of University of Chicago Press.

By 400,000 years ago, populations of *Homo* innovated with fire to alter food, tools, and the day-night cycle. Around this time the hunting of large game became more common across populations, as did more and more complex stone, bone, and wooden tool kits. By ~120,000–200,000 years ago, our ancestors created composite materials from plants, animals, and minerals in increasingly complex new forms (such as glues and pigments) and painted themselves with ochres, made jewelry from shells, and moved, successfully, into the more northern landscapes. By at least 65,000 years ago *Homo* met and surpassed the challenges of crossing expanses of open water and developed a habit of creating images and using color on cave walls to tell stories about the world.

In the past ~15,000 years, the magnitude, diversity and impact of humans' relationships with an increasing array of other species and ecologies transformed bodies, societies, and the global ecosystem at a pace outstripping the accumulation of capacities across the previous two million years. It is in this period that more complex human communities, economies, polities, and belief systems emerge. Domestication, sedentism, storage, and new forms of societies facilitate the appearance and the rise of practices of property and identity that, in combination with expanding patterns of inequality, create novel landscapes in the human experience. These landscapes set the stage for what some now term "the Anthropocene"—the contemporary niche in which we exist (Fuentes 2017a, 2017b).

Today, *Homo sapiens* is incredibly widespread for a large-bodied mammal and extremely numerically successful. There are more than 7.8 billion members of our species, likely the largest number for a mammalian species of our body size (and at the very top of the distribution for mammals of any size). Those numbers reflect radical increases in population size in the past few centuries. That is, humanity hit one billion individuals—humans on the planet—approximately 200 years ago. Subsequently, our global population has grown more than 700%. That is an unprecedented growth trajectory for a long-lived, large-bodied mammal (or any vertebrate except those coevolving with humanity). This most recent population explosion follows on the heels of another radical expansion, probably in the past 5,000–10,000 years, from approximately a few hundred million humans globally to the one billion mark by the 1800s.

From a simplistic evolutionary assessment of whether humans are numerically robust and growing, there is no doubt that we are successful. However, in understanding the success of the genus *Homo* in a more complex and more accurate evolutionary sense, there are a couple of elements that we have to consider. First, we have good evidence that in the terminal Pliocene and early Pleistocene, there were at least two—if not three—genera of hominins (*Australopithecus*, *Homo*, and *Paranthropus*; Wood and Boyle 2016). But by one million years ago all non-*Homo* hominin lineages were extinct. *Homo*, then, is also highly successful relative to the other hominin lineages. And we are more generally successful than other large-bodied mammalian lineages. We have more than two million years of lineage history, and at many points in that period there were multiple, diverse populations or subspecies, or even species, in the genus, and eventually a global geographic distribution. We also know that, throughout the Pleistocene, there was much variation in regional, geographic, and temporal populations in the genus *Homo* (see Antón, this volume). Some scholars tend to lump many of the Pleistocene *Homo* into a general *Homo erectus* category, but others are more likely to divide the various temporal and morphological groupings into multiple *Homo* lineages that are different at the species or subspecies levels (see Antón 2013 and this volume; Wood and Boyle 2016; Scerri et al. 2018). Regardless of whether we call these species or populations or subspecies, it is clear that the genus *Homo* begins to radiate geographically and to diversify morphologically, behaviorally, and technologically across the Pleistocene (Kissel and Fuentes 2021). This seems to reflect a capacity to effectively engage very diverse environmental challenges, a pattern often referred to as "successful" in a basic evolutionary sense.

From such a perspective, one could say that the genus *Homo* is incredibly successful as a migratory species that expands and is able to live successfully in a variety of different ecologies. In fact, the genus *Homo* is probably one of, if not *the*, most diverse habitat-type occupiers in the order Primates—and likely has had success in more diverse ecologies than any mammal (except those we've brought along for the ride). This, too, is probably important as a measure of success.

The genus *Homo* itself is incredibly successful in the sense of populational diversity (that is, multiple populations exhibiting a range of phenotypic and behavioral diversity; Ackermann, Mackay, and Arnold 2015; Antón, Potts, and Aiello 2014; Kissel and Fuentes 2018; Wood and Boyle 2016). While most individual populations of the genus *Homo* across the Pleistocene went extinct, enough populations in diverse habitats survived, and we—all 7.8 billion of us—are their descendants.

By the terminal Pleistocene/early Holocene (circa 10,000 years ago), *Homo sapiens*, the only surviving member of the hominin lineage, had moved successfully into the vast majority of niches on the planet and to all of the continental areas except for Antarctica. Such an accomplishment is impressive, making us, in a traditional sense, the most successful of any large-bodied mammal and of any primate.

4. Making "Evolutionary Success" in Humans a Bit More Complicated

A closer look at how we might expand or alter the definitions of evolutionary success in humans might be of relevance in the 21st century. Under the basic assessment model (e.g., SET), we can say we are an incredibly successful species. We can note this success at the demographic level (overall species population) and at the environmental adaptation level (currently successfully existing and reproducing in a majority of terrestrial environments on the planet). In a general evolutionary sense one would say that humans are very high in average fitness as measured by reproduction (for example, we have a positive overall species growth rate for the past 10,000 years; Roser, Ritchie, and Ortiz-Ospina 2019).

However, there are now over 7.8 billion human beings on the planet. When we are theorizing about human evolutionary success, should we be looking *only* to these basic measures of population maintenance and/or expansion? Or, in the contemporary landscape of what some term the Anthropocene, might we need to develop new measures of evolutionary success given the distinctive nature and history of our lineage and our species (see the Hourdequin, Buchanan and Powell, and Desmond chapters in this volume)? Have we created such a distinctive niche on this planet that true assessment of our evolutionary success should be seen through the context of the human niche and not just via a generalized comparison with other forms of life, such as barnacles or salmon or chickens?

In fact, chickens offer us a good reason for thinking that pure demography might not be the best indicator of evolutionary success in the Anthropocene, neither for humans nor for many others. There have been approximately 20 billion domestic chickens (*Gallus gallus domesticus*) on the planet at any given moment since 2009 (and because of our consumption patterns as many as three times that number will be born and eaten over the course of the next few years). Chickens, with their billions-strong representation on this planet, by traditional measures are incredibly evolutionarily successful.

But can we actually say they've achieved any evolutionary success as a species given that, without massive anthropogenic sustenance and structuring, they, in their contemporary biological forms, could not maintain their current demographic patterns? In fact, Bennett et al. (2018) recently argued that the chicken can act as a robust signal of a human reconfigured biosphere, noting that "chickens, now unable to survive without human intervention, have a combined mass exceeding that of all other birds on Earth; this novel morphotype symbolizes the unprecedented human reconfiguration of the Earth's biosphere." Chickens did not evolve this rate of reproductive success (or excess) on their own.

Humans have created incredible numeric success for themselves and for other species (and cataclysmic decreases for yet others), and such patterns appear to be inextricably tied to anthropogenic and anthropocentric processes—outcomes of the human niche (Estrada et al. 2017; Ripple et al. 2017). Even though chickens are more widespread than they have been at any point in their evolutionary histories, can we say they're evolutionarily successful? The fact that this is a human-caused alteration resulting in more chickens than humans and is tied to both human and chicken demographic "success" (and to radical, and possibly catastrophic, habitat alterations globally) suggests that simply assessing the numeric representation of a given species may not be a sufficient measure of evolutionary success in the Anthropocene.

This brings us back to humans. What measures should we be using? I suggest that if for some other species pure demographic success no longer reflects a basic evolutionary success measurement because that success is wholly contingent upon human intervention or human structuring, then we could argue that evolutionary success in the Anthropocene might need to be measured against, or with, a certain set of environmental, historical, or structural contexts.

If overall population numbers are not sufficient, what is? Coming back to the context of humans, might we say that there is a difference between measuring human evolutionary success across the Pleistocene and human evolutionary success in the Anthropocene? How do we move beyond traditional notions of success based on populational representation, maintenance, and heritability of the genetic basis for traits and processes leading to such success? It makes sense to explore a novel understanding of what it means to be evolutionarily successful in the Anthropocene, especially if such a measure involves more than just populations maintaining themselves or expanding.

5. What Might Success in the Anthropocene Look Like?

In asking questions about evolutionary success in the Anthropocene, it becomes important that we remember the potential differences between SET and the EES. Under SET, measurement of success is ultimately associated with genotypic success, that is, the fitness of particular genotypes or portions of genotypes. Success is measured in terms of the relative proliferation of units of selection. Under the EES, we can recognize that modes of success or fitness at the organismal or populational levels are not exclusively tied to genotypic patterns. That is, there is one type of ultimate unit of selection in SET (allele), and many more in the EES (allele, organism, cultural variant). Thus, epigenomic, behavioral, and cultural or symbolic processes may also facilitate evolutionary success as measured by the maintenance of viable demographic patterns, or a population's capacities to deal with challenges across one or more environments (Demps and Richerson, this volume).

If that is the case, then we can consider including assessments of these other targets of selection, or even evolutionary processes not directly relying on selection, in our measures of success (see Uller and Laland 2019; Baedke, Fábregas-Tejeda, and Vergara-Silva 2020). Given this, in the contemporary context of the Anthropocene, we then might want to ask *What is characterizing human populations, and how do populations vary internally and interpopulationally in the facets of evolutionary relevance beyond the genotype?* That is, what do the epigenomic, the behavioral, and the cultural or symbolic landscapes look like, and are they impacting human populations in evolutionarily significant manners?

If it is the case that certain structuring components of human societies, and the populations that make them up, create highly differential processes of transmission of inheritance of particular cultural/symbolic, behavioral, or epigenomic patterns, and those patterns have differential impact on the individuals or subsections of populations (e.g., Krieger 1994; Kuijper et al. 2019; Lock 2017), then we might want to alter our notions or our measurements of success to include such patterns. In doing so we can move beyond demographic maintenance or expansion (based on the cumulative fitness of individuals in those populations) as the *only* indicator of evolutionary success and think about the ways in which populations themselves are structured and if that impacts the evolutionarily relevant processes in those populations and on the individuals within them. In this context, I suggest that it might be worth assessing elements such as health, morbidity, and development across the life stages, in addition to quality and frequency of access to those aspects in societies (social ecologies) that enable/structure outcomes in each of these areas. In short, I think we should ask: *Does the structuring of a society impact epigenetic, behavioral, and cultural patterns such that the capacities for success are limited for subsets of individuals in the population or society?*

If populational structuring is radically unequal, that is, if populations are maintaining themselves numerically (or even growing) overall, but the quality of what is being transmitted in regard to overall health and well-being is highly unequal, then we might want to query the standard model of maintenance or numeric expansion, and the population level, as the sole measure of success. Rather than just using mean overall reproductive output as our primary measure of evolutionary fitness, what if we also include assessment of the relative capacity for individuals within a population to achieve developmental and social performance at levels concomitant with their biological potential?

If the passing on of epigenomic, behavioral, and cultural/symbolic inheritance can influence evolutionarily relevant outcomes, we can attempt to measure microscale or local evolutionary success with regard to the capacity for achieving physiological and psychological functioning in a given environment by individuals. Rather than just taking the traditional mode of success, the passing on of the genotype, we can also include the epigenome or particular behavioral or cultural or even perceptual (symbolic) characteristics that influence development, health, and morbidity and assess how successfully passing those from generation to generation impacts populations.

For example, let's say within a given population we have a range of capacities in both genomic and epigenomic patterns for physiological functioning and overall health. For a majority of individuals in the population, their variant genetic sequences offer a potentially high level of physiological functioning that facilitates the possibility of successful reproductive output. However, the epigenomic range includes a suite of components that deleteriously affect developmental processes and morbidity. And these deleterious elements are facilitated/augmented via behavioral and cultural structures and patterns structuring the social ecology of that population. These might not be severe enough to significantly impact overall average reproductive output, but they do affect systemic functioning across segments of the population and impinge on their ability to achieve phenotypic results concomitant with underlying capacities (e.g., Krieger 2020). If, because of differential patterns in the structuring of societies, there are substantive differences in the ability to navigate challenges in a given population, resulting in differential success of individuals across generations in achieving physiological and other forms of functioning in this population, one could argue that there are differential levels of success, despite the lack of underlying genetic determinants for such differentials or of significant differentials in reproductive output. That is, these differences in outcomes are not due to fitness differences but rather to differences in the structuring of epigenetic, behavioral, and cultural/symbolic processes and patterns in the population. Obviously, I am referring to political, economic, and environmental structuring in human societies.

The medical anthropologist Margaret Lock (2017), integrating genetics, epigenetics, epidemiology, physiology, and ethnography, coined the term "local biologies," referring to the manner in which biological and social processes are entangled throughout life, producing the diversity of human bodily experiences across the globe. There is no simple relationship between local biologies and nation-states, social groups, ethnicities, communities, or even families, but it is clear that societal, historical, political, and economic processes—all deeply social elements in the human experience—have substantive biological effects on the functioning of human bodies. These effects are not only on the individuals directly affected; they also include intergenerational transmission of toxins and other deleterious cellular invaders, of epigenetic changes leading to proclivities toward a variety of physiological and endocrine disorders, of physiologies of reactivity, and of perceptual notions of positionality and identity associated with negative psychoneuroendocrine physiologies (see studies reviewed in Lock 2017 and Krieger 2020; see also for cancers Tiffon 2018, and for war-related trauma Clarkin 2019).

Social/cultural lives and experiences matter in human biology. The anthropologist Clarence Gravlee (2020; Gravlee, Non, and Mulligan 2009) and the social epidemiologist Nancy Krieger (2020) document in great detail how the patterns and structures of racism and sexism (among many other "isms") interface with health pathways, societal power hierarchies, physiologies, and epigenomes to clarify how health inequities create biological expressions of injustice and constraint. Put simply, perceptions, ideologies, societal experiences, and beliefs can be as relevant as genes, bones, and muscle in human evolutionary processes. The EES, with multiple modes of evolutionarily relevant inheritance, with niche construction, and with a deep emphasis on reciprocal causation, offers modes of evolutionary analysis that enable a more nuanced engagement with this complexity than does SET.

Let me be clear here: I am not arguing for a simple cultural evolution model wherein some individuals and groups do "better" in a cultural context because their "variants" of that culture are a better fit. This is what Herbert Spencer (1851) and Darwin ([1871] 1981) suggested and what many contemporary racists and nationalists argue (see Marks 2017; Saini 2019). Cultures do not evolve in the same manner as populations of organisms, and cultural "units" are not the same as "genes." Success in economic or political terms in a given society is not simply due to the possession and capacity to transmit more "fit" cultural variants in a system akin to natural selection on genetic variation. Rather, inequity and outcomes in societies are due (largely, but not exclusively) to the structures and processes, the economic and political institutions and governance, in addition to the social histories of inequity and oppression and/or facilitation. This process is not a parallel to natural selection. However, cultural processes, while not directly comparable to genomic evolutionary processes, are certainly intertwined with biological evolutionary processes (see Lewens 2015; Han 2017; Henrich 2016, 2020; Demps and Richerson, this volume).

In SET, the lived details of local biologies, especially those that do not directly impact the potential for reproduction, and the contexts that produce them, are not included in ultimate measures of evolutionary success; only reproductive success (fitness) is included. Even if an organism suffers, physiologically and psychologically, across its life but is successful at passing along its genotype at a reasonable rate relative to other individuals in the population, then we can say, in a SET sense, it has achieved evolutionary success.

But is this an accurate, or effective, measure for humans in the Anthropocene? We know that, for the past 3,000–5,000 years, and especially the past few centuries, there has emerged a disconnect between most genotypes and successful reproduction in humans. That is, in the contemporary landscape of the Anthropocene, nearly all human phenotypes—and thus genotypes—are successful at reproducing. Achieving some level of reproductive success has become a limited constraint on most humans, in large part because of medical technology in recent centuries but also because of the nutritional, social, and cultural landscape that has emerged over the most recent aspects of history in the human niche. Humans have created a distinctive niche, a context in which the connection between many (most?) genotypic and phenotypic patterns and reproductive success is at least somewhat decoupled. This does not mean that natural selection and the range of other evolutionary processes are not acting on human populations—they are. Rather, it means that the patterns and outcomes of those processes are mediated through and structured by a broader and more dynamic set of interlacing processes/variables than in most other organisms.

If this is the case, how do we measure human evolutionary success today? I would like to suggest that it might be necessary for us to think specifically about human evolutionary success in the Anthropocene using the multiple modes of evolutionary processes offered in the EES (in addition to selection), alongside an anthropological approach to understanding cultural complexity and history (e.g., Fuentes 2016, 2018; Fuentes and Wiessner 2016). Evolutionary success within the contemporary human niche of the Anthropocene could be measured by more than just numeric representation (population demography) or reproductive success (at the individual and populational levels), or the traditional calculation of the fitness of particular traits. All of these remain important as factors in understanding evolutionary processes in humans, but they are not the whole of our assessment possibilities (or requisites) for evolutionary success. Rather, evolutionary success for humans in the Anthropocene should also include measures of the capacities and the realities of achieving physiological and psychological outcomes relative to societal structuring—"local biologies" and the human experiences co-constructing them.

This is not simply an argument for gene-culture coevolution where both culture and genomic processes are moving along more or less in modes explained by fitness and reproductive success, that is, genes reproducing themselves and cultural memes or cultural units reproducing themselves. What I am suggesting is that, over evolutionary time, particularly over the Pleistocene, and particularly the terminal Pleistocene and Holocene, *Homo sapiens* developed a niche wherein patterns for relatively high levels of physiological and psychological functioning emerged across populations of the species. I would add that most recently in the Anthropocene, the human niche has become structured such that the ability to achieve effective physiological and psychological functioning has been restricted for some segments of some populations. In other words, human suffering and institutional and structural violence (Farmer et al. 2006; Galtung 1969; Krieger 1994, 2020; Lock 2017) are relevant to the measurement of human evolutionary success in the Anthropocene (Fuentes 2016).

The argument to add human suffering and the assessment of local biologies as a measure of evolutionary success in the Anthropocene goes something like this: *Homo sapiens*, by the 21st century, has created a niche wherein most of the genotypic and phenotypic variation existing in this species achieves some sufficient level of reproductive success. While there is variance in how many offspring are produced across individuals, that variance is very rarely tied, causally, to specific genomic characteristics or processes, or even specific phenotypic traits. Most of the genotypic and phenotypic variants in contemporary *Homo sapiens* achieve at least sufficient reproductive success to maintain themselves in the population. If this is the case, the traditional measurement of success, fitness as measured via representation of certain genetic variants or phenotypes in a population, is no longer sufficient as the *single measure* (but it does remain important in an overall assessment). This is because the vast majority of variants are able to maintain themselves in the population across time, and their relative frequency at the species and population level is often not directly connected to the specific DNA sequence function or its relationships to a specific phenotypic outcome. In short, successful reproduction in a given context (fitness) does not capture all of, or even most of, the biologically relevant aspects that can be involved in shaping the patterns and processes of the human niche and thus human evolution.

Obviously, this is not true for a range of serious genomic diseases and disorders, but those variants are rare at the species level, and even then cultural processes can enable their maintenance in contemporary populations to some extent (e.g., Tay Sachs). Humans have created a niche where, by being born in the 21st century, one is more likely than not to have a sufficient SET fitness outcome, measured as the passing on of distinct genetic sequences (personally or via close genetic relatives) to the next generation. But does this, by itself, capture sufficiently the relevant evolutionary dynamics for humans in the Anthropocene? Drawing on the research I have reviewed above, I suggest that it does not.

Thus, I propose that we might look at some measure of local biologies, such as the quality of development, health, and morbidity, as an evolutionary measure as well. Here I define "quality" as the level of equity[2] in the capacity to effectively achieve the range of physiological and psychological potential within individuals and populations. Contemporary human genomes and epigenomes most likely reflect a very well-structured suite of variation for achieving physiological and psychological success across myriad environments/contexts. We can see evidence of this in the incredible demographic success of our species across the past few millennia (Roser, Ritchie, and Ortiz-Ospina 2019). We have created a niche in which most of our genotypic and phenotypic variation has sufficient chances of successfully reproducing and being passed down to the next generation (with little evidence of substantial variance across the majority of individuals in most populations). If that is the case, we have to ask what it means to be successful in an evolutionary sense for humans. I suggest adopting a form of a human inequity index as an additional aspect of our assessments of evolutionary success in the Anthropocene—that is, an index of human inequity or constraints on our capacity to flourish physiologically and psychologically as per the potential of our genotypes and phenotypes.

6. Conceptualizing Success and Assessing Sustainability

It is a challenge to outline specifically what this *Homo sapiens* distinctive measurement of success in the Anthropocene might look like. To date there are a number of indices and composite measures that purport to reflect aspects of inequity in societies and the amount/grade of suffering that they cause. There is the International Human Suffering Index (Camp and Speidel 1987), the Human Development Index (Stanton 2007), and multiple other indices by the United Nations and a range of other global NGOs, national governments, and research institutes. However, they mostly focus on national-scale measures of economic and political structures and outcomes, educational attainment, and general measures of health and well-being.

At the theoretical scale, this sort of work is currently done largely in global development, epidemiology, economics, and sociology, and to a lesser extent in anthropology—but such assessments are rarely connected to evolutionary contexts/processes (again, see Krieger 1994, 2020; Farmer et al. 2006) or make serious efforts to engage with the concept of local biologies (Lock 2017). I suggest that an evolutionarily relevant suite of measures would need to connect patterns of structural inequity to specific patterns of morbidity and health challenges and to the developmental hindrances/insults they cause. This would then have to be connected to transgenerational inheritances of such patterns in order to develop evolutionarily situated models of such processes.

There are copious studies demonstrating the specifics of structural inequity's negative impacts on developmental and health outcomes (e.g., Mendenhall et al. 2017; Noble et al. 2015; see also Lock 2017; Krieger 2020), but there is not any current robust theoretical context with which to link them to broader evolutionary narratives. However, connecting studies of inequity and its deleterious social and biological impacts to an evolutionary context is inherently difficult due to the history of simplistic, racist, and nationalist attempts to ascribe the causality of poverty, discrimination, and health inequity, erroneously, to inherent (read "biological") differences in human groups, races, and classes (see Marks 2017; Roberts 2012; Saini 2019).

Since the 19th century, there has been substantial misuse of standard evolutionary contexts and facile (and erroneous) arguments about natural selection and human cultural groups. This is why it is critical to approach this via the lens of the EES and in the anthropological context I have tried to outline above. These assessments are explicitly *not* about intrinsic differences at the genetic level between individuals or groups, but rather about how differences in societal structural processes affect the potential for effective development and expression of the species-wide capacities for health and social well-being. Increased attention to the cross-generational transmission of epigenetic, behavioral, and cultural/symbolic patterns related to societal structuring may be helpful in developing such an approach.

One final area where such expanded evolutionary thinking might also be beneficial is in the broader context of *Homo sapiens*'s relations to local and planetary ecosystems. The impacts of the human niche have differential deleterious effects not only within the human species but also on the global environment. In 1992, 1,700 scientists published a letter calling on humankind to curtail environmental destruction and cautioned that "a great change in our stewardship of the Earth and the life on it is required, if vast human misery is to be avoided" (Union of Concerned Scientists 1992). In 2017 a larger group of scientists (over 16,000) published a second letter, noting that "since 1992, with the exception of stabilizing the stratospheric ozone layer, humanity has failed to make sufficient progress in generally solving these foreseen environmental challenges, and alarmingly, most of them are getting far worse" (Ripple et al. 2017). The letter (and much affiliated research) goes on to illustrate the key crises across the planet and issues an evidence-based dire warning that our species' current patterns of growth and use of the planet are not sustainable.

That is, we are facing massive ecosystem collapse and massive extinction events if we continue the demographic expansion and resource-use trajectory established over the past few centuries. The COVID-19 pandemic is an excellent example of this. It is becoming clear that gross demographic success and capacity to maximize extraction and use of environmental resources are likely deleterious to long-term (millennia-scale) success not just for our species but also for a majority of ecosystems and other species on the planet. A meaningful assessment of human evolutionary success, in this context, should likely include the sustainability of anthropogenic ecologies and their trajectories as a key component. In such a context the development and transmission of cultural/symbolic and behavioral variation (as included in the EES) will likely play a key role in structuring future ecologies for the planet and are thus directly relevant to a model of human evolutionary success.

Our species' ecological and geographic range, population growth, and adaptability in the face of micro- and macro-environmental challenges are substantial. We have such an impact at a global scale that the most recent epoch of the Quaternary Period has been dubbed by many "the Anthropocene." There is no question that under a SET *Homo sapiens* is among the most successful, if not the most successful, large-bodied mammal. However, as I have argued here, I do not think that SET offers the most comprehensive assessment framework for measuring or conceptualizing evolutionary success in the 21st century. In this chapter, I have offered the suggestion that engaging the EES in dialogue with a dynamic anthropological (or more broadly social scientific) approach enables a frame of reference that is more in tune with the specifics of the entire range of evolutionarily relevant processes at play in the contemporary human niche. While there is no formal model in this argument, I do hope that it at least inspires a broader discussion across the range of scholars interested in assessing what we might mean by "evolutionary success" when focusing on a species as evolutionarily dynamic as *Homo sapiens*.

Notes

1. There is the substantive question of whether fitness is a property of traits or of organisms. For example, Pence and Ramsey (2015) argue that while it is tempting to think of fitness as a property of traits, organismic fitness is what is ultimately assessed and modeled. However, given the focus of this chapter on forms of success other than fitness, I do not delve into these aspects of the fitness debate here.
2. Equity is not equality. Equity is the establishment of broad-scale (equitable) landscapes with the potential for flourishing/expression of capabilities. Equality is treating everyone the same. I am not arguing for equality, as that negates the substantive individual variation in humans at all levels.

References

Ackermann, R. R., A. Mackay, and M. L. Arnold. 2015. "The Hybrid Origin of 'Modern' Humans." *Evolutionary Biology* 43: 1–11. doi:10.1007/s11692-015-9348-1.

Antón, S. C. 2013. "Homo erectus and Related Taxa." In *A Companion to Paleoanthropology*, edited by D. R. Begun, 497–516. Oxford: Wiley-Blackwell.

Antón, S. C., R. Potts, and L. C. Aiello. 2014. "Evolution of Early Homo: An Integrated Biological Perspective." *Science* 345 (6192): 45–57. doi:10.1126/science.1236828

Baedke, J., A. Fábregas-Tejeda, and F. Vergara-Silva. 2020. "Does the Extended Evolutionary Synthesis Entail Extended Explanatory Power?" *Biology and Philosophy* 35 (1): 1–22.

Bennett C. E., R. Thomas, M. Williams, J. Zalasiewicz, M. Edgeworth, H. Miller, B. Coles, A. Foster, E. J. Burton, and U. Marume. 2018. "The Broiler Chicken as a Signal of a Human Reconfigured Biosphere." *Royal Society Open Science* 5 (12): 180325. http://doi.org/10.1098/rsos.180325

Camp, S. L., and J. J. Speidel. 1987. "The International Human Suffering Index." *Washington: The Population Crisis Committee.* https://www.worldcat.org/title/21186688.

Clarkin, P. 2019. "The Embodiment of War: Growth, Development, and Armed Conflict." *Annual Review of Anthropology* 48: 423–442.

Darwin, C. (1871) 1981. *The Descent of Man and Selection in Relation to Sex*. Princeton, NJ: Princeton University Press.

Estrada, A., P. A. Garber, A. B. Rylands, C. Roos, E. Fernandez-Duque, A. Di Fiore, K. A.-I. Nekaris, et al. 2017. "Impending Extinction Crisis of the World's Primates: Why Primates Matter." *Science Advances* 3 (1): e1600946. doi: 10.1126/sciadv.1600946.

Farmer, P. E., B. Nizeye, S. Stulac, and S. Keshavjee. 2006. "Structural Violence and Clinical Medicine." *PLoS Medicine* 3 (10): 1686–1691. doi:10.1371/journal.pmed.0030449.

Foley, R. A. 2016. "Mosaic Evolution and the Pattern of Transitions in the Hominin Lineage." *Philosophical Transactions Royal Society B* 371: 20150244. doi:10.1098/rstb.2015.0244.

Fuentes, A. 2016. "The Extended Evolutionary Synthesis, Ethnography, and the Human Niche: Toward an Integrated Anthropology." *Current Anthropology* 57 (Suppl 13): 13–26. doi:10.1086/685684.

Fuentes, A. 2017a. *The Creative Spark: How Imagination Made Humans Exceptional*. New York: Dutton/Penguin.

Fuentes, A. 2017b. "Human Niche, Human Behaviour, Human Nature." *Interface Focus* 7: 20160136. http://dx.doi.org/10.1098/rsfs.2016.0136.

Fuentes, A. 2018. "How Humans and Apes Are Different, and Why It Matters." *Journal of Anthropological Research* 74 (2): 151–167. doi.org/10.1086/697150.

Fuentes, A. 2019. *Why We Believe: Evolution and the Human Way of Being*. New Haven, CT: Yale University Press/Templeton Press.

Fuentes, A., and P. Wiessner. 2016. "Reintegrating Anthropology: From Inside Out." *Current Anthropology* 57 (Suppl 13): 3–12.

Futuyma, D. J. 2017. "Evolutionary Biology Today and the Call for an Extended Synthesis." *Interface Focus* 7 (5): 20160145. https://doi.org/10.1098/rsfs.2016.0145.

Galtung, J. 1969. "Violence, Peace, and Peace Research." *Journal of Peace Research* 6 (3): 167–191.

Gravlee, C. 2020. "Systemic Racism, Chronic Health Inequities, and COVID-19: A Syndemic in the Making?" *American Journal of Human Biology* 32 (5): e23482.

Gravlee, C., A. L. Non, and C. J. Mulligan. 2009. "Genetic Ancestry, Social Classification, and Racial Inequalities in Blood Pressure in Southeastern Puerto Rico." *PLoS ONE* 4 (9): e6821. doi:10.1371/journal.pone.0006821.

Han, S. 2017. *The Sociocultural Brain*. Oxford: Oxford University Press.

Henrich, J. 2016. *The Secret of Our Success*. Princeton, NJ: Princeton University Press.

Henrich, J. 2020. *The Weirdest People in the World: How the West Became Psychologically Peculiar and Particularly Prosperous*. New York: Farrar, Straus and Giroux.

Kissel, M., and A. Fuentes. 2018. "'Behavioral Modernity' as a Process, Not an Event, in the Human Niche." *Time and Mind* 11 (2): 163–183. doi:10.1080/1751696X.2018.1469230.

Kissel, M., and A. Fuentes. 2021. "The Ripples of Modernity: How We Can Extend Paleoanthropology with the Extended Evolutionary Synthesis." *Evolutionary Anthropology* 18 (5): 773–778.

Krieger, N. 1994. "Epidemiology and the Web of Causation: Has Anyone Seen the Spider?" *Social Science & Medicine* 39 (7): 887–903.

Krieger, N. 2020. "Measures of Racism, Sexism, Heterosexism, and Gender Binarism for Health Equity Research: From Structural Injustice to Embodied Harm—an Ecosocial Analysis." *Annual Review of Public Health* 41: 4.1–4.26.

Kuijper, B., M. A. Hanson, E. I. K. Vitikainen, H. H. Marshall, S. E. Ozanne, and M. A. Cant. 2019. "Developing Differences: Early-Life Effects and Evolutionary Medicine." *Philosophical Transactions of the Royal Society B* 374: 20190039. http://dx.doi.org/10.1098/rstb.2019.0039.

Laland, K. N. 2017. *Darwin's Unfinished Symphony: How Culture Made the Human Mind*. Princeton, NJ: Princeton University Press.

Laland, K. N., T. Uller, M. Feldman, K. Sterelny, G. B. Müller, A. Moczek, E. Jabonka, and J. Odling-Smee. 2014. "Does Evolutionary Theory Need a Rethink? Yes, Urgently." *Nature* 514 (7521): 161–164.

Laland, K. N., T. Uller, M. Feldman, K. Sterelny, G. B. Müller, A. Moczek, E. Jablonka, and J. Odling-Smee. 2015. "The Extended Evolutionary Synthesis: Its Structure, Assumptions and Predictions." *Proceedings of the Royal Society B* 282: 20151019. http://dx.doi.org/10.1098/rspb.2015.1019.

Lewens, T. 2015. *Cultural Evolution: Conceptual Challenges*. Oxford: Oxford University Press.

Lock, M. 2017. "Recovering the Body." *Annual Review of Anthropology* 46: 1–14.

Marks, J. 2017. *Is Science Racist?* New York: Polity Press.

Mendenhall, E., B. A. Kohrt, S. A. Norris, D. Ndetei, and D. Prabhakaran. 2017. "Non-communicable Disease Syndemics: Poverty, Depression, and Diabetes among Low-Income Populations." *Lancet* 389: 951–963.

Noble, Kimberly G., Suzanne M. Houston, Natalie H. Brito, Hauke Bartsch, Eric Kan, Joshua M. Kuperman, Natacha Akshoomoff, et al. 2015. "Family Income, Parental Education and Brain Structure in Children and Adolescents." *Nature Neuroscience* 18 (5): 773–778. doi:10.1038/nn.3983.

Pence, C., and G. Ramsey. 2015. "Is Organismic Fitness at the Basis of Evolutionary Theory?" *Philosophy of Science* 82: 1081–1091.

Roser, M., H. Ritchie, and E. Ortiz-Ospina. 2019. "World Population Growth." Our World in Data. https://ourworldindata.org/world-population-growth.

Ripple, W. J., and 15,364 scientist signatories from 184 countries. 2017. "World Scientists' Warning to Humanity: A Second Notice." *BioScience* 67 (12): 1026–1028. https://doi.org/10.1093/biosci/bix125.

Roberts, D. 2012. *Fatal Invention: How Science, Politics, and Big Business Re-create Race in the Twenty-First Century*. New York: New Press.

Saini, A. 2019. *Superior: The Return of Race Science*. New York: Penguin/Random House.

Scerri, E. M. L., L. Chikhi, and M. G. Thomas. 2019. "Beyond Multiregional and Simple Out-of-Africa Models of Human Evolution." *Nature Ecology and Evolution* 3 (10): 7–9. doi:10.1038/s41559-019-0992-1.

Scerri, E. M. L., M. G. Thomas, A. Manica, P. Gunz, J. T. Stock, C. Stringer, M. Grove, et al. 2018. "Did Our Species Evolve in Subdivided Populations across Africa, and Why Does It Matter?" *Trends in Ecology & Evolution* 33: 582–594.

Spencer, H. 1851. *Social Statics: or The Conditions Essential to Happiness Specified and the First of Them Developed*. London: John Chapman. https://oll.libertyfund.org/title/spencer-social-statics-1851.

Stanton, E. A. 2007. "The Human Development Index: A History." *Political Economy Research Institute Working Papers* 127: 14–15.

Tiffon, C. 2018. "The Impact of Nutrition and Environmental Epigenetics on Human Health and Disease." *International Journal of Molecular Sciences* 19 (11): 3425. https://doi.org/10.3390/ijms19113425.

Uller, T., and K. N. Laland. 2019. *Evolutionary Causation: Biological and Philosophical Reflections*. Cambridge, MA: MIT Press.

Union of Concerned Scientists. 1992. "World Scientists' Warning to Humanity." https://www.ucsusa.org/resources/1992-world-scientists-warning-humanity.

van Schaik, Carel P. 2016. *The Primate Origins of Human Nature*. Oxford: Wiley-Blackwell.

Whitehead, H., K. N. Laland, L. Rendell, R. Thorogood, and A. Whiten. 2019. "The Reach of Gene-Culture Coevolution in Animals." *Nature Communications* 10: 2405.

Wilson, E. O. 1975. *Sociobiology: The New Synthesis*. Cambridge, MA: Harvard University Press.

Wood, B., and E. Boyle. 2016. "Hominin Taxic Diversity: Fact or Fantasy?" *Yearbook of Physical Anthropology* 159: S37–S78.

11

Adaptability and the Continuation of Human Origins

Richard Potts

1. Starting Point

A species is ideally identified by a unique suite of morphological and genomic characteristics. It may also possess distinctive behavioral and ecological adaptations that underlie how its member organisms interact with their biotic, resource, and social surroundings over time and space. It is appropriate, therefore, that a study of human success takes into account the conditions in which the behavioral and ecological adaptations of *Homo sapiens* arose. Following this line of inquiry requires us to consider the evolutionary contingencies—the intersection of survival challenges that led humans to exist, to become distinguishable in the first place, which may then allow insight into the nature of our species' success.

To this end, I begin by describing an example from personal research, truly the work of an interdisciplinary team of "ologists"—paleontologists, archaeologists, geologists, paleoecologists—along with environmental scientists, whose data have combined to compose the most detailed historical narrative we presently possess about the emergence of behaviors near the time of the oldest known members of *Homo sapiens* in Africa. The example begins by considering a prehistoric era of extraordinary endurance known as the Acheulean—a way of life defined by the production of large stone tools such as handaxes. The example then tracks this technology's demise at the expense of a new suite of tools and behaviors known as the African Middle Stone Age, the technology associated with the emergence of our species (Brooks et al. 2018).

Acheulean technology persisted for more than one million years, a regularity of tool production nearly unfathomable to most of us, given our own experiences in the rapid ride from slide rule to laptop, from aerogram

Richard Potts, *Adaptability and the Continuation of Human Origins* In: *Human Success*. Edited by: Hugh Desmond and Grant Ramsey, Oxford University Press. © Oxford University Press 2023. DOI: 10.1093/oso/9780190096168.003.0011 to internet, or the massive change in a mere two centuries from energy dependence on solar radiation (agriculture) to the resurrection of ancient solar power (fossils fuels in the Industrial Revolution). The material signal of the highly durable Acheulean is the production of ovoid-shaped large cutting tools, which archaeologists group into categories called handaxes, cleavers, picks, and knives (Isaac 1977) (Figure 11.1). The oldest current date for the Acheulean is around 1.76 Ma (Lepre et al. 2011; Beyene et al. 2013), and in some areas of Africa and Eurasia, Acheulean tool kits lasted to less than 200,000 years ago.

The demise of the Acheulean's extraordinary run—and the conditions under which this occurred—has significance for any reasonable understanding of the concept of human "success." In the sedimentary basin in southern Kenya called Olorgesailie, one of the oldest examples of this technology's termination and replacement is recorded. There, our research team has unearthed layer upon layer of Acheulean archaeological sites dated between 1.2 million years old (the oldest known layers of sediment in that region) and 500,000 years old (Deino and Potts 1990; Potts, Behrensmeyer, and Ditchfield 1999; Behrensmeyer et al. 2002; Potts 2013). This stretch of 700,000 years contrasts starkly with a period of far more rapid change that took place between 500,000 and about 320,000 years ago (Brooks et al. 2018;

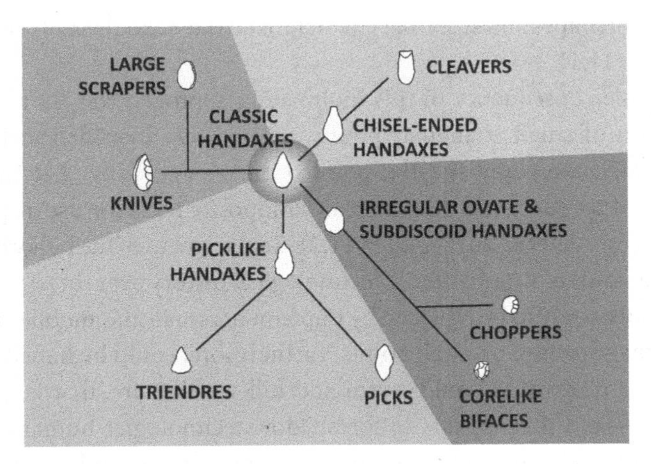

Figure 11.1 The gradual continuum of stone tools technology that typifies the Acheulean technology of Africa. Classic handaxe shapes and sizes of the Acheulean are connected in a fluid spectrum with other large cutting tools and tool types, including choppers and scrapers.
Source: based on Isaac 1977.

Deino et al. 2018; Potts et al. 2018). During this latter interval, handaxes disappeared from the southern Kenya rift and were replaced by tool innovations and behaviors previously unknown in the prehistoric record.

The new technology—the Middle Stone Age (MSA)—included a set of behavioral adaptations that ended up typifying early African *Homo sapiens*. In our excavations, these new adaptations included a remarkable trio: the repetitive, planned manufacture of small flake tools; the exchange, or trading, of obsidian stone over unprecedented distances; and the use of coloring materials, red and black pigments, which are widely considered to indicate the development of human symbolic communication. Let's take these three matters one at a time.

First, the manufacture of tools from carefully prepared cores required thinking many steps ahead. Through a careful, intentional series of strikes, a diverging array of flake scars around the stone core was produced, enabling the toolmakers to create, with one blow, flakes of predictable size and thinness. A flake from a prepared core was like the last step in a chess match, with the final predicted rocky sliver equivalent to checkmate. Although the development of prepared core methods had its roots in earlier forms of toolmaking, their regular application characterizes MSA technologies (McBrearty and Brooks 2000; Tryon, McBrearty, and Texier 2006). As a result, a wider diversity of distinguishable tool types replaced the smooth spectrum of tool shapes and sizes that characterized the Acheulean of Olorgesailie (see Figure 11.1).

A notable consequence of this toolmaking approach was the intentional production of small triangular points, which were then delicately nicked at the wide bases opposite the pointed end—suggesting that the points were hafted to wooden shafts to create composite tools for use as projectile weapons (Brooks et al. 2018; Figure 11.2). Replacement of the bulky handaxes by smaller tools is reminiscent of technological history ever since: tools begin big and bulky only to be replaced by implements small and mobile. The newfangled manufacture of small points, furthermore, could be hafted to shafts (composite tools) that could maim and kill via the air—that is, projectile weaponry—and this appears to foreshadow technologies humans have recently fashioned that enormously intensify the primate tendency to bluff or, otherwise, to inflict harm.

Evidence of this remarkable transition in technology is preserved in buried archaeological sites dated to 320,000 years old, at least in the southern Kenya rift valley. Yet there is more to the story: The projectile points and

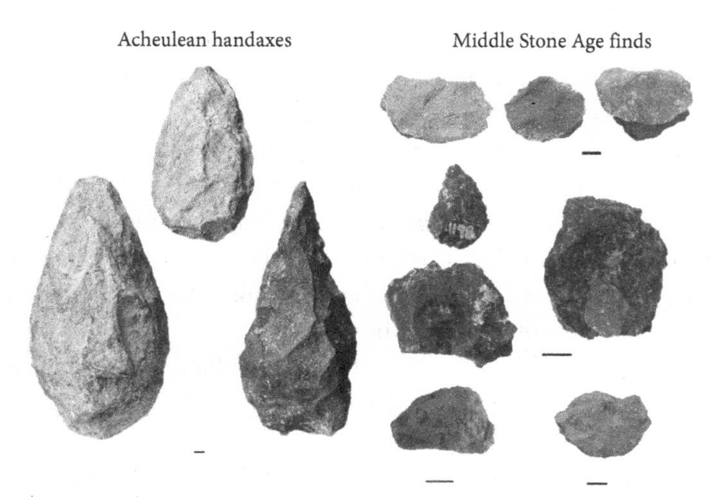

Figure 11.2 A dramatic transition in African hominin behavior between 500,000 and 320,000 years ago, as recorded in the Olorgesailie basin, southern Kenya rift valley. Acheulean tools (left side) were made throughout the time from 1.2 million to 500,000 years ago, but disappeared by 320,000 years ago, when the Acheulean technology was replaced by new kinds of tools and behaviors typical of the African Middle Stone Age (right side). The early Middle Stone Age at Olorgesailie was characterized by smaller, more diverse tool kits (upper right), including innovations such as projectile points; the acquisition of stone material, including black obsidian (middle right), from multiple, distant sources and directions; and the use of coloring material, including reddish-orange pigments (bottom right).
Sources: Brooks et al. 2018; Deino et al. 2018; Potts et al. 2018.

other early MSA tools at Olorgesailie were made on a rare kind of rock in this part of the rift valley: the sharp, black shiny volcanic glass known as obsidian. This extraordinary rock type, which yields the sharpest cutting edges when it is fractured, is virtually unknown (~0.05% of stone artifacts) in the Acheulean archaeological sites at Olorgesailie. By contrast, obsidian made up from 4% to more than 42% of all the tools and flaking debris in the oldest MSA sites (Brooks et al. 2018; Potts et al. 2018). The amount of obsidian the Olorgesailie hominins brought to and flaked at these sites indicates that large, heavy pieces of this rock were brought from distant sources into the region.

By testing the distinct chemical signatures of obsidian sources across the Kenyan rift, our research team determined that this type of rock was transferred from at least eight distinct sources located in different directions over distances that ranged from at least 25 to 95 kilometers away from the Olorgesailie MSA archaeological sites (Brooks et al. 2018). Our calculations, moreover, reflect straight-line distances. The actual distances across the high relief and rugged terrain of the rift valley, up and down geological faults, and around volcanic mountains, were easily four times greater by our estimates. In earlier stone technologies, a group of early human foragers would be expected to have continually flaked and used the rock as they traversed the landscape. This is indeed what the Acheulean toolmakers did at Olorgesailie, even across distances of just a few kilometers.

In the MSA, however, the transfer of obsidian across hundreds of kilometers of estimated walking distance from multiple directions across the rugged rift valley terrain did not lead to complete consumption of the obsidian but rather the movement of considerable volumes of this material and its flaking at the Olorgesailie sites. For this reason, we hypothesize that the valued obsidian was conserved and transferred from group to group in a network of resource exchange—essentially, early evidence of trade across multiple groups spread out over the multiple directions where the obsidian sources occurred (Brooks et al. 2018). Indeed, mobile foraging peoples today (hunter-gatherers of sub-Saharan Africa) rarely exceed a 25-kilometer radius of annual movement while maintaining alliances among networked social groups up to 100 kilometers away as resource exchange partners; multiple social groups can thus enlarge the network and thereby exchange and gain access to resources over very large areas (Yellen and Harpending 1972). In short, when vital resources such as food, water, and toolmaking materials become scarce in one place, the supplies of such things may be more abundant and available elsewhere. A social network that enables distant exchange helps mitigate the risks, creating partners both practical and symbolic who can rely on each other. In the prehistoric record of Olorgesailie, we are limited in answering certain interesting details, such as how many social groups were involved in these networks (we only know the locations of the eight distant obsidian source locations); how far any one individual might have traveled between one social group and another; or whether the individuals who brought the black stone to Olorgesailie were the most highly skilled toolmakers or the most skilled socially, or both.

Our team was surprised to find such indicators of social networks of resource exchange as far back as 320,000 years ago. For one thing, it implies that the earliest MSA groups in the region had developed some capacity to cognitively model and potentially pass on sophisticated maps of the social and resource landscape. This approach to life stands in stark contrast to the Acheulean way of life in the same region, where 98% of all stone used in making handaxes and other large cutting tools was obtained within a distance of only 5 kilometers, well within sight of the places on the landscape where we find the tools.

With a high degree of certainty, we can infer that the manner in which MSA hominins understood and modeled their social, physical, and ecological landscape became vastly enlarged sometime between 500,000 and 320,000 years ago—that is, in a very brief time relative to the impressive endurance of the Acheulean way of life. What had changed so dramatically was not only the mental mapping of rock sources but also the social connections that allowed access to this most valued type of rock for making tools, including ones that facilitated projectile hunting and energy-efficient food acquisition. Our MSA predecessors had become socially connected with others well beyond their immediate social vicinity, aware and mindful of their fellow hominins even when unseen.

This raises an interesting question: How is it that distant obsidian locations, and the distant groups of hominins who possessed the stone, could be kept in mind by those who likely valued obsidian for its sharpness and delicate flaking quality? How were exchanges of this resource conveyed between individuals or groups? People today possess complex symbolic systems, especially language, alliances, and symbols that facilitate exchange, which make such transactions possible. Symbols are essential to how we communicate abstractions, plans, and possibilities that are displaced in time and place. As it turns out, at these same digging sites where abundant MSA tools and obsidian were found, we also discovered coloring material—lumps of red and black rock that, at least in some cases, were drilled and chopped to produce powdered pigment (Brooks et al. 2018).

Our research team wondered, of course, what these pigments might imply. Color is symbolically loaded in the life of all people—in dress, flags, and all manner of signaling our personal and societal identity. Of course, coloration, sound, and other signals are generated by other vertebrate species through their anatomy and movement. In the case of the MSA hominins at Olorgesailie, however, red and black pigments were brought to and aggregated at the sites, not through some natural geological process but rather as purposeful behavior from some distance well beyond the limits of the Olorgesailie Basin (Brooks et al. 2018; Potts et al. 2020). Meaningful use of color is fundamental to humans and the groups to which they belong. It seems quite possible and perhaps very likely, then, that the amplified use of symbolic communication, with color as a basic expression, was beneficial and ultimately essential in assigning value to any distant resource and in developing, negotiating, and sustaining a network of social resource exchange partners in multiple directions across the southern and central Kenya rift (Brooks et al. 2018; see also Smithsonian Human Origins Program 2022).

The onset of this novel suite of behaviors—technological innovation, which included efficient hunting tools; intergroup exchange of valued material; and intensified symbolic ability—all took place near the time of the earliest fossil evidence of African *Homo sapiens*. The oldest currently dated fossils assigned to our species, although more advanced in the face than in the braincase, are from the site of Jebel Irhoud, Morocco, at 300,000 years old (± 30,000 years; Hublin et al. 2017; Richter et al. 2017). At the site of Omo Kibish, Ethiopia, two fossils typically assigned to our species have been dated to around 233,000 years old (McDougall, Brown, and Fleagle 2005; Vidal et al. 2022), although dating uncertainty indicates a minimum age of between 211,000 and 255,000 years old. These two fossil skulls from Omo show somewhat different appearances. One (Kibish I) possesses the distinctive small face and vertically arching forehead of *H. sapiens*, and the other (Kibish II) more archaic facial and cranial traits. To the south, a partial cranium from Florisbad, South Africa, dated 281,000 ± 73,000 years old, also exhibits traits indicative of *H. sapiens* (Grün et al. 1996; Kuman, Inbar, and Clarke 1999). Together, the three widely separated discovery sites may indicate that the cranial characteristics of our species were assembled over a period of 100,000 years by gene flow across much of the continent (Stringer 2016; Galway-Witham and Stringer 2018; Scerri et al. 2018; for updated geological ages, see Deino et al. 2018). It is possible therefore that the replacement of Acheulean by MSA behaviors in southern Kenya reflects a contemporaneous transition in behavior, the emergence of an adaptive repertoire that may have been a critical wedge between the African gene pool of *H. sapiens* and that of our immediate ancestor.

For the purpose of this chapter, we face a provocative question: What if the Acheulean toolmakers had found the means to persist? What if they had found the means to maintain their system of behavior, continuing their technological tradition and dampening any tendency of individuals to roam beyond their local range and risk interacting with distant social groups? Although this did not happen, one might wonder *Why not?* What exactly destabilized the system? What perturbation shook the tenacity of the Acheulean, leading to its demise, the end of its great run?

Our recent investigations illustrate an unexpected finding. According to prior studies of how environmental change may have influenced human evolution, basic changes in African climate—increasing aridity or greater precipitation or a rising degree of climate variability—have been thought to have instigated the major changes in African hominin adaptation (e.g., deMenocal 1995; Vrba 1995; Potts 1996, 2013; Trauth et al. 2007). Our recent study, however, of the most precisely dated environmental record for the past one million years, obtained by drilling a sedimentary core ~24 kilometers south of the Olorgesailie Basin, offers a more complicated story. A conjunction of four factors appears to have pushed the Acheulean over the edge (Potts et al. 2020). In a very brief synopsis, first, tectonic rifting divided the landscape into high and low terrain, resulting in a series of separate basins. Second, fluctuations in water supply, governed by climate oscillation, also took place. Due to these first two factors, the vegetation, water, and food supplies in this region of the Kenya rift became fragmented; in other words, some basins were more arid, others wetter, and each distinct area of the complex terrain developed its own uncertainties and dynamics of available resources over time. The implications point to a third factor: breaking up previously extensive grasslands, large tracts of short grass that ecologists call "grazing lawns," which are maintained by large-bodied, grass-loving herbivores and which sustain ecological communities comprised of these herbivores. Thus, the spatial repackaging of vital food sources and water supply across the region set the stage for a fourth critical factor by placing the previously dominant huge herbivore grazers at a disadvantage. What we know is that a distinctive suite of large-bodied grazing elephants, baboons, hippos, pigs, and zebras—dominant members of the ecosystem inhabited by Acheulean toolmakers—began to unravel. An estimated 85% replacement of mammal herbivore species took place, contemporaneous with the Acheulean's replacement by the MSA (Potts et al. 2018).

As a result, by the time MSA technology had developed and flourished in this region, the community of herbivore species and their feeding ecology had fundamentally changed as vegetation and food resources fluctuated widely and a fresh (potable) water supply became less predictable. The rough concurrence of these factors between 500,000 and 320,000 years ago appears to have destabilized the region ecologically, ultimately favoring not only the turnover of mammal species but also the demise of the Acheulean way of life in this region at the expense of adaptations typical of the MSA (Potts et al. 2020). The success of the Acheulean had come to an end in a time of ecological foment.

2. Adaptability: Reconfiguring the Notion of Adaptive Success

What can be lost in assumptions concerning adaptive success and progress in human evolutionary history is the demise of entire ways of life. In our starting example, the suite of behaviors associated with Acheulean technology in the Kenya rift valley represented one such way of life. We may wonder, however, whether the Acheulean might be considered more successful than the technologies and behaviors that replaced it. Lasting more than four times longer than our species is known to have existed, the Acheulean (by measure of longevity) and the toolmakers who made it represented a highly persistent approach to hominin adaptation. We may ask, though, whether the ways of life that replaced the Acheulean—behaviors that typify our own species—qualify as more successful (see also McShea in this volume; Hourdequin in this volume).

My thesis here is that evolutionary success is not only contingent and ephemeral but also most meaningful in circumstances that are limited when compared to the highly dynamic evolutionary environments evident in virtually all long-term paleoenvironmental records studied over the past quarter-century pertaining to human evolutionary history (e.g., deMenocal and Bloementhal 1995; Potts 1996; Kingston 2007; Trauth et al., 2007; deMenocal 2011; Potts and Faith 2015). Ecological dynamics of the late Cenozoic, the era encompassing human evolution, have been volatile—unstable to such a degree that adaptive states and latent capacities to adjust to environmental conditions ultimately fail. Biologists have formulated how the processes of selection obtain a match between an organism and its typical environment, which reflect the specific history of survival conditions the organism has encountered. We can enlarge this understanding by considering that an organism's characteristics can also evolve in response to the environmental dynamics its genetic lineage has endured. The result is what I refer to as evolutionary adaptability, or adaptation to change itself. The outcomes assist in buffering an organism from environmental instability and risks that might otherwise prove fatal if prior adaptations had not been abandoned.

The idea that adaptability can evolve in response to increased unpredictability or instability of ecological conditions (Potts 1996) implies that boundary conditions exist for any set of adaptations that characterize a species. Evolutionary adaptability arises from novel and unpredictable conjunctions of environmental variables, conjunctions never previously experienced by a population of organisms. The evolution of adaptability results in the ways by which an organism can overcome such a highly variable fitness landscape, ultimately affecting the ability to adjust to unprecedented circumstances that impact its survival and reproduction (Potts 1998). A strong and prolonged increase in environmental variability may thus influence whether and how adaptability is expanded. According to this variability selection hypothesis, variable fitness outcomes, when integrated over time and space, may favor genetic, developmental, behavioral, and ecological characteristics that make a difference in an organism's ability to adjust to changing and even novel situations (Potts and Faith 2015; see also Antón in this volume; Grove in this volume).

Evolutionary adaptability thus can entail an increase in the ecological versatility or flexibility of a population of organisms in response to dynamic surroundings. This evolved versatility improves a population's (or species') chances of (1) enduring change in the environment, (2) thriving in novel environments, (3) spreading to new habitats, and (4) responding in new ways to the surroundings, characteristics that are observable in a wide variety of organisms and distinctive of the genus *Homo* (e.g., Levins 1968; Solé and Goodwin 2000; Kirschner and Gerhart 2005; Potts and Faith 2015). It differs from a long-standing concept of plasticity in which particular phenotypes, including behaviors, are elicited by the specific range of environments encountered in an organism's evolutionary past. This limited concept of plasticity has been critiqued and replaced by views that recognize dynamic, including novel, evolutionary environments (e.g., Schlichting and Pigliucci 1998; West-Eberhard 2003; Oomen and Hutchings 2020).

Our starting point in this chapter suggests that the Acheulean's earliest demise in the Kenya rift, and its replacement by elements in our species' capacity for cultural innovation, were contingent on a new conjunction of geological, climatic, and ecological events. These events were unprecedented relative to the lengthy prior period when the Acheulean prevailed. The environments that accompanied this transition most likely varied from one region to another across the African homeland of *H. sapiens*. However, the Olorgesailie evidence points to the oldest set of circumstances currently known in which technological innovation, social exchange networks, and symbolic behavior took root and endured in this part of Africa. These substantive developments occurred at the expense of an ancient suite of behaviors that characterized the genus *Homo* throughout most of the Pleistocene.

Broadening this idea, I postulate that an increase in environmental unpredictability, or uncertainty, is necessary for adaptability to evolve. It is useful to think of uncertainty as a measure of change in the tempo, amplitude, and diversity of environmental configurations that an organism and its gene pool encounter over time and space. Intensified uncertainty results from an increase in the number of environments, the range and rapidity of fluctuation, and the combined temporal and spatial complexity of change in those environments as experienced by a population of organisms. Any rise in these variables signifies an increase in ambiguity in how, when, and where an organism can find food and water, for example—that is, how efficiently it can absorb and use energy as it carries out activities related to its survival and reproduction.

When adaptability evolves, uncertainty is necessarily reduced, moderated, or avoided by a change in the amount and quality of information the organism can produce, communicate, and rely upon. In any population of organisms, its storehouse of genomic information serves as a path to adaptability in dynamic surroundings. In social settings, on the other hand, observation and behavioral signaling can enhance the information available to organisms that are dependent on cognition. In human cultural settings, a symbolic system of information transmission makes possible an advanced degree of spatial displacement (e.g., distant resource exchange) and temporal displacement (e.g., prediction, anticipation, evaluation of conditionality) as means of enhancing adaptability in response to uncertainty. Genetic information, then, has a potentially stabilizing effect on external uncertainty, enabling the evolution of novel degrees of biological organization. The stabilizing effect of learned information in the social realm promotes interactions among individuals that permit the formation of multi-individual groups and social organizational complexity. And in the cultural realm, symbolic culture provides a novel means of coding, storing, and accumulating information that can serve as a source of adaptability in the face of social and ecological instability.

We can formulate these concepts as *the vowels of evolutionary adaptability*: A, E, I, O, and U: Adaptability, Energy, Information, Organization, and Uncertainty. Inspired by the starting example in this chapter, uncertainty (a rise in the number and range of environmental configurations that hominins and the contemporary biota encountered over time and space) disrupted the dominant systems of evolutionary adaptation at that time. The Acheulean was an incredibly persistent success. Yet the conjunction of landscape fragmentation due to faulting, the rise in the complexity of the rift valley terrain, an increase in climate variability, and the demise of large tracts of grazing land that girded a relatively stable ecosystem of megaherbivores, all had a destabilizing effect.

The Acheulean dissolved during a rise in ecological resource uncertainty, causing the previously durable adaptive system to become less effective. It was displaced by new means of energy acquisition (one of which was probably increased efficiency in hunting through projectile technology), expanded information transmission via symbolic means (represented by the use of color as a proxy for symbolic behavior), and organization into multigroup, distant resource-exchange partners. The result represented a new combination of behaviors that enhanced the adaptability of the population.

To change the order of the vowels slightly, we may represent this hypothesis in the following way: U →E + I + O →A. Environmental disruption (U, increased uncertainty) favored those behavioral variations that happened to (E) enhance energy/food acquisition, (I) improve the means and regularity of transmitting information symbolically among individuals, and (O) reorganize social interactions to include resource-exchange partners across distance. These aspects of behavior provided the behavioral currency by which adaptability (A) to environmental and ecological dynamics evolved. There is no valid reason I can think of to argue that the novel suite of MSA behaviors can be judged more successful than the Acheulean. It was merely adaptable in the context of a selective milieu instigated by environmental instability and increased uncertainty.

3. Living in the Anthropocene

In this era of the human-altered planet, the Anthropocene, the adaptability of humans worldwide will continue to be challenged. Will the presumed success of Western society, or any number of the varied ways of life manifested around the world, end up like the Acheulean, although obviously without the same temporal endurance? The deep-time perspective afforded by hominin evolutionary history lays out some clues as to the consequences of our current unprecedented challenges.

To step back for a moment, the evolutionary group of bipedal species to which *Homo sapiens* belongs has been diverse, consisting of some two dozen or more distinct lineages, as estimated from the current fossil record. This diversity came about during the past six million years in a messy process of population isolation, divergence, range expansion, and genetic admixture, followed by further divergence and genetic separation. Although these evolutionary cousins and ancestors of ours possessed ways of life resembling our own, the unique adaptations that furthered the survival and success of those particular lineages are gone. The demise of their species leaves us paleoanthropologists grasping at and arguing over slim lines of evidence as to how they lived. *Homo sapiens* is the last biped standing, owing to having evolved a magnified degree of adaptability relative to our now-extinct evolutionary predecessors and enabling us to become a global species. We are not more successful, but our adaptability is impressive. *H. sapiens* evolved the capacity to transform its behaviors, mating systems, and means of adaptation into many diverse cultural and other phenotypic expressions.

It is fair to say that a fundamental dimension of our species' adaptability is to live in the world by altering it. We have surpassed making tools by using tools to assemble ever more complex things that do more work. We have surpassed problem-solving by accumulating extraordinary storehouses of information that assist in buffering risks, avoiding inconveniences, and constructing physical cocoons and highly varied cultural understandings. People have also surpassed the quest for food to rely on energy sources barely known two centuries ago. Modifying our surroundings by direct and indirect means is the mainstay of human life, increasingly evident in industrial societies and all societies touched by industrial culture's global tendrils.

Well documented are the data regarding "the great acceleration"—the marked rise in world temperature, atmospheric CO_2, population, water use, motor vehicles, and fossil fuel consumption, along with the loss of forest and species—all traceable to the period between 1750 and 1950, with a steepening rate of change to the present (e.g., Steffen et al. 2004; Zalasiewicz, Williams, and Waters in this volume). The displacement of free-ranging mammal populations across the world by humans and our livestock and pets provides but one astonishing example of the takeover of the planet by our species (Smith et al. 2016).

The human-altered world, now unprecedented in degree and rate, presents enormous hurdles to societal adaptability and the resilience of prevailing ecological foundations. To the Pleistocene ancestors we have discussed here, environmental disruption was invisible and unfelt until it was inescapable, tearing down and replacing the most enduring way of prehistoric human life. It could happen to us. Humans survive in the world by altering it—thus, our species' evolutionary adaptability (A in the formula) instigated by uncertainty (U) in our ancestors' surroundings has come back around to instigate further environmental instability.

What happens next? Fortunately, it is hard to stem the human impulse to try to figure things out. While the Anthropocene is defined by the alteration of our surroundings, our species is defined, in part, by the tendency to alter ourselves—our repertoires of behavior and environmental adaptations. There is simply no other way to account for the grand expressions of human cultural diversity. Although humans survive by altering the environment, the capacity to alter our ways of life, borne out through history, is also a distinguishing aspect of the anthropogenic world.

Therefore, it is reasonable to ask what we can, or should, do in response to this new era of uncertainty. If our thesis here has value, environmental instability has the potential to shake current states of adaptation—that is, the "success" of human life as we currently know it and live it. As imperfect and faulty a guide as it is, the past is all we have to go on. Understanding how our species' adaptability came about may offer some clues about how the Anthropocene may play out. As already noted, the groundwork of adaptability in *H. sapiens* emerged during a time of rising risks and unpredictability and coincided with a new framework of hominin energy acquisition, means of communicating about the world, and social organization involving distant groups. How do these three interacting variables in the equation help explain our future adaptability?

We are an energy-hungry species, and thus much will depend on whether we can transform how energy is obtained, stored, and managed. There is nothing new or particularly insightful about this statement. We know too much to abide any further delusion that the crutch of fossil fuels will continue to enhance our bipedal, cognitive, and cultural way of life, or that land and water degradation and ecological decimation can be considered "unintended." It's gotten to the point that these activities can be considered intentional in that they are carried out knowingly, even though they are perhaps not an express purpose of our actions.

How do the pathways of information currently available to us and the ways we organize our social relationships come into play? As electronic messaging increasingly supersedes face-to-face interaction, practical criteria and ethical principles have yet to be developed that encourage trust, avoid delusion, and kindle reliability in the information we share and the relationships we forge across the spectrum from individuals to nation-states, from communities to international conglomerates. New beliefs and values concerning how we communicate and engage with one another will likely need to emerge in this rapidly shifting age of humans if people across local to global scales are ever to construct common goals and standards to mitigate our current energy use, waste accumulation, and biotic disruption. It is well acknowledged that communities are now created at a distance; by no means are they exclusively local. To evoke our prehistoric example as metaphor, much depends on whether norms of social networking will magnify divisiveness and isolation, or (as in the MSA transformation, apparently) will stimulate interdependence, exchange, and reciprocity. In the latter part of the spectrum, social networks provide safety nets that activate when uncertainty throws societies around the world into peril.

Discussions of the Anthropocene have tended to lean toward complaint, protest, and doom over the unintended consequences of the human global footprint. The question is how humanity may create purposeful outcomes that enhance our adaptability. A focus on *intended consequences* requires us to question the social, economic, political, and other assumptions that influence how we currently conduct our lives. How do we live responsibly in the human-altered world, where dynamic natural systems and traditions of cultural values and practices are intimately knitted together? Where will our adaptability come from? The argument here is that adaptability may evolve as existing, status quo ecological relations are dissolved and replaced as new means of energy use, information transmission, and social relationships (organization) are developed. For there to be "success"—that is, adaptability—the status of our economies, political practices, and mental and moral frameworks will evolve to look different from our present expectations, hopes, and underlying assumptions.

In light of international agreements concerning carbon emission limits aimed at stemming the tide of climate disruption, nations are the most obvious, overarching aspect of human social arrangements that may generate an enduring, positive response to Anthropocene challenges. Existing institutions involved in governance, however, include a wide variety of cultural systems and levels of social organization, including geographic and online communities, corporations, religious systems, and activist movements (Allenby 2005, chapter 1). At present, there is relatively little interdigitation of effort. Yet the interplay of these institutions and scales of social organization, requiring multicultural cooperation that transcends national boundaries and other sources of bias, will be crucial if we are to develop new ideational and intellectual frameworks, value-driven and pragmatic ways of life, and if the desired outcome is to create interconnected, empathetic, and altruistic paths in the future. Such paths of unprecedented organizational networking, I believe, will be central to our future adaptability.

Practicable moral frameworks are bound to be deeply entwined with this vision of future adaptability. Such frameworks are integral to human relationships—that is, how people view and interact with one another. Adopting meaningful ethical precepts arises, in part, from the cultural narratives we create and pass on, which then give rise to particular mental realities that address complex relationships toward risk, hardship, and demise, by seeking sources of security, observing rites of passage, and celebrating birth. These are obviously not all that moral systems address. The point here is that moral frameworks in practice affect how people treat one another and how we consider ourselves as actors and agents in the world. Ethical stances, I believe, will thus deeply influence how well people will navigate the nexus of ecological disorder, technological innovation (including energy acquisition), cultural narratives, and novel forms of social interaction. It is inevitable that moral systems will contribute vitally to human adaptability in the Anthropocene.

I have summarized elsewhere the ethical choices that, to me, make sense as people everywhere stumble into the future and adjust meaningfully to adversity and the prospects ahead. These choices ultimately revolve around whether people will be able to act collectively (Potts 2017). People are suspicious of others whom they consider to be unlike themselves. Yet it is equally true that human society requires the propensity to trust strangers; we see and interact trustfully all the time with persons we have never previously encountered. Suspicion may well arise from experience but is perhaps more often hidden in stereotypes and virulent assumptions about others. The geographic displacement of people and entire cultural groups due to flooding, intense heat, sea-level rise, and disputes over food, water, and land, all will force larger numbers of unfamiliar people into contact with one another. This intersection of people never before encountered may well present the biggest moral question of the Anthropocene: How will we treat each other and act collectively? Empathy, reciprocity, and inclusion are among the focal responsibilities in the face of such new extremes and future uncertainty. Although it is naïve to imagine that xenophobia is eradicable, adaptability in the age of humans will ultimately ride on the possibility of burying suspicion in favor of interdependence. By this I mean abandoning destructive stereotypes in favor of pragmatic exchange and reciprocity. Personally felt narratives about the universality of the human condition could help, as moral codes are best absorbed not theoretically, in the abstract, but by stories, music, art, and all avenues of communication that create purpose. New narratives and the means by which meaningful information is passed along will likely make ethically informed contributions to our future adaptability.

4. Conclusion

As founders of an anthropogenic planet, our species' adaptability—replacing traditional concepts of success—will require persistent decision-making and acts that reshape human life in the face of the novel environments that we create. The adaptability hypothesis proposed here derives from prehistoric evidence near the time of the origin of *Homo sapiens*, and it emphasizes the contingencies that underlie evolutionary history. In our starting example, the longest-enduring hominin way of life became extinct at the expense of technologies and behaviors associated with the earliest members of our species. An intersection of landscape disruption, climatic instability, and ecological transition—which began sometime between 500,000 and 400,000 years ago—led to the possibility that adaptability to change itself might expand, although by no means guaranteeing that hominins would persist. The overall effect of this cross fire of environmental events was an increase in adaptive uncertainty, which undermined the most enduring standard of Pleistocene human life up to that point. Those hominins that emerged from this period of unpredictable environment made a smaller, more mobile technology with innovations that included more effective tools of predation. They depended increasingly on symbolic behavior with its transformative implications, and they formed alliances of survival with distant groups. This trio of changes—greater technological innovation, enhanced communication of information, and reorganization of social awareness and group interaction—all now reside at the heart of the human social experience.

The strategy of the Acheulean was focused on local adaptation by separate groups—that is, group independence, perhaps isolation, compared with the MSA in which intergroup connectivity and resource sharing provided a vital difference. One might imagine how tenacious the Acheulean way of life was in confrontation with new levels of environmental instability.

Likewise, the headwinds against change today are fierce. Resolving Anthropocene disruptions—by which I mean the ecological and existential threats to the status quo, which include powerful nation-state isolationism, ideological walls, and ethnic prejudices—may come down to a similar strategic difference across the anthropogenic planet. Facing the unpredictability of climate and ecosystem disruption currently underway, our current concepts of "owned" and "mine" may well break down, whereas cultural ideals of "shared" and "ours" may prove beneficial. Can we dissolve our status quo acceptance of "us versus them" in favor of a new belief founded on interdependence? Shall we replace our fallback posture of "true to one's own people" with a new sense of belonging to humanity, with kinship that extends beyond the standard definitional boundaries of humanity? Change never arises de novo. The cultural values and behavioral shifts implied by these questions already exist as centers of gravity that may one day draw people together in new possibilities that will enhance life in an uncertain world.

In this chapter I have posited that uncertainty is the engine of adaptability. What are the broader implications? Long-standing philosophies, religious doctrines, and a variety of worldviews have been biased toward the desire for and pursuit of certainty, secure and stable knowledge that lacks ambiguity. I propose here the pervasiveness of uncertainty in the natural world as an instigating force in evolution, including the processes by which humanity and cultures have come into being. Indifferent to human goals, uncertainty appears to drive our relationship with the universe. The evolution of new information systems emanates from ambiguity, born of an ever-present insufficiency of information. Here, I venture deeply into my own ignorance, perhaps naïveté. Because I am no philosopher, I have no defensible proposal as to what a philosophy of uncertainty would look like or how to formulate one. I do suggest, however, that a dedicated philosophical inquiry into uncertainty and the inherent ambiguity of information would be consonant with how adaptability originates and evolves. This effort would make sense if it is reasonable to substitute the concept of adaptability for the supposition of evolutionary success.

The practical significance of this line of thinking will likely become obvious in the face of future uncertainties. Global warming means climate disruption; the sharp release of heat energy from burning fossil fuels has fragmented and redirected the interwoven circulatory system of the planet, the atmosphere, and ocean. Our unprecedented handprint redistributes continental water and revises plant biomes, animal diversity, and biomass. The disruptions are not experienced simultaneously or in the same manner everywhere; the impacts might not be predictable in a pattern that can be easily anticipated. If the pattern of change could be anticipated, our existing penchant for innovation and adaptation could potentially meet and prepare for such predictability. If solving future challenges required only an accommodation to a progressive environmental trend, even severe and unprecedented in rate, our existing capabilities for biocultural adaptation could well be sufficient.

Our future instead requires adjusting to uncertainty, a problem that may well instigate the repackaging of human behavior and revising the social, economic, and political strategies that result. The argument here is that adaptability may evolve as existing, status quo ecological relations are dissolved and replaced as new means of energy use, information transmission, and social relationships are developed. The status of our mental and moral frameworks would thus look vastly different from our present expectations, hopes, and underlying assumptions and will require the development of systems of meaning by which we create new cultural narratives, understandings, and mental realities that address the complex relationships we humans have with an uncertain world.

Acknowledgments

My great appreciation to Hugh Desmond and Grant Ramsey for the invitation to contribute this chapter and for the transdisciplinary mix of authors they have attracted here. Comments from two anonymous reviewers also helped sharpen the clarity of this chapter. Thanks to Jennifer Clark for her assistance on Figures 11.1 and 11.2, and I acknowledge the amazing skills of Kenyan excavators and the dedication of scientific collaborators who have worked with me at Olorgesailie for more than three decades. The Kenyan research reported here was supported logistically by the National Museums of Kenya and funded by the Peter Buck Fund for Research in Human Origins (Smithsonian).

References

Allenby, B. 2005. *Reconstructing Earth: Technology and Environment in the Age of Humans.* Washington, DC: Island Press.

Behrensmeyer, A. K., R. Potts, A. Deino, and P. Ditchfield. 2002. "Olorgesailie, Kenya: A Million Years in the Life of a Rift Basin." In *Sedimentation in Continental Rifts*, edited by R. W. Renaut and G. M. Ashley, 97–106. SEPM (Society for Sedimentary Geology) Special Publication 73.

Beyene, Y., S. Katoh, G. WoldeGabriel, W. K. Hart, K. Uto, M. Sudo, M. Kondo, et al. 2013. "The Characteristics and Chronology of the Earliest Acheulean at Konso, Ethiopia." *Proceedings of the National Academy of Sciences* 110: 1584–1591.

Brooks, A. S., J. E. Yellen, R. Potts, A. K. Behrensmeyer, A. L. Deino, D. E. Leslie, S. H. Ambrose, et al. 2018. "Long-Distance Stone Transport and Pigment Use in the Earliest Middle Stone Age." *Science* 360: 90–94.

Deino, A. L., A. K. Behrensmeyer, A. S. Brooks, J. E. Yellen, W. D. Sharp, and R. Potts. 2018. "Chronology of the Acheulean to Middle Stone Age Transition in Eastern Africa." *Science* 360: 95–98.

Deino, A., and R. Potts. 1990. "Single Crystal $^{40}Ar/^{39}Ar$ Dating of the Olorgesailie Formation, Southern Kenya Rift." *Journal of Geophysical Research* 95 (B6): 8453–8470.

deMenocal, P. B. 1995. "Plio-Pleistocene African Climate." *Science* 270: 53–59.

deMenocal, P. B. 2011. "Climate and Human Evolution." *Science* 331: 540–541.

deMenocal, P. B., and J. Bloemendal. 1995. "Plio-Pleistocene Climatic Variability in Subtropical Africa and the Paleoenvironment of Hominid Evolution." In *Paleoclimate and Evolution, with Emphasis on Human Origins*, edited by E. S. Vrba, G. H. Denton, T. C. Partridge, and L. H. Burkle, 262–288. New Haven, CT: Yale University Press.

Galway-Witham, J., and C. Stringer. 2018. "How Did *Homo sapiens* Evolve?" *Science* 360: 1296–1298.

Grün, R., J. S. Brink, N. A. Spooner, L. Taylor, C. B. Stringer, R. G. Franciscus, and A. S. Murray. 1996. "Direct Dating of Florisbad Hominid." *Nature* 382: 500–501.

Hublin, J.-J., A. Ben-Ncer, S. E. Bailey, S. E. Freidline, S. Neubauer, M. M. Skinner, I. Bergmann, et al. 2017. "New Fossils from Jebel Irhoud, Morocco and the Pan-African Origin of *Homo sapiens*." *Nature* 546: 289–292.

Isaac, G. L. 1977. *Olorgesailie: Archeological Studies of a Mid-Pleistocene Lake Basin in Kenya*. Chicago: University of Chicago Press.

Kingston, J. D. 2007. "Shifting Adaptive Landscapes: Progress and Challenges in Reconstructing Early Hominid Environments." *Yearbook of Physical Anthropology* 50: 20–58.

Kirschner, M., and J. Gerhart. 2005. *The Plausibility of Life*. New Haven, CT: Yale University Press.

Kuman, K., M. Inbar, and R. J. Clarke. 1999. "Palaeoenvironments and Cultural Sequence of the Florisbad Middle Stone Age Hominid Site, South Africa." *Journal of Archaeological Science* 26: 1409–1425.

Lepre, C. J., H. Roche, D. V. Kent, S. Harmand, R. L. Quinn, J.-P. Brugal, J.-P. Texier, A. Lenoble, and C. S. Feibel. 2011. "An Earlier Origin for the Acheulean." *Nature* 477: 82–85.

Levins, R. 1968. *Evolution in Changing Environments.* Princeton, NJ: Princeton University Press.

McBrearty, S., and A. S. Brooks. 2000. "The Revolution That Wasn't: A New Interpretation of the Origin of Modern Human Behavior." *Journal of Human Evolution* 39: 453–563.

McDougall, I., F. H. Brown, and J. G. Fleagle. 2005. "Stratigraphic Placement and Age of Modern Humans from Kibish, Ethiopia." *Nature* 433: 733–736.

Oomen, R. A., and J. A. Hutchings. 2020. "Evolution of Reaction Norms." In *Oxford Bibliographies in Evolutionary Biology*, 1–23. Oxford: Oxford University Press. doi:10.1093/OBO/9780199941728-0130.

Potts, R. 1996. *Humanity's Descent: The Consequences of Ecological Instability*. New York: William Morrow.

Potts, R. 1998. "Variability Selection in Hominid Evolution." *Evolutionary Anthropology* 7: 81–96.

Potts, R. 2013. "Hominin Evolution in Settings of Strong Environmental Variability." *Quaternary Science Reviews* 73: 1–13.

Potts, R. 2017. "What Will It Mean to Be Human?" In *Living in the Anthropocene: Humanity in the Age of Humans*, edited by W. J. Kress and J. K. Stine, 27–31. Washington, DC: Smithsonian Press.

Potts, R., A. K. Behrensmeyer, and P. Ditchfield. 1999. "Paleolandscape Variation in Early Pleistocene Hominid Activities: Members 1 and 7, Olorgesailie Formation, Kenya." *Journal of Human Evolution* 37: 747–788.

Potts, R., A. K. Behrensmeyer, J. T. Faith, C. A. Tryon, A. S. Brooks, J. E. Yellen, A. L. Deino, et al. 2018. "Environmental Dynamics during the Onset of the Middle Stone Age in Eastern Africa." *Science* 360: 86–90.

Potts R., R. Dommain, J. W. Moerman, A. K. Behrensmeyer, A. L. Deino, S. Riedl, E. J. Beverly, et al. 2020. "Increased Ecological Resource Variability during a Critical Transition in Hominin Evolution." *Science Advances* 6: eabc8975. doi:10.1126/sciadv.abc8975.

Potts, R., and J. T. Faith. 2015. "Alternating High and Low Climate Variability: The Context of Natural Selection and Speciation in Plio-Pleistocene Hominin Evolution." *Journal of Human Evolution* 87: 5–20.

Richter, D., R. Grün, R. Joannes-Boyau, T. E. Steele, F. Amani, M. Rué, P. Fernandes, et al. 2017. "The Age of the Hominin Fossils from Jebel Irhoud, Morocco, and the Origins of the Middle Stone Age." *Nature* 546: 293–296.

Scerri, E. M. L., M. G. Thomas, A. Manica, P. Gunz, J. T. Stock, C. Stringer, M. Grove, et al. 2018. "Did Our Species Evolve in Subdivided Populations across Africa, and Why Does It Matter?" *Trends in Ecology & Evolution* 33: 582–594.

Schlichting, C. D., and M. Pigliucci. 1998. *Phenotypic Evolution: A Reaction Norm Perspective*. Sunderland, MA: Sinauer.

Smith, F. A., C. E. Doughty, Y. Malhi, J.-C. Svenning, and J. Terborgh. 2016. "Megafauna in the Earth System." *Ecography* 39: 99–108.

Smithsonian Human Origins Program. 2022. "Evolution of Human Innovation." June 28. http://humanorigins.si.edu/research/east-african-research-projects/evolution-human-innovation.

Solé, R., and B. Goodwin. 2000. *Signs of Life*. New York: Basic Books.

Steffen, W., A. Sanderson, P. D. Tyson, J. Jäger, P. A. Matson, B. Moore III, F. Oldfield, et al. 2004. *Global Change and the Earth System*. Berlin: Springer.

Stringer, C. 2016. "The Origin and Evolution of Homo sapiens." *Philosophical Transactions of the Royal Society B* 371: 20150237. http://dx.doi.org/10.1098/rstb.2015.0237.

Trauth, M. H., M. A. Maslin, A. L. Deino, M. R. Strecker, A. G. N. Bergner, and M. Dühnforth. 2007. "High- and Low-Latitude Forcing of Plio-Pleistocene East African Climate and Human Evolution." *Journal of Human Evolution* 53: 475–486.

Tryon, C. A., S. McBrearty, and P.-J. Texier. 2006. "Levallois Lithic Technology from the Kapthurin Formation, Kenya: Acheulian Origin and Middle Stone Age Diversity." *African Archaeological Review* 22: 199–229.

Vidal, C. M., C. S. Lane, A. Asrat, D. N. Barfod, D. F. Mark, E. L. Tomlinson, A. Z. Tadesse, et al. 2022. "Age of the Oldest Known *Homo sapiens* from Eastern Africa." *Nature* 601: 579–583.

Vrba, E. S. 1995. "On the Connections between Paleoclimate and Evolution." In *Paleoclimate and Evolution, with Emphasis on Human Origins*, edited by E. S. Vrba, G. H. Denton, T. C. Partridge, and L. H. Burckle, 24–45. New Haven, CT: Yale University Press.

West-Eberhard, M. J. 2003. *Developmental Plasticity and Evolution.* Oxford: Oxford University Press.

Yellen, J., and H. Harpending. 1972. "Hunter-Gatherer Populations and Archaeological Inference." *World Archaeology* 4: 244–253.

12

Evolving Measures of Moral Success

Allen Buchanan and Rachell Powell

1. Introduction

The aim of this volume is to reflect upon and problematize different ways of understanding the concept of "human success." Any comprehensive theory of human success should consider achievements in the moral domain, and any naturalistic theory of success in the moral domain should explain how these achievements are possible given the kinds of evolved beings that we are. Of course, we can talk about moral progress without invoking evolutionary concepts at all—the productive history of moral philosophy is a testament to this fact. However, an evolutionary approach to the question of moral success may be especially enlightening, since human evaluative capacities and norms have been shaped by natural selection for specific roles in human ecology, and theorists of moral progress ignore these evolutionary roles at their peril. Any evolutionary account of success in the moral domain must not only show *that* moral progress is compatible with certain evolved features of human psychology, but also it must explain *how* moral progress has been achieved, given these facts.

For the overwhelming majority of human natural history, moral success and evolutionary fitness were closely intertwined. According to the received evolutionary account, morality arose and spread in early human cultural groups because it enhanced within-group cooperation in an arena of intergroup competition over scarce and scattered resources. Moralities that diverged too far from this adaptive role would have had significant evolutionary costs (including existential risks), which would have translated into prohibitive normative costs. During this harsh period of human history, the superiority of one morality over another might well have been gauged by how well it fulfilled these strategic evolutionary functions.

Recently, however, the premium of evolutionary fitness has given way to a truly normative project of human betterment. Human moral success, in

Allen Buchanan and Rachell Powell, *Evolving Measures of Moral Success* In: *Human Success*. Edited by: Hugh Desmond and Grant Ramsey, Oxford University Press.
DOI: 10.1093/oso/9780190096168.003.0012

the thickly normative sense that we mean here, does not refer to the greater differential reproduction of certain cultural moral variants over others. The mere fact that certain moral norms have proliferated over long stretches of human history (e.g., those underpinning institutions of slavery) has no bearing whatsoever on whether their proliferation constitutes moral success as we understand it here.

In this chapter, we summarize our own view of the evolutionary origins of morality, which we use as a springboard for understanding the decoupling of human morality from its long-standing evolutionary roots. Appreciating the ways that moral success and evolutionary fitness have come apart allows us to expose several defects in adaptationist accounts of morality and moral progress. We then go on to show that these flaws prevent standard evolutionary accounts of morality from serving as the bulwark of an argument for ethical pluralism—or the notion that there can be multiple valid moralities. The standard evolutionary argument for ethical pluralism fails, we argue, because there is a yawning gap between a morality's validity and its ability to solve problems of social living. We go on to offer a better argument for ethical pluralism based on two kinds of evolutionary constraints: the developmental dynamics of individual moral ontogeny, and path dependency in the evolution of moral systems. We conclude that human success in its moral dimension can assume many different forms.

2. Evolutionary Accounts of Morality

2.1 The Brighter Side of Morality

According to the received view among evolutionary psychologists and naturalistic philosophers, morality is an adaptation for coping with cooperation problems that early human groups faced during the trying times of the middle to late Pleistocene (~1.9–0.4 million years ago). "Morality" means different things to different research communities; as the term is used in the evolutionary literature, it refers to the interlocking capacities for moral sentiments and moral judgments and the acquisition of social norms that figure into culturally transmitted systems of evaluative attitudes, beliefs, and behaviors.

The prevailing view has it that morality arose and spread in early humans because it increased the fitness (expected reproductive success) of groups and/or their constituent members by transforming *Homo* into a lineage of strategic "super-cooperators" (Tomasello 2016; Henrich 2015; Boehm 2000; Kitcher 2011; Kumar and Campbell 2022), rivaled in their collaborative feats perhaps only by the comparably inflexible eusocial insects. Morality did this, so the empirically constrained adaptationist story goes, by coordinating collective action, managing conflict within groups, and mitigating free riding and other selfish tendencies that undermine cooperation in closely related hominins (such as the extant chimpanzees and the extinct australopithecines and habilines). The vigorously enforced egalitarian norms of hunter-gatherer societies sit in sharp contrast with the hierarchical, dominance-structured societies of chimpanzees and other great apes (Boehm 2000).

According to the standard view, fairness norms and other moral virtues and sentiments are pervasive in hunter-gatherer societies because they enabled human moral groups to compete successfully with nonmoral or weakly moral groups for control of scarce and patchy resources. In conjunction with other cognitive capacities such as joint intentionality and a theory of mind, and along with physiological adaptations that support long-distance running, morality coordinated the successful hunting of large dangerous prehistoric game and it ensured that the fruits of this cooperation were shared among all those who participated. Morality was arguably a key evolutionary innovation that enabled *Homo* (either *erectus* in Africa or *heidelbergensis* in Europe) to ascend to an apex predatory niche, with major ramifications for Earth's terrestrial megafauna (Sandom et al. 2014) and eventually the entire biosphere.

Crucial to the ecological success of *Homo sapiens* was the capacity for cumulative culture, which allowed for the stepwise improvement of technologies, such as the manufacture of weapons, fire (Wrangham 2009), gathering implements, clothing, and other tools, skill sets, and natural history information that were vital to human flourishing in the late Pleistocene world. The transmission of cumulative technological traditions would have been possible only with scaffolded learning structures, in particular apprenticeship-like institutions. Philosopher of science Kim Sterelny (2012) argues that the transmission of moral competences, like any other complex skill set, would have required a similar system of apprenticeship. However, the apprenticeship relationship, whether technological or moral, is an inherently normative institution. The social categories of "teacher" and "student" are infused with standards of appropriate conduct, such as respect, deference, care, and loyalty—creating expectations on the part of those entering this mutually beneficial relationship (Powell 2020). Sophisticated learning environments like apprenticeships thus presuppose, and likely coevolved with, sophisticated normative competences.

In sum, a robust scientific case is building for the idea that morality was adaptively fine-tuned to solve various "problems of social living." Morality allowed early humans to realize otherwise inaccessible cooperative goods, and it laid the groundwork for cumulative culture and hence the rise of *Homo* to ecological preeminence. If this evolutionary story is right in broad strokes, then the "ultimate" reason morality exists, and why it has certain contentful features across cultures—such as the punishment of free riding, the extolling of altruism, and norms of fairness—is because it was shaped by a history of biocultural selection for these strategic cooperative effects.

2.2 The Darker Side of Morality

As told so far, the story of morality has been quite sanguine. Evolutionary theorists of morality have generally focused on the prosocial aspects of moral psychology and culture, namely how evaluative capacities made certain cooperative ways of life possible for human beings. But the received evolutionary explanation of morality also has a sinister side.

A growing body of ethnographic, archaeological, psychological, and mathematical modeling work indicates that in order for altruistic norms to stabilize, there must be institutionalized (third-party) punishment of free riders. The rub is that the evolution of institutionalized punishment, assuming it is costly to punishers, creates a higher-order altruism problem that only group-level selection can solve. And for group-level selection to be strong enough to solve the punishment free-rider problem, there must be competition among discrete human groups resulting in frequent and frequently lethal intergroup conflict.

Only in such an ecological setting would groups be capable of sustaining more egalitarians, altruists, and moralizing punishers, which in turn would enable them to outperform and replace groups with less effective moralities in economic, military, and cultural contests between groups (Bowles 2008, 2009; Sober and Wilson 1998). Traces of this dark human evolutionary past can be found in human psychology: pan-cultural studies have shown in-group favoritism combined in many cases with out-group antagonism (Aboud 2003). In short, according to the received view, our evolutionary legacy is not altruism tout court but rather a highly parochial altruism which underlies an inherently, and one might think unalterably, tribal moral core.

For most of the duration of the human species, the cost of a given morality (a specific, contentful system of moral norms and beliefs) diverging from its adaptive, tribalistic role was evolutionarily—and perhaps normatively—prohibitive. The ability of human members and cultural variants to move fluidly between groups was relatively restricted, and the continued existence of human groups, let alone their flourishing beyond a state of bare persistence, hinged on whether morality sustained these strategic functions. Poorly configured moralities would have left groups unable to engage in effective hunting, gathering, territorial acquisition, defense, and other collaborative activities, and would have made it difficult for them to retain and transmit the technology, skill sets, and natural history information necessary for humans to survive and even thrive in the late Pleistocene.

Poorly configured moralities would also have left groups susceptible to exploitation, absorption, or elimination by competing groups that were blessed with higher-functioning moralities. One way that moralities could fail in this sense is by becoming too inclusive in their delineation of the moral community. Extending robust moral consideration to out-group members, especially when there was no reciprocal strategic advantage to doing so, would have reduced a group's fitness and imperiled its members. At the same time, hunter-gatherer groups that became morally hierarchical *internally* would have suffered severe disruptions of in-group cooperation. Cultural moral groups that reserved meaningful moral consideration to in-group members and interacted with out-groups strategically, warily, and (where advantageous) exploitatively would have systematically outcompeted moral groups that were more "promiscuous" in their extensions of moral regard.

Although in-group favoritism is not necessarily linked to out-group antagonism (Brewer 1999), xenophobic biases are expected to evolve where outgroups pose significant threats or can safely be exploited (Haselton and Nettle 2006). Out-group threats included competition over scarce resources, warfare and raiding, the transmission of infectious disease, the potential for social parasitism through free riding in the absence of effective institutional punishment across groups, and the disruption of the "normative glue" of society through the introduction of foreign attitudes, norms, or behaviors. This ecological setting would have provided fertile grounds for the biocultural evolution of intergroup aggression, antipathy, and distrust. The fact that a certain trait would have been adaptive in the distant past does not imply that it was in fact selected for. Nevertheless, we believe that the totality of the evidence from across the social and psychological sciences supports this group-modulated picture of human morality.

During this harsh period of human history, the superiority of one morality over another might well have been gauged solely by how well it fulfilled these adaptive intergroup roles. It was not until the Neolithic revolution and, in particular the industrial and scientific revolutions, that measures of moral progress and evolutionary fitness began to part ways. The demands of fitness now impose only minimal strictures on the shape of human morality, thanks to the unprecedented surplus, transactional efficiency, technological sophistication, and complex sociopolitical institutions that characterize the modern human world. The relaxing of evolutionary constraints has opened new spaces of normative possibility, allowing for the proliferation of moral goals that are orthogonal or even contrary to evolutionary fitness. In brief, the premium of evolutionary success has been relaxed, carving out space for a truly normative project of human betterment.

3. From Fitness to Flourishing: The Great Decoupling

Without a functional morality, countless cooperative social goods would not be produced and human well-being would undoubtedly suffer for it. Moralities that perform their cooperative functions well are undoubtedly a critical component of human flourishing and well-being to this day. At the same time, it is equally clear to us that there is now a wide gap between moral norms that are conducive to evolutionary fitness (whether indexed to individuals or to cultural groups) and our considered conceptions of what morality prescribes.

Much of morality today sits in stark contrast with the group-centered moralities that dominated the vast majority of human history, calling into question the idea that morality as it now exists is limited to its strategic evolutionary function (Buchanan and Powell 2015). These anomalous features cannot be dismissed as mere aspirations or unobtainable ideals, as they are reflected in transformations in institutional design and large-scale patterns of human behavior. Human moral success, as we currently understand it, involves much more than a tendency to promote fitness-enhancing cooperation, or even to obtain reciprocal benefits by expanding our circle of cooperators.

Moral norms adhered to by large swaths of the human population are now far more inclusive than one would predict if group-beneficial co-operation strongly constrained the possibility space of morality. These "inclusivist" elements of morality, as we call them, are best illustrated by two great expansions of the moral circle (Buchanan and Powell 2018; Buchanan 2020). The first great expansion is the legal and institutional realization (albeit imperfect) of inalienable or human rights that recognize the basic moral equality of all persons regardless of their group membership, race, gender, ethnicity, religion, and so on. The second great expansion involves the attribution of moral standing to some nonhuman animals, or the recognition that they count morally in their own right irrespective of their species membership and that consequently there are moral limits on how such beings may be treated.

Each of these great expansions is underwritten by a conceptual shift away from a strategic conception of the grounds of moral standing and toward a "subject-centered" view of what makes a being a member of the moral community whose interests cannot be arbitrarily disregarded. Delineations of the moral community are now commonly drawn on the basis of subject-centered considerations, such as the possession of certain cognitive capacities. This foundational approach to moral standing cannot easily be explained in strategic terms.

Although many would argue that recent expansions of the moral circle are due to ecological conditions that make such expansions fitness-enhancing or advantageous, we do not find this line of argument persuasive for reasons we discuss at length in our book *The Evolution of Moral Progress* (Buchanan and Powell 2018). In a nutshell, the strategic approach cannot account for the growing recognition that there are moral constraints on how we may treat nonhuman animals—beings which lack strategic capacities and hence are vulnerable to unrestrained exploitation. Further, the costs of social practices that discount or disregard the interests of the world's worst-off people fall disproportionately on the world's worst off. While trade, economic development, and other mutually beneficial arrangements are sometimes in the self-interest of states, in many cases there are no significant costs to nations for their support of a deeply inegalitarian global order or their exploitation of strategically weaker peoples. The motivations for constraining such behaviors are fundamentally other-regarding, not strategic.

How, then, can we make sense of this dramatic decoupling of morality from its strict, tribalistic foundations, and how can we explain regressions in recent times to more "exclusivist" moralities? In a recent book (Buchanan and Powell 2018), we proposed that these trends can be understood in terms of the operation of an adaptively plastic moral psychological mechanism that toggles between exclusivist and inclusivist moral orientations depending on whether out-group threat cues are detected in individual moral ontogeny. These detections can include culturally constructed out-group threats irrespective of their veracity, and in the context of the ethnically diverse modern state, the out-group placeholder can be filled in by internal "others."

For instance, historical eugenics programs falsely cast minority groups as free-riding social parasites with diseased genes and a penchant for prolific reproduction that would lead to the deterioration of the human gene pool and ensuing collapse of Western civilization. The skillful deployment of out-group threat cues in the service of tribalistic morality can also be seen today in the rhetoric of right-wing propaganda outlets in the United States, which warn of an ongoing invasion of physically dangerous, free-riding, disease-ridden aliens pouring across the southern border or taking refuge in urban areas. Similar rhetorical tactics have been used to preserve the supposed purity and security of the European Union. The familiar social moral epistemic tool kit of nativists and would-be genocidaires has deep evolutionary roots.

Importantly, such regressions do not amount to the disbanding or "disengagement" of moral capacities, as some psychologists have suggested (e.g., Bandura 2002, 2016). Rather, they are reversions to thick exclusivist moralities. The promulgation of out-group threat cues can rapidly shrink the moral circle, whereas extensive institutional scaffolding has been necessary to expand it over the past several centuries. The project of inclusion is particularly difficult given the hierarchical structure of post-Neolithic populations, which, unlike pre-state societies, are vulnerable to resource monopolization, free riding, and exploitation by powerful elites and skilled demagogues (Buchanan and Powell 2017). The vulnerability to demagoguery has arguably been exacerbated by the social media revolution (Floridi 2021), which is ironic given that social media was initially touted as a revolutionary force for democracy in the context of the Arab Spring in the 2010s.

In sum, although morality originated as an adaptation to solve certain problems of social living, it has since come to encompass many inclusivist features that are not explicable in functionalist terms. These inclusivist features of morality are difficult to construct and sustain, and they are fragile in the face of demagoguery and deteriorating social conditions, but they nevertheless play an important role in human moral ecology. Even if moral norms contribute to reproductive success of individuals or groups or did so in the past, it is a mistake to think that this is all they do or that this is the only criterion of their success. Once humans achieved surplus reproductive success, morality was no longer tied to the demands of fitness, and more demanding standards for the validity of moralities became applicable. In the remainder of this chapter, we consider what these additional criteria might be and develop an evolutionary argument for ethical pluralism, or the proposition that there is more than one way for humans to succeed on the moral front.

4. Evolutionary Arguments for Ethical Pluralism

4.1 A Simple Adaptationist Argument for Ethical Pluralism

Although the standard adaptationist explanation of morality is purely descriptive, it has been invoked to support the normative thesis of ethical pluralism, the view that there is more than one valid (or reasonable or justifiable or rationally defensible) morality (Wong 2009; Flanagan 2016). Ethical pluralism is distinct from cultural relativism, in that it accepts there are objective, cross-cultural standards for assessing moral validity, but it argues that these objective standards indicate there is more than one valid morality.

While the thesis that there is more than one way for humans to succeed in a truly normative sense may be intuitively appealing, intuitions are not enough; an argument in support of ethical pluralism is needed. It will not suffice, of course, to point to the anthropological fact that there is a plurality of moralities in our world, because that in itself does nothing to establish that there is a plurality of *valid* moralities. Appealing to the standard evolutionary account of morality is one way of filling out the argument for ethical pluralism (though obviously not the only way).[1]

As discussed above, the received view in evolutionary anthropology is that human morality originated as an adaptation for solving problems of social living that early humans faced in the late Pleistocene. If one makes the further assumption that the validity of a morality is indexed to how well it performs this evolutionary function—and as we shall see, this is a big "if"—then it follows that moralities which perform this evolutionary function effectively are better than moralities that do not. In this way, functional efficacy can provide vectors for moral progress and regression. If the validity of a morality is determined by how well it performs this evolutionary function, then moralities could be more or less valid depending upon how well they performed the function. In other words, validity would be a matter of degree. Note that we use the term "validity" here as a stand-in for various concepts that may be used for evaluating moralities, including reasonableness, truth (whether understood in a robust realist or a constructivist manner), and rational justifiability. We need not choose among these for purposes of our argument here.

Some prominent evolutionary theorists of morality appear to identify moral progress with the increased performance of morality's sociobiological function. For instance, Kitcher (2011) argues that morality simply is a social technology for coping with what he calls "altruism failures," or situations in which the interests of some individuals are discounted rather than bringing them into the cooperative fold, resulting in social disruption and conflict. For Kitcher, altruism failures are failures because they fail to capitalize on an ecological opportunity for cooperation, and thus they fail to perform morality's evolutionary function, or they do so less effectively than other moralities.

The view that morality is constituted by its strategic evolutionary function is widely held by philosophers of science (e.g., Sterelny and Fraser 2016; Campbell and Kumar 2013; Kitcher 2011; Railton 1986), evolutionary anthropologists (e.g., Curry, Mullins, and Whitehouse 2019; Tomasello 2016), and empirical moral psychologists (e.g., Greene 2015; Haidt 2012). In a recent paper, for example, Curry, Mullins, and Whitehouse (2019, 48) claim to offer cross-cultural data in support of the view that "morality consists of a collection of biological and cultural solutions to the problems of cooperation recurrent in human social life." If one does hold this sort of "constitutivist" view—and we will offer reasons for rejecting it—then it follows that the validity of a moral system depends on how well it performs these cooperative functions. And if different cultural and ecological contexts call for different functional moralities, or if there is more than one morality that can perform equally well the functions that all societies require, then at least a moderate version of ethical pluralism may be true.

We say "moderate" ethical pluralism because all valid moralities may share some basic norms even if they diverge reasonably in others. Indeed, the received evolutionary account makes this "partial moral overlap thesis" a reasonable assumption: if all human groups faced the same basic design problems in social living, *and* if there are significant constraints on the space of solutions to those problems, then one would expect there to be some commonality in adaptive responses among cultural moral groups. By the same token, if human groups face the same basic design problems, but numerous alternative solutions to these design problems are readily accessible to selection, then there is no reason to expect significant convergence on the shape of moralities, even if there is convergence on their cooperative *function*.

Work in ethnography and evolutionary anthropology indicates that some basic norms, such as "Keep promises," "Reciprocate when others provide aid," "Be brave," and "Do not kill members of your own group," tend to be included in all or most moralities, presumably because they underpin crucial social functions without which cooperative frameworks would diminish or disintegrate. Let us call these overlapping norms "core moral norms." Even if core moral overlap is substantial, this is compatible with reasonable cross-cultural variation in the content of other norms, in the interpretation of overlapping basic norms, and in the weightings accorded to the latter when they conflict with one another (Nisbett 2004). These divergent weighting schemes could reflect functionally equivalent solutions to common social design problems, or they could reflect solutions to entirely different design problems. Either way, it is unlikely there will be only one evolutionarily "correct" way to assign weights to core values.

As we descend into more fine-grained social ecological details, the number of evolutionarily effective moralities begins to multiply. The set of moral characteristics that are evolutionarily optimal for a hunter-gatherer group in an arena of intense intergroup competition may be very different from those well suited for the challenges of sedentary agricultural communities, which in turn differ from those that fit the ecology of rugged livestock herding, which again will diverge from those that are appropriate for life in modern hierarchically differentiated states, and so on.

If we add to this descriptive functionalist analysis the normative premise that the validity of moral systems depends on their ability to solve problems of social living, then it follows that there will be a plurality of valid moralities, so long as there are differing social design problems or shared problems of social living that can be solved by more than one morality. This is not to say that the plausibility of the adaptationist explanation of morality *entails* ethical pluralism. After all, ethical pluralism is a normative thesis, and consequently there can be no straightforward inference from any descriptive-explanatory account to its truth or its falsity. But the adaptationist explanation could provide a naturalistic framework for ethical pluralism that puts pressure on ethical monism, or the thesis that there is only one valid morality or set of moral principles against which all moral progress should be judged.

4.2 Why the Adaptationist Evolutionary Argument for Ethical Pluralism Fails

Moral philosophers who draw on evolutionary thinking to bolster the case for ethical pluralism have adopted a cogent argumentative strategy. Unlike the logically flawed (if well-meaning) anthropological assault on moral objectivity, they do not infer from the fact that there are multiple moralities that there are multiple valid moralities, nor are they committed to the indefensible relativist idea that no morality can be better than any other. And unlike facile attempts to derive ethics from evolution, the adaptationist case for ethical pluralism does not commit a gross is-ought logical violation or rely on problematic appeals to human nature. But in attempting to elide the naturalistic fallacy and essentialist conceptions of human nature, the adaptationist argument for ethical pluralism stumbles into other serious logical difficulties.

First, there is a yawning gap between a morality's validity and its success in "solving problems of social living," whether these functional solutions are indexed to fitness or to some other causal role. It is true that morality is closely associated with cooperative functions, which in turn explain its origin and proliferation. But this does not imply that human morality as it exists today is exhausted by this function. The point is not that we can always ask whether functional performance is in fact good, à la Moore's "open question" argument. Rather, it is that the very justification that evolutionary functionalist accounts of morality give in their own defense—such as that they unify or systematize our understanding of morality and moral progress—fails on its own terms, or at least has significant limitations.

Recall that if the received evolutionary account of morality is right, then the evolutionary function of morality is not to support cooperation per se but rather *within the group*—while *exacerbating* intergroup altruism failures in an arena of competing groups. If this is so, then the strategic functionalist view cannot encompass crucial cases of moral progress that involve extending altruism to strangers, to members of foreign cultural groups, and to individuals with no strategic capacities at all. For instance, the two great expansions of the moral community discussed earlier are paradigmatic cases of moral progress, and yet they are not explicable, let alone normatively defensible, in evolutionary functionalist terms. Because the standard evolutionary account is a hypothesis about the *origins* of human morality—and not about its present constitution or effects—it is compatible with human morality becoming something more than a social technology. The function of the rotationally flexible human shoulder joint may be to permit the skillful throwing of projectiles, but it is now used for all sorts of functions. The shoulder is no more "a type of projectile thrower" than morality is just "a type of cooperation."

Moreover, there are many ways of achieving social coordination, managing conflicts, and distributing resources, and some of these ways are thought to be fairer or more humane than others. From the standpoint of selection, however, whether it occurs at the individual or the group level, all functionally equivalent solutions are identical—none has an edge over the others. But *functional* equivalence clearly does not imply *moral* equivalence.

Indeed, the link between biological fitness and psychological well-being is contingent, and the two often come apart. It is true that being adequately nourished, being disease-free, and having positive relations with members of one's social group can figure in both evolutionary fitness and moral success. But there are severe constraints on the evolutionary process when it comes to producing outcomes that line up with human good (Powell and Buchanan 2011; Buchanan 2011; Powell 2010; cf. Lewens 2012). Although natural selection created the capacity for welfare (Mikhalevich and Powell 2020), it does not aim at well-being; the latter is simply an instrument or by-product of an amoral evolutionary process driven solely by differences in reproductive success. Natural selection has managed to produce creatures who care even though the evolutionary process itself does not. Just as the well-being of individual organisms is often sacrificed in the name of fitness (Williams 1993),[2] so too is the evolutionary success of groups sometimes promoted by massive reductions in the well-being of a subset of their members. Institutions of slavery and military conscription in wars of conquest are obvious examples.

For human organisms, which not only *have* a life but *lead* a life, the non-overlap between fitness and flourishing is especially pronounced. Many individuals in developed societies choose to forgo behavior that maximizes their own (or their own group's) reproductive success in the pursuit of other personal goods, such as career development or the ability to invest more in fewer offspring. The strong divergence between the performance of an evolutionary function, on the one hand, and human health and well-being, on the other, has led some theorists to abandon biofunctionalist accounts of disease in favor of value-based conceptions (see, e.g., Ereshefsky 2009; for discussions, see Powell and Scarffe 2019a, 2019b).

The best one can say is that under conditions in which solving problems of social living promotes valid moral goals or values, a morality's contribution to fitness counts in favor of its validity (other things being equal). But in that case, it is value, not fitness, that is doing the work of moral justification. Thus, a morality's ability to fulfill a narrow understanding of morality's function should provide at most tentative, defeasible evidence that it is valid. In response, the evolutionary functionalist could simply concede that their working conceptions of "altruism failure" and "solving problems of social living" have inherent normative content. The problem with this move, however, is that this normative content cannot be derived from or justified by any evolutionary facts, and so it is unclear why such an account should be characterized as functionalist in the first place.

Note that there is an ambiguity in what one might mean in saying that the evolutionary function of morality is to "solve problems of social living." One might, for instance, conceive of this phrase in a broad sense that is not limited to enhancing the reproductive success of individuals, groups, or even cultural variants themselves. If one insists on characterizing morality as functional, then there are good reasons, given the multifaceted role of morality in contemporary human affairs, to assume a more expansive notion of function. However, untethering function from evolutionary mechanisms raises thorny conceptual and methodological questions: What do social functions consist in? How do they come about? And how can we identify them in practice? It is not clear that such problems have ever been answered satisfactorily outside of an evolutionary framework.

Another compelling reason to reject the view that morality simply is a cultural solution to recurrent problems in social living is that some elements of morality are not social in any significant sense. They are instead ideals of character or value commitments that have to do with individual moral excellence or integrity, irrespective of any relationship to solving social design problems. It is true that some intrapersonal moral virtues (generosity, bravery, altruism, honesty, friendship, etc.) may enhance group or individual performance and could thus be explained by their instrumental contributions to individual or group fitness. In some cases, however, these intrapersonal aspects of morality are not only independent of but also in tension with fitness. Included here would be various ascetic ideals according to which the highest moral excellence is to withdraw from society and to remain celibate.

All of this is not to discount the significant moral progress that solving intragroup social design problems has achieved, both in our hunter-gatherer past and in the contemporary world of developed states. Like other social animals, the conditions for individual human flourishing are undoubtedly tied, to some degree, to the evolutionary fate of one's group. But there is no reason to think that morality consists solely in these intragroup effects.

According to what criteria, then, could one adjudicate between various moralities, including those that succeed equally in solving problems of social living? This is the million-dollar question that we will not attempt to resolve here. But one obvious answer is their success in promoting human flourishing or well-being, where this is a more demanding feat than merely increasing reproductive success or successfully achieving cooperation. However, there is no reason to assume that conduciveness to human flourishing or well-being is the only relevant criterion by which to evaluate rival moralities. One could argue that any viable human morality must include norms of fair distribution as well. Alternatively, or in addition, one might hold that a necessary condition for the validity of a morality is that it includes norms that acknowledge the equal basic moral status of persons or the recognition of moral standing. The point is that even if more than one morality succeeds in fulfilling social functions, it does not follow that there is more than one valid morality. Indeed, there might not be any valid moralities at all.

5. A Better Evolutionary Argument for Ethical Pluralism

Another reason that evolutionists have not made a strong case for ethical pluralism is that their borrowings from evolutionary theory have been too meager, confined almost exclusively to adaptationist thinking. Merely focusing on the idea that morality originated as an adaptation that performed the function of coping with recurrent problems of human social living, and then noting that there is a plurality of ways of successfully performing that function, is insufficient to establish ethical pluralism. We will now develop a richer evolutionary argument for ethical pluralism that does not make the mistake of equating moral validity with fitness enhancement or even with the fulfillment of any functions at all.

For purposes of this argument, we will proceed on the reasonable assumption that whatever the exhaustive set of criteria for evaluation may turn out to be, it is likely to include flourishing or well-being. There are likely to be other criteria for a morality's validity, such as its overall logical coherence, whether it reposes on false empirical beliefs, and the extent to which it is capable of guiding practical action. However, our aim here is to provide an evolutionary argument for ethical pluralism that is consistent with accounts of moral success that take flourishing or well-being to be an important factor.

We will appeal to two evolutionary dynamics to support this argument: (1) early plasticity paired with later rigidity in individual moral development and (2) path-dependent constraints on the evolution of moral systems. Taken together, these two phenomena suggest that there is more than one way of flourishing for human beings—and if conduciveness to flourishing or well-being is one plausible criterion for evaluating the success of moralities, then there is likely to be more than one valid morality (if there is at least one valid morality).

5.1 Moral Psychological Malleability

Generally speaking, traits (including behaviors) are more plastic or sensitive to environmental inputs in earlier phases of developmental ontogeny (West-Eberhard 2003). Precisely why developmental systems exhibit this plasticity-rigidity asymmetry is not fully understood, and its causes are likely to vary across species. In the case of moral psychology, explanations are likely to include a combination of adaptive canalization of early cultural learning, combined with the intrinsic structure of developmental systems—with earlier nodes in the developmental cascade being connected with many distal nodes such that the alteration of earlier nodes tends to have cascading deleterious consequences. Whatever its cause, early malleability paired with later relative inflexibility is likely to be a pervasive feature of cognitive-social-moral development in humans.[3]

The second key step in our moral malleability argument for ethical pluralism is to recognize that the sociomoral environment shapes one's capacity for flourishing. Whether an individual flourishes by living a particular type of life generally depends upon whether their sociomoral environment favors, privileges, and rewards that type of life, especially during the period in which the individual is developing their capacities for flourishing. In other words, flourishing requires a certain fit between one's moral psychology and the possibilities offered by one's sociomoral environment.

For example, if an individual's moral development unfolds in a society that provides rich opportunities for engagement in family life and features moral norms and practices that reward such engagement as well as esteemed role models who exhibit it, then they are likely to develop a capacity for finding fulfillment in family life, and their social environment will make it easier for them to pursue this end without serious psychological conflicts or frustrations. If the social environment makes the rewards of family life high and the costs low, the capacity for flourishing as an individual for whom family life is central will develop accordingly.

In essence, that individual will become *psychologically attuned* to flourishing in that mode of life. If they were transported into an extremely individualistic social environment, they may not achieve the kind of fit between their psychology and their life prospects that flourishing requires, given the recalcitrance of early moral development, as discussed above. And if conduciveness to flourishing is an important criterion of validity for moralities, it will also be true that for individuals who have traversed that path of moral development, a morality that privileges family life will be more likely to be valid (ceteris paribus) than one that does not. And conversely for the individual whose moral development unfolds in a robustly individualistic social-moral environment.

The psychological distress caused by cultural moral mismatch is examined by Jennifer Morton (2014) in her excellent political philosophical analysis of code-switching in multicultural societies. "Code-switching" refers to the ability of low-income minorities to adapt their behavior, language, and presentation in order to take advantage of economic and educational opportunities available to the middle class while remaining meaningfully engaged with the collectivist values of their families and local communities. The result is that code-switchers must straddle "considerably different and often conflicting ways of being" (279). Individuals who are unable or unwilling to code-switch are forced to choose between moral alienation from their communities and the pursuit of economic opportunity.

So long as conduciveness to human flourishing is a significant contributor to the validity of a morality, then even if all human beings share the same core moral foundations (Graham et al. 2013) or features of human flourishing, there is likely to be a plurality of valid moralities, if there are any valid moralities.[4]

This view contrasts sharply with the traditional, conservative idea that universal features of human nature strongly constrain the shape that successful moralities can take (Buchanan 2009). Dressed in modern evolutionary garb, such views see evolved human nature as a source of substantive moral rules (e.g., Ruse 1986). If that view were correct—if human moral nature were so contentful and fixed as to prescribe a single way of flourishing—then one would be forced to conclude that there is in fact only one "correct" path of moral development. Nevertheless, the traditional view has little to commend it. Most contemporary moral philosophers, surveying the wreckage of attempts to derive a single, comprehensive, substantive morality from an understanding of human nature, have largely abandoned the project (Hull 1986; Lewens 2012). In contrast, our proposal simply argues that flourishing and well-being can be attuned to different moral developmental environments, which in turn give rise to a plurality of valid moralities.

5.2 Path Dependency and the Moral Validity Landscape

Even if there is only one form of human life that is morally optimal, and hence one uniquely valid morality for humans, that would still be compatible with ethical pluralism. Making this counterintuitive point requires two shifts of focus: first, a shift from individual moral development to societal moral evolution, and second, a shift from thinking about moral progress in terms of moral *destinations* to thinking in terms of moral *trajectories.*

Societies undergo changes in moral norms and practices over time. In some cases, these changes are responses to new problems, and solutions to new problems sometimes create further problems, which in turn stimulate further changes. The different moral trajectories that societies take can constrain their ability to transition to a globally optimal or ideal moral state (if such a state exists), or even to locally optimal states that are in the moral neighborhood. The result is that moral evolution is *path-dependent* in the following sense: whether a morally better state, judged from some ideal perspective, is accessible to a moral system depends upon previous states of the system. By way of a simple analogy: in the early building phases, a house can take on many different configurations; halfway through construction, however, the possible outcome space has been narrowed considerably and certain designs are now off-limits unless one scraps the entire project and starts over.

Path-dependent constraints, understood as biases that earlier outcomes place on the probability distribution of subsequent outcomes, are a ubiquitous feature of biological evolution (Szathmary 2006). Natural selection cannot simply demolish an organism and go back to the drawing board; it must work with the developmental platform available in all of its convoluted interdependencies, and it must fold in useful variations in the contingent and cobbled order in which they arise. Thus, path dependence involves a degree of irreversibility: certain outcomes in possibility space are rendered highly unlikely (or effectively inaccessible) as new "steps" are taken along an evolutionary trajectory. We will call these sorts of constraints "causal path-dependent constraints."

There is good reason to think that significant parts of human morality are subject to path-dependent constraints of the causal variety. To say that the history of a moral community imposes causal constraints on its possibilities for further moral development is to say that, given the current location of the community's norms and practices in the space of possible human moralities—a location that is strongly influenced by the contingent evolutionary developments that the community has already undergone—certain further changes are not plausible (i.e., are extremely unlikely to occur) given the evolutionary laws and causal structures that govern those systems. It is plausible that the same generic causes that make morphological evolution path-dependent (and, in some cases, irreversible) also make cultural evolution path-dependent (and, in some cases, irreversible).

One reason to think this may be so is that cultures exhibit patterns of evolutionary stasis much like species do (for a discussion, see Powell and Shea 2014). This indicates that human cultures, including moral cultures, may be subject to "homeostatic" evolutionary mechanisms akin to those that keep species within certain bounds of variation for the duration of their existence. Cultural systems, like morphological systems, are likely to have interconnections among components that make modifying "entrenched" central nodes extremely difficult without radically reengineering the entire platform (Wimsatt 2014; but see Carruthers 2006 on modularity in complex systems). Unlike evolution by natural selection, however, cultural moral change can be directed by the forward-looking decisions of human agents. Yet, even if human beings came to have both sufficient knowledge of the complex causal structure of biocultural moral systems and the technical wherewithal to effect desired "cultural macromutations," this still leaves a further path-dependent constraint: normativity.

Even if navigating to morally optimal peaks is causally possible, doing so can be normatively prohibitive. Depending upon the current features of a given moral community—which in turn depend on the contingent path that community has already traversed—certain changes, including ones that would otherwise be morally progressive, may be achievable only at excessive human costs, including moral costs. For example, a causally feasible state of society might be morally superior to the status quo, but due to path dependency there might be no feasible way of getting to it without massive violations of basic rights, including the wrongful deaths of many people. Part of what makes the morality of revolution so problematic, for instance, is that path-dependent constraints on moral evolution will often entail the imposition of morally repugnant costs on the present generation if progressive improvements for later generations are to be secured.

Normative path-dependent constraints on moral evolution can be illustrated by borrowing another key idea from evolutionary thinking: the notion of suboptimality traps. Most readers of this volume will be familiar with "fitness landscapes": visual representations of different spaces that a lineage could occupy, where space reflects the developmental distance of particular assemblages of genes/traits and vertical dimensions of the landscape (peaks, valleys, and plains) reflect the fitnesses of those genes/traits. The higher the peak, the greater the reproductive fitness of organisms having those traits, and vice versa.

Now suppose that the current generation of some lineage occupies a particular peak on the landscape. Suppose also that there is a higher peak that members of the lineage could occupy—a different assemblage of genes or traits that would confer greater reproductive fitness if that assemblage could be biologically realized. But there's a catch: the only feasible causal path from the lower fitness peak to the higher one traverses one or many valleys. In other words, the path involves one or more reductions in fitness in one of its segments. Because natural selection will not permit a reduction in fitness in order to achieve a higher gain at some later time, the lineage cannot navigate to the higher peak—or, at least, it cannot do so under the sole influence of natural selection. So long as the adaptive landscape is rugged, globally optimal fitness peaks, if they exist, will be causally inaccessible to selection.

Consider now a *moral* landscape, in which the peaks, valleys, and plateaus are a function of moral morphologies and their validity, with distance in the landscape a function of cultural distance. A given moral community now occupies a local, morally suboptimal peak. There is a higher moral peak in reasonable proximity that is visible to the community, but it may be impossible, either causally or normatively (or both), for the community to navigate to that peak. It will be causally impossible if, given the initial developmental and social conditions that characterize the status quo, there is no causal sequence that is likely to begin with those initial conditions and culminate in the morally superior state of affairs. For instance, perhaps making this transition would require the systematic, population-level alteration of basic human motivations—an outcome that we have no idea how to effect, with even well-intentioned attempts to do so at risk of pushing the community into a morally worse position on the landscape.

Even where the transition to a higher moral peak in a rugged landscape is causally possible—say, through radical forced social engineering or the systematic biomedical alteration of human capacities—it may be morally impossible: that is to say, not achievable through morally acceptable means or processes. Of course, if a moral community could instantaneously reconfigure itself so as to teleport to a higher moral validity peak, it would avoid any normative costs associated with the transition. However, human cultural change, though often faster than biological evolution, is still largely incremental, and so it will inevitably pass through valleys to reach distant peaks in a rugged moral landscape. Wishing for cultural teleportation is as futile as wishing for a hopeful genetic monster.

Transitions to a higher moral validity peak might require unacceptable human rights violations or intolerable intergenerational injustice, as might be the case if the path to the higher peak spanned several generations with the greatest burdens falling on the earlier generations and the greater benefits accruing to the later ones. Or it might require the systematic, forcible redistribution of legitimately acquired wealth or other severe violations of individual autonomy or liberty in the service of bringing about other social goods. Or it could require violent social struggles and the moral risks they entail.

It is plausible that in any or all of these cases, the moral "valleys" that a community would need to traverse would not offset, or could not acceptably be traded off with, the moral benefits that come from climbing a higher moral peak. Even a strict utilitarian could weigh the expected utility costs associated with traversing moral valleys and judge that a transition to the single optimal peak is not, on balance, worth the arduous trip. Deontologists could simply find the violations of moral norms that would be required to reach the higher peak unconscionable. The point is that even if there is one globally optimal morality, path-dependent constraints on cultural moral evolution are likely to result in multiple valid moralities that have no transverseable path to the global peak. In other words, a morality can be valid if, inter alia, it promotes human flourishing and can be feasibly obtained given the path-dependent constraints to which it is liable. And given path dependency, there is no reason to think that any one morality has or will more closely approximate the ideal or best human morality, if there is such a thing.

6. Conclusion

We began this chapter by emphasizing the distinction between the original evolutionary function of human moralities and what moralities have now become. We argued that when human cooperation became sufficiently successful, and various cultural innovations effectively buffered human beings from the hostile environments that shaped early moralities, the character of moral success became decoupled from its long-standing connection to fitness. In particular, a shift toward more inclusive moralities could occur without unacceptable evolutionary costs, and additional criteria apart from conduciveness to fitness, such as human flourishing and well-being, could become crucial for evaluating moral success. We further concluded that two evolutionary dynamics—the structure of individual moral ontogeny and the path dependency of cultural moral phylogeny—make it unlikely that there is only one valid morality for human beings. Hopefully, our analysis has shown how evolutionary thought should and should not factor into conceptions of moral success.

Acknowledgments

The authors are indebted to Jacob Barrett, Daniel Kelly, Victor Kumar, Alexander Rosenberg, Hanno Sauer, audiences at Oxford University, University of Arizona, and Boston University, and the editors of this volume for very helpful comments on an earlier draft of this manuscript.

Notes

1. In normative ethical theory, "ethical pluralism" is sometimes used to refer to moral theories that have more than one basic value or principle, whereas "ethical monism" refers to moral theories that feature only one basic value or principle. Here, we take

ethical pluralism to refer not to the number of basic principles in a sound ethical theory but rather to the notion that there are many valid moralities, as understood above.

2. For instance, selection for less tissue maintenance, reduced lifespan, and large numbers of low-quality offspring (so-called *r-selection*) can occur even when this is not in the best interests of the individual organisms that comprise a lineage, because those interests have no bearing on fitness.
3. To claim there is greater malleability in early development paired with greater rigidity later on does not commit one to the discredited thesis that human moral psychology is a "blank slate" upon which the social environment inscribes the whole of morality, or even to more plausible antinativist views. Even if it is true, as some moral psychologists have speculated, that there is a set of innate "moral foundations" present in all human beings who are capable of developing a morality (Graham et al. 2013), these are fully compatible with it being the case that the specification of "foundations" that occur earlier in the developmental sequence are more causally potent than those that occur later.
4. The proviso "if there are any valid moralities" is important. It is at least possible that even if there is a plurality of moralities that promote human flourishing (because there are different ways of flourishing), it might still be the case that there is no morality that satisfies the entire suite of requirements for validity, whatever they should turn out to be. That possibility cannot be ignored, especially if there are likely to be other criteria for validity in addition to conduciveness to human flourishing. Further, if validity, as applied to moralities, is a scalar concept (if moralities can be more or less valid), then it is possible that even if two or more moralities did equally well in promoting human flourishing, one might be "more" valid than the others. Alternatively, moral validity could be conceived as a threshold concept, such that any morality that meets a certain threshold of validity criteria is valid; on this view, even if some moralities that exceed the validity threshold are superior to other moralities that exceed the validity threshold, ethical pluralism would still be true.

References

Aboud, F. E. 2003. "The Formation of In-Group Favoritism and Out-Group Prejudice in Young Children: Are They Distinct Attitudes?" *Developmental Psychology* 39 (1): 48–60.

Bandura, A. 2002. "Selective Moral Disengagement in the Exercise of Moral Agency." *Journal of Moral Education* 31: 101–119.

Bandura, A. 2016. *Moral Disengagement: How People Do Harm and Live with Themselves*. New York: Worth.

Baumard, N. 2018. "Increased Affluence, Life History Theory, and the Decline of Shamanism." *Behavioral and Brain Sciences* 41: e67.

Boehm, C. 2000. *Hierarchy in the Forest: The Evolution of Egalitarian Behavior*. Cambridge, MA: Harvard University Press.

Boehm, C. 2011. *Moral Origins: The Evolution of Virtue, Altruism, and Shame*. New York: Basic Books.

Bowles, S. 2008. "Conflict: Altruism's Midwife?" *Nature* 456 (7220): 326–327.
Bowles, S. 2009. "Did Warfare among Ancestral Hunter-Gatherers Affect the Evolution of Human Behaviors?" *Science* 324 (5932): 1293–1298.
Brewer, M. 1999. "The Psychology of Prejudice: Ingroup Love or Outgroup Hate?" *Journal of Social Issues* 55 (3): 429–444.
Buchanan, A. 2009. "Human Nature and Enhancement." *Bioethics* 23 (3): 141–150.
Buchanan, A. 2011. *Better Than Human: The Promise and Perils of Enhancing Ourselves*. Oxford: Oxford University Press.
Buchanan, A. 2020. *Our Moral Fate: Evolution and the Escape from Tribalism*. Cambridge, MA: MIT Press.
Buchanan, A. and R. Powell. 2015. "The Limits of Evolutionary Explanations of Morality and Their Implications for Moral Progress." *Ethics* 126 (1): 37–67.
Buchanan, A., and R. Powell. 2017. "De-moralization as Emancipation: The Evolution of Invalid Moral Norms." *Social Philosophy and Policy* 34 (2): 108–135.
Buchanan, A., and R. Powell. 2018. *The Evolution of Moral Progress: A Biocultural Theory*. New York: Oxford University Press.
Campbell, R., and V. Kumar. 2012. "Moral Reasoning on the Ground." *Ethics* 122 (2): 273–312.
Campbell, R., and V. Kumar. 2013. "Pragmatic Naturalism and Moral Objectivity." *Analysis* 73 (3): 446–455.
Carruthers, P. 2006. *The Architecture of the Mind*. New York: Oxford University Press.
Curry, O. S., D. A. Mullins, and H. Whitehouse. 2019. "Is It Good to Cooperate? Testing the Theory of Morality-as-Cooperation in 60 Societies." *Current Anthropology* 60 (1): 47–69.
Ereshefsky, M. 2009. "Defining 'Health' and 'Disease.'" *Studies in History and Philosophy of Biological and Biomedical Sciences* 40 (3): 221–227.
Flanagan, O. 2016. *The Geography of Morals: Varieties of Moral Possibility*. New York: Oxford University Press.
Floridi, L. 2021. "Trump, Parler, and Regulating the Infosphere as Our Commons." *Philosophy & Technology* 34 (1): 1–5.
Graham, J., J. Haidt, S. Koleva, M. Motyl, R. Iyer, S. P. Wojcik, and P. H. Ditto. 2013. "Moral Foundations Theory: The Pragmatic Validity of Moral Pluralism." *Advances in Experimental Social Psychology* 47: 55–130.
Greene, J. D. 2015. "The Rise of Moral Cognition." *Cognition* 135: 39–42.
Haidt, J. 2012. *The Righteous Mind*. New York: Vintage Press.
Haselton, M. G., and D. Nettle. 2006. "The Paranoid Optimist: An Integrative Evolutionary Model of Cognitive Biases." *Personality and Social Psychology* 10: 47–66.
Henrich, J. 2015. *The Secret of Our Success: How Culture Is Driving Human Evolution, Domesticating Our Species, and Making Us Smarter*. Princeton, NJ: Princeton University Press.
Hull, D. L. 1986. "On Human Nature." *Proceedings of the Philosophy of Science Association* 2: 3–13.
Kitcher, P. 2011. *The Ethical Project*. Cambridge, MA: Harvard University Press.
Kumar, V., and R. Campbell. 2022. *A Better Ape: The Evolution of the Moral Mind and How It Made Us Human*. New York: Oxford University Press.
Lewens, T. 2012. "Human Nature: The Very Idea." *Philosophy and Technology* 25: 459–474.
Mikhalevich, I., and R. Powell. 2020. "Minds without Spines: Evolutionarily Inclusive Animal Ethics." *Animal Sentience* 29 (1). https://www.wellbeingintlstudiesrepository.org/cgi/viewcontent.cgi?article=1527&context=animsent.

Morton, J. M. 2014. "Cultural Code-Switching: Straddling the Achievement Gap." *Journal of Political Philosophy* 22 (3): 259–281.

Nisbett, R. 2004. *The Geography of Thought: How Asians and Westerners Think Differently . . . and Why*. New York: Simon and Schuster.

Powell, R. 2010. "What's the Harm? An Evolutionary Theoretical Critique of the Precautionary Principle." *Kennedy Institute of Ethics Journal* 20 (2): 181–206.

Powell, R. 2020. *Contingency and Convergence: Toward a Cosmic Biology of Body and Mind*. Cambridge, MA: MIT Press.

Powell, R., and A. Buchanan. 2011. "Breaking Evolution's Chains: The Prospect of Deliberate Genetic Modification in Humans." *Journal of Medicine and Philosophy* 36 (1): 6–27.

Powell, R., and A. Buchanan. 2016. "The Evolution of Moral Enhancement." In *The Ethics of Human Enhancement: Understanding the Debate*, edited by S. Clarke, J. Savulescu, C. A. J. Coady, A. Giuibilini, and S. Sanyal, 239–260. Oxford: Oxford University Press.

Powell, R., and E. Scarffe. 2019a. "Rehabilitating 'Disease': Function, Value, and Objectivity in Medicine." *Philosophy of Science* 86 (5): 1168–1178.

Powell, R., and E. Scarffe. 2019b. "Feature Article: Rethinking Disease: A Fresh Diagnosis and New Conceptual Treatment." *Journal of Medical Ethics* 45 (9): 579–588.

Powell, R., and N. Shea. 2014. "Homology across Inheritance Systems." *Biology and Philosophy* 29 (6): 781–806.

Railton, P. 1986. "Moral Realism." *Philosophical Review* 95 (2): 163–207.

Ruse, M. 1986. "Evolutionary Ethics: A Phoenix Arisen." *Zygon* 21 (1): 95–112.

Sandom, C., S. Faurby, B. Sandel, and J.-C. Svenning. 2014. "Global Late Quaternary Megafauna Extinctions Linked to Humans, Not Climate Change." *Proceedings of the Royal Society B: Biological Sciences*: 281 (1787): 20133254.

Sober, E., and D. S. Wilson. 1998. *Unto Others*. Cambridge, MA: Harvard University Press.

Sterelny, K. 2012. *The Evolved Apprentice: How Evolution Made Humans Unique*. Cambridge, MA: MIT Press.

Sterelny, K., and B. Fraser. 2016. "Evolution and Moral Realism." *British Journal for the Philosophy of Science* 68 (4): 981–1006.

Szathmáry, E. 2006. "Path Dependence and Historical Contingency in Biology." In *Understanding Change: Methods, Methodologies and Metaphors*, edited by A. Wimmer and R. Kössler, 140–157. London: Palgrave Macmillan.

Tomasello, M. 2016. *A Natural History of Human Morality*. Cambridge, MA: Harvard University Press.

West-Eberhard, M. J. 2003. *Developmental Plasticity and Evolution*. Oxford: Oxford University Press.

Williams, G. C. 1993. "Mother Nature Is a Wicked Old Witch." In *Evolutionary Ethics*, edited by M. H. Nitecki and D. V. Nitecki, 217–231. Albany, NY: SUNY Press.

Wimsatt, W. C. 2014. "Entrenchment and Scaffolding: An Architecture for a Theory of Cultural Change." In *Developing Scaffolds in Evolution, Culture, and Cognition*, edited by L. R. Caporael, J. R. Griesemer, and W. C. Wimsatt, 77–105. Cambridge, MA: MIT Press.

Wong, D. B. 2009. *Natural Moralities: A Defense of Pluralistic Relativism*. New York: Oxford University Press.

Wrangham, R. 2009. *Catching Fire: How Cooking Made Us Human*. New York: Basic Books.

13

Future Human Success

Beyond Techno-Libertarianism

Hugh Desmond

Some of today's salient environmental challenges, such as climate change and ecosystem depletion, suggest that metrics such as population size or ecological dominance seem inappropriate ways to define "human success" in the 21st century. An increase in population size and rate of resource consumption seems neither desirable nor sustainable. How, then, should we reconceptualize success moving forward? What principles should guide our reflection about this question?

One striking vision of human success is one where technological progress delivers a desirable and sustainable state of affairs. This need not imply some dystopian technocracy. At least in the literature on enhancement ethics (e.g., Harris 2007; Persson and Savulescu 2012; Bostrom 2014; Sandberg 2014), an optimism in technology is typically combined with an embrace of Millian "experiments in living." The enhancing of our bodies and minds should not be coordinated by governments or collectives but should be conducted by individuals in pursuit of personal well-being. For this reason, this vision's concept of human success can be labeled "techno-libertarian success":

> Techno-libertarian success is the realization of maximal individual choice by means of progress in enhancement technology.

The core concept here is choice. It is what grounds the teleological nature of techno-libertarian success (see Hourdequin in this volume): the preferences or choices of the individual determine what "success" means. Thus techno-libertarian success is very pluralistic. It could, in principle, refer to any outcome or state as long as it is the object of an autonomous wish. Common

Hugh Desmond, *Future Human Success* In: *Human Success*. Edited by: Hugh Desmond and Grant Ramsey, Oxford University Press. © Oxford University Press 2023. DOI: 10.1093/oso/9780190096168.003.0013 desires would be for enhanced cognitive/athletic performance or for the enhanced health or talents of one's children (e.g., Savulescu 2005; Sandberg 2014). In fact, *any* valued aspect of human life can be the target of enhancement. Does an individual desire intimacy and healthy relationships? These can be enhanced through "love drugs" (Earp and Savulescu 2020). Are biodiversity and animal well-being to be valued? Then human self-centeredness, speciesism, or parochial altruism should be enhanced by targeting human moral cognition (Persson and Savulescu 2012).

Is techno-libertarian success plausible? One important consideration in its favor is that bets against human technological creativity have often been on the losing side. In fact, Arthur C. Clarke (1964, 14) even deemed it a "law" that "when [an elderly but distinguished scientist] states that something is impossible, he is very probably wrong." And one need not share Clarke's level of optimism to agree that a creative culture focused more on what may work will generate more technological progress than a culture focused on what cannot work. An optimism in technology may be a useful belief, even if it is not always justified.

However, one should not be too quick to downplay technology's potential. Even our basic anatomical features—bipedalism, opposable thumbs, a small gut, and a large skull—reflect how humans evolved in symbiosis with various technologies, including Oldowan/Acheulean stone technologies (see Vermeij or Potts in this volume) and the control of fire (Wrangham 2010; but see also Vermeij, Grove, or Demps and Richerson in this volume). Nobody can deny how important technology has been in the human evolutionary story.

There seems to be no strong reason to doubt this importance will diminish in the future. On the contrary, today it seems possible that we will be able to directly intervene on human genotypes or on brain function. Who can say what the limits are to this symbiosis between humans and technology? Human minds have become so intertwined with technology that for some time now it does not seem absurd to claim that information technologies are simply extensions of the mind (Clark and Chalmers 1998). From a broader evolutionary perspective, powerful technology represents the way in which human evolution could be shaped according to the goals of human agency.

Formulated in terms of more fundamental concepts, this vision of techno-libertarian success would correspond to an unprecedented empowerment of human agency: our beliefs, intentions, or ideals would control how our minds and bodies develop. It is not that human agency has not played a role in human evolution thus far. (The role of technology attests to that.) However, there have not been attempts to intentionally direct human evolution, except perhaps for eugenics, which relied on mistaken beliefs about heredity. Eugenics also entailed the ethically problematic asymmetry between the empowering of the agency of some humans (policymakers) and the disempowering of others (e.g., those undergoing forcible sterilization). By contrast, techno-libertarian success does not seem—at least at first sight—to involve this asymmetry and, moreover, allows, at least in principle, a direct shaping of individual genotypes and phenotypes. If the vision of techno-libertarian success were to come about, it would represent a significant discontinuity in human evolutionary history, reducing the causal role played by several "natural" processes that have been historically important for the human species: natural selection, development, and—because of the futuristic potential of cognitive enhancement—even forms of social learning such as education.

Even if we should not be quick to dismiss techno-libertarian success, we come back to the question: Is it plausible, given what we know and understand about human evolution? The danger in thinking about future human success is that it quickly becomes overly speculative. Hence, as a first step in bringing this discussion onto a more secure footing, I will introduce a framework by which success concepts can be evaluated: namely, how well they resolve the "problem of human success." Roughly, this is the problem that the human species has become "too" successful in terms of population size and ecological dominance. The problem of human success serves as a benchmark by which to judge candidate success concepts: How sustainable and desirable is a given success metric?

In these terms, my main argument will be that techno-libertarian success is not desirable nor likely to be sustainable. And the main idea is relatively straightforward: when individuals are left to pursue their personal well-being in any way they see fit, they end up competing over a small range of scarce resources all relating to social status, whether nice houses, good education, well-paying jobs, or attractive romantic partners. By focusing on individual-level metrics, techno-libertarian success downplays the role of the social environment and the community. These criticisms, I argue, constitute desiderata for concepts of future human success.

1. The Problem of Human Success

The question of what future human success might be may seem speculative—akin to asking historians to make predictions about where humanity will be in 100 years. However, a more principled treatment can be achieved by focusing on why precisely eco-evolutionary metrics of success are unsatisfactory. For this chapter we will limit the discussion to two important metrics: population size and ecological dominance.

These metrics are quite common when evaluating the evolutionary fate of other species. For instance, general overviews of ant evolution explicitly adopt success talk: ants are said to be "one of evolution's great success stories" (Ward 2006, R152) or, alternatively, "arguably the greatest success story in the history of terrestrial metazoa" (Schultz 2000, 14028). This success talk can and has been transposed to the human species (most recently and notably in Henrich 2016).

However, there are at least two reasons to judge eco-evolutionary metrics insufficient when it comes to defining "human success." The first is that the ecological dominance of the human species is currently paired with unsustainable rates of habitat destruction for other species, threatening future ecological collapse. The second refers to the role of human agency in defining what counts as "success" (McShea and Hourdequin each emphasize this point in this volume). For instance, a continued explosive growth in the human population size would seem, in Malthusian fashion, quite compatible with increasing misery and hardly a scenario one would be tempted to call "one of evolution's great success stories."

Taken together, these two reasons constitute a challenge for any success concept—a challenge that eco-evolutionary success seemingly cannot meet. The failure of the eco-evolutionary success metrics to define a satisfactory success concept can be dubbed "the problem of human success":

> The problem of human success. If success is defined according to the eco-evolutionary metrics of ecological dominance and population size, then human success seems neither sustainable nor desirable, and thus not really a concept of success.

It is not self-evident why eco-evolutionary success should not be sustainable or desirable. For instance, Thomas Malthus worried about overpopulation, and yet the human population has increased about eightfold since the 18th century. Moreover, while increased rates of resource extraction cause pollution and climate change, they have also allowed for the flourishing of human culture in the past 10,000 years. Nonetheless, here is some further argumentation for why genuine "human success" cannot involve maximal levels of eco-evolutionary success.

Current trends in eco-evolutionary success are unsustainable. The prospect of environmental depletion due to human ecological dominance is not new. For instance, during the hegemony of the Roman Empire, large-scale deforestation and extinction of large fauna occurred (Hughes 2014). However, the scale of threatened depletion is so high today that humans cannot ignore it in the way they could 2,000 years ago. First, no longer do humans merely dominate their evolutionary rivals (i.e., large predatory mammals that require large habitats); now they also threaten to cause the extinction of species across a wide range of phylogenies (Barnosky et al. 2011). Consumption of other species is not the main cause of this, but rather the destruction of habitats, for instance by converting forests into agricultural land or infrastructure. A recent study found that the appropriation of the net primary production of biomass by humans rose from 13% in 1910 to 25% in 2005 (Krausmann et al. 2013).

However, contrary to depictions in the popular media, one should caution against some overly pessimistic conclusions. First, this trend may not continue. Krausmann et al. (2013) note that while biomass appropriation doubled, economic output—a measure for efficiency—rose 17-fold between 1910 and 2005. Hence, future economic growth may well be possible without large increases in biomass appropriation. Second, it is uncertain what precise level of maximal ecological dominance is also sustainable. Through better environmental management and more efficient energy and food production, it may be possible to extract more resources without any existential risk (Ord 2020) or even a cascading reduction in biodiversity. Nonetheless, in the assumption that investment in biomass production requires less than maximal biomass appropriation (i.e., to burn wood one must plant trees), it is safe to say that maximal biomass appropriation is unsustainable.

A similar argument can be made for the metric of population size. The size of the global population has increased dramatically in the past three to four centuries, engendering great uncertainty as to whether the human population size would exceed the carrying capacity of the environment. In the 18th century, Malthus was worried about collapse; almost two centuries later, Paul Ehrlich (1968) predicted imminent population collapse. Both underestimated the importance of technological progress for agricultural yields (Trewavas 2002). Nonetheless, here too it is safe to say that continued increases in population size and ecological dominance cannot be sustained indefinitely and that the maximal possible population size would not be the sustainable population size.

The lack of sustainability means that "eco-evolutionary success" is not an entirely consistent notion: when population size or biomass appropriation increases beyond a certain point, that increase implies a future *decrease* in population size or biomass appropriation. Beyond a certain point, increased eco-evolutionary success actually corresponds to human failure.

Eco-evolutionary success is not what matters most. Not only do population size and biomass appropriation seem to fail on their own terms, but, for varying reasons, they are not what human agency aims at. We seem to care for biodiversity and animal welfare for their own sakes: environmental ethicists typically think of biodiversity and animal welfare as intrinsic values. Moreover, in a different sign that such value judgments are widespread, the scenario of a future human state where the human population size is astronomical and biodiversity all but destroyed is a staple of dystopian science fiction.

Similarly, in the hypothetical trade-off between global population size and our own flourishing and well-being, we seem to squarely side with the latter. The Parfitian argument illustrates this value judgment: if further increases in the human population size would correlate with the misery of individual humans, then it would be "repugnant" to aim at increasing human population, even if utility calculus dictated it (Parfit 1984). In fact, the vast growth of a population consisting of valueless individuals calls to mind images of parasites or cancers, to which the human species is sometimes compared (Hern 1993).

Of course, these ethical intuitions are not universal: humans once believed that having a large number of future descendants was an ethical priority.[1] Perhaps the trade-off then between large population size and other values was not as sharp as it seems to be today. Nonetheless, we can safely assume for the purpose of this chapter that indefinite increases in eco-evolutionary success, defined in terms of population size and ecological dominance, do not capture what humans actually care about.

2. Evaluating Success Concepts: Eugenicist Success

The problem of human success allows candidate success concepts to be evaluated with some systematicity, by means of two questions: Are indefinite increases in the associated metrics of success sustainable? And are indefinite increases desirable? If not, then the success concept does not identify a long-term direction for human evolution.

As an illustration of such an evaluation, consider the concept of human success that late 19th- and early to mid-20th century eugenicists implicitly operated with. Eugenicists were worried about how natural selection was disrupted by the improved circumstances in modern societies, whether through nutrition, healthcare, or vaccination. The problematic consequence was that those who *would have been* fitter in a "natural" environment no longer out-reproduced those who *would have been* less fit—and in particular, the lower socioeconomic classes were out-reproducing the upper ones. These lower socioeconomic classes were believed to be characterized by hereditary traits such as "pauperism," "feeble-mindedness," or "imbecility" (Kevles 1985, 20–21).[2] In fact, it was this eugenicist concern with heritability that spurred the biostatistical research that was to become the modern synthesis (see Desmond 2022 for a further exploration of this). In any case, this difference in reproductive output, together with the hereditary nature of their undesirable traits, was believed to be leading to the "degradation" of the human species. Action was needed to stop and reverse this trend. This was the rationale, as is well known, for deliberate artificial selection by means of antimiscegenation laws, forcible sterilization, and worse.

The concept of human success implicit in this eugenicist vision of human evolution is a species-level concept: success is achieved not by the species as a whole but by individuals alone. Moreover, eugenicist success is not pluralistic but is spelled out in terms such as "intelligence" or "giftedness" or "virtue." For instance, the state of human success then would have been, in the words of Galton (1869, 1), a "highly gifted race of men."

In its own way, the eugenic concept of human success can be interpreted as a response to the problem of human success. Increases in population size were viewed as unsustainable and undesirable, not so much because of the threat of ecosystem collapse but rather because they created a differential reproduction rate between "high-quality" and "low-quality" humans. The eugenicists were the progressives of their era: they were critical of hereditary aristocracy,[3] and their main opposition, at least in the United States, came from religious (Catholic) corners. In this way, eugenics is an illustration about how a certain understanding of the problem of human success gave rise to a concept of future human success, which in turn was heavily laden with nefarious ethical and political implications.

3. Techno-Libertarian Success and Its Evolutionary Rationale

Of main concern to this chapter is what sometimes is termed contemporary "liberal eugenics," as distinguished from "population eugenics" of the late 19th and early to mid-20th centuries (Agar 2005). Liberal eugenics leverages developments in gene-editing technology and holds that the state should neither actively intervene on human phenotypes or genotypes nor prohibit individuals from enhancing themselves, as long as others are not harmed by doing so. In this way, liberal eugenics typically has a distinctively libertarian flavor (Sparrow 2011).[4]

The associated success concept is techno-libertarian success: the future success of the human species lies in putting more and more powerful enhancement technologies at the disposal of individuals, so that they can live longer, healthier, and more satisfied lives. In this way, techno-libertarian success seems to offer a relatively straightforward response to the problem of human success. First, it straightforwardly answers the problem of desirability: if one assumes that individuals are the best judge of what they want in life, then by maximizing individual choice individuals can achieve what they desire. Usually this is greater flourishing and/or happiness. In Harari's (2017) view, it will mean eternal youth, a permanent state of happiness, and the possession of "super-abilities." In Harris's (2007, 9) view, it will be the "healthier, longer-lived, and altogether 'better'" lives mentioned earlier.[5] At no point would it seem like further increases in these goods would be a bad thing: it seems desirable that techno-libertarian success should increase indefinitely.

Second, techno-libertarian success addresses the issue of sustainability through cognitive and especially moral enhancement. The underlying idea here is that our self-centeredness and lack of concern for both non-human species and future human generations play a significant role in the unsustainability of eco-evolutionary success. There is in fact some evidence that this is the case. For instance, habitat destruction correlates with economic inequality within a country (Mikkelson, Gonzalez, and Peterson 2007). In other words, the more relative poverty there is in a society, the more willing individuals are to exploit environmental resources to further their own economic standing (and, likely, the more willing regulators are to turn a blind eye). Hence our self-centeredness and parochial altruism are a prime target for cognitive enhancement technology.[6] In sum, through increased techno-libertarian success, both dimensions of the problem of human success can be resolved.

Techno-libertarian success could be read as a purely ethical notion: a value judgment of what humans should aim for, affirming the principle of autonomy and indirectly other values such as biodiversity or the well-being of animals or that of future human generations. If that were the only reading possible, it would not be of interest in the context of this volume. However, what is perhaps more fascinating—and problematic—about techno-libertarian success is how it is closely linked to a narrative or quasi-prediction of what will happen if we *do not* enhance. The success concepts involve an evolutionary rationale: without the promoting of enhancement technologies, the current path of human evolution will lead to suboptimal outcomes, if not catastrophe. Thus, techno-libertarian success is a predictive notion as well and is another illustration of how success concepts often uneasily straddle the divide between the ethical-normative and the causal-descriptive (see Desmond and Ramsey or McShea, this volume).

Why is techno-libertarian success, at least as it is commonly used, a hybrid concept in this way? First, it often involves referring to evolution by natural selection as a blind, directionless, and contingent process. This view of evolutionary history has deep roots and tends to be used to argue that some human state of affairs is contingent and likely maladaptive (Street 2006). In other work I argue that it is problematic (Desmond 2018, 2021a). However, using it as if it were as a "meme," the view has been taken up in enhancement ethics, where it is combined with the mismatch hypothesis in evolutionary psychology. As an illustration, consider the following:

> After all, our brains are products of evolution, which is a blind process that hardly seeks to maximize the good, or make us morally best. Evolution "cares" only about reproductive success. Moreover, even if the evolutionary process somehow led to what is in one sense an optimal result, this result may be optimal only in the environment in which our very distant ancestors

> lived. It is very unlikely to be optimal in our utterly different modern environment. (There was, for example, no police in the primeval savannas, nor were there planes or hijackers. . . .) But if the current level isn't optimal, and we now have means of improving it (in whichever direction), then surely we have strong reasons to do so—including by biomedical means. (Kahane and Savulescu 2015, 138)

This is the "stone-age brains in modern skulls" logic most readers of this volume will be familiar with (Barkow, Cosmides, and Tooby 1992). The difference is perhaps that it entails a pessimism about our adaptiveness: our brains and bodies are not designed for the challenges of the Anthropocene, and our maladaptedness is so deep that it needs direct intervention by means of enhancement technology (Harris 2007; Harari 2017; Persson and Savulescu 2012). Our inherited genotypes and phenotypes are barriers to our well-being and must be overcome.

It would require a different type of work than is possible here to ascertain just *how* frequently this view pops up in enhancement ethics, but even a superficial review reveals other instances.[7] It seems fair to say that the scientific presuppositions of the view are seldomly made explicit, and that the mismatch hypothesis is applied far beyond relatively circumscribed psychological mechanisms like mate preferences or parental care (cf. Buss 2019) to all aspects of the human phenotype—including feelings of intimacy in long-term relationships (Earp and Savulescu 2020) as well as even our capacities of reasoning about what is valuable (Schaefer, Kahane, and Savulescu 2014).

At this point one can wonder why evolution is invoked in order to make an ethical argument. Why not keep ethics and evolution separate? And why connect ethics and evolution in this specific way? After all, the mismatch hypothesis need not imply any specific ethical conclusion. Mismatch could be viewed as warranting a collectivist or totalitarian approach to enhancement, where the state mandates enhancement regardless of individual choice. Alternatively, the mismatch could even be judged ethically unproblematic, or yet again, as a source of pain that should be tolerated as part of the human condition.

I believe that the best answer to this question is multifaceted, involving both logic and rhetoric, as it were. Regarding the former, a view of the evolution and causal nature of humans supports a utilitarian calculus about consequences. In particular, the emphasis on maladaptedness not only implies that enhancement is desirable but also that it involves a focus on individuals. Proponents of techno-libertarian success tend not to acknowledge that maladaptedness can be a group-level property, but tend to emphasize individual-level properties: our brains are incapable of dealing optimally with evolutionary novelties in our environment, whether the presence of police, planes, or hijackers. Well-being is also defined in terms of individual properties, as a "state of an [*individual*] person's biology or psychology" (Savulescu, Sandberg, and Kahane 2014, 7, my emphasis). In this conceptual framework, when the utility calculus is carried out, only individual states and properties are taken into consideration, and thus group-level considerations are ignored. (This will be crucial for my argument later in the chapter.)

One consequence of this focus on the individual is that technologically mediated enhancement appears to be the only plausible way to overcome humans' deep maladaptedness. In this respect, techno-libertarian success should be distinguished from rival social liberal approaches to enhancement ethics, where the importance of policy and/or education are emphasized (see Buchanan and Powell, this volume). Given our inherited limitations, especially regarding our tribal moral psychology, the question arises as to whether policy and education are powerful enough.[8] We need cognitive enhancement, ranging from pharmaceuticals that manipulate neurochemicals (such as oxytocin) to computer-brain interfaces. Thus, interestingly, techno-libertarian success seems to depend on some implicit human nature concept: namely that it lies within "human nature" to be maladaptive to such an extent it can be meliorated only by means of technological phenotypic or genotypic interventions (see Kaebnick 2014 for how the concept "nature" is difficult to entirely eliminate).

In this way, the evolutionary considerations help rig the utilitarian calculus in the direction of the necessity of technological enhancement. However, a rhetorical element (if it can be called that) is involved as well, in that the evolutionary rationale helps position techno-libertarianism within a broader intellectual-political landscape. First, techno-libertarians champion change and progress and seek to overcome the status quo. Some have argued for this explicitly: cautionary arguments have been recast as irrational aversions caused by cognitive biases (Bostrom and Ord 2006; Caviola et al. 2014). At other times the association with Enlightenment and progress are more implicit. Consider, for instance, the following:

> This possibility of a new phase of evolution in which Darwinian evolution, by natural selection, will be replaced by a *deliberately* chosen process of selection, the results of which, instead of having to wait the millions of years over which Darwinian evolutionary change has taken place, will be seen

> and felt almost immediately. This new process of evolutionary change will replace natural selection with deliberate selection, Darwinian evolution with "enhancement evolution." (Harris 2007, 3–4)

Here is a dichotomous understanding of human evolution: a "natural" part where human evolution is driven by blind and chancy natural selection, and a "rational" part where humans intentionally shape their own evolutionary destiny in the image of moral values. Add to this the fact that prominent cautionary approaches use quasi-religious language, such as respect for givenness or the sacredness of life (L. Kass 2003; Sandel 2007), and science versus religious overtones are added to the concept of techno-libertarian success.

This politicization of human evolution is fascinating, even though an attempt to analyze it would bring us beyond the scope of this chapter. However, it illustrates a mixing of fact and value that is strongly rejected by others (see McShea or Powell and Buchanan, this volume) and that yet seems to be more than fleetingly fallacious thinking. For it is striking how techno-libertarian success at least chimes with some of the eugenicist arguments. Moreover, the old eugenics was portrayed as a progressive movement overcoming human obscurantism and/or irrationality. This fact is often misrepresented today in political discourse, but Ordover (2003, 53) reminds us, "Eugenics meant, to its proponents, the victory of rationality over shortsighted altruism . . . science over sentimentalism." Or yet again, in the words of Galton (1909, 42), "[W]hat Nature does blindly, slowly, and ruthlessly, man may do providently, quickly, and kindly."

Much more could be said about techno-libertarian success, but what can be concluded here is that it is much more than a cleanly ethical concept. It is also supported by an evolutionary rationale: humans are deeply maladapted to the challenges of the Anthropocene and hence are in need of enhancement. In this way, techno-libertarian success seems to offer a prima facie resolution to the problem of human success by sketching a vision of future human evolution involving sustainable and desirable increases in liberty through technological progress.

4. Liberty and Status Competition

The "stone-age brains in modern skulls" logic that techno-libertarianism relies on has its strengths; for instance, it helps to explain why many sexual, mating, parenting, and kinship behaviors seem cross-cultural and recalcitrant to changes in the social environment (Buss 2019). However, as a sweeping view of human evolution, it is quite limited. First, the very concept of a single type of ancestral environment characterizing the majority of human evolution is suspect (e.g., Foley 2005). Second, and more important, many aspects of human cognition are highly influenceable by social learning and thus are very adaptable to the cultural environment (Boyd and Richerson 1985; Henrich and Boyd 1998). Humans respond to social norms, whether status or sexual norms, and in fact, sensitivity to norms seems to be hardwired via various cognitive biases, such as the conformity bias (Baron, Vandello, and Brunsman 1996), which is adaptive whenever social learning is adaptive (Henrich and Boyd 1998). Thus, portraying human phenotypes as primarily shaped by natural selection in the ancestral environment ignores the role played by changing social structures. What happens to techno-libertarian success—the maximization of individual choice through technological means—when social structure is taken into consideration?

Let us return to the quote by Kahane and Savulescu (2015), where, to illustrate the challenges of environmental novelty, they give examples of police, planes, and hijackers. Do these factors constitute evolutionary novelties? Consider police: obviously there were no individuals in hunter-gatherer societies enforcing social norms with the backing of batons, judges, and prisons. But the function of enforcing social norms is hardly a novelty. In fact, many view enforcement as a necessary counterpart to the spread of altruistic social norms (see, e.g., Tomasello 2016). What counts as an evolutionary novelty or as a mere variation of past environmental circumstances is a difficult question (Desmond 2022). Perhaps planes are more plausible evolutionary novelties, but then the further question arises: Do evolutionary novelties always warrant cognitive enhancement?

Here it is instructive to look at the history of technology, and in particular at how people sometimes panicked at technological change. For instance, rather amusingly today, the advent of train travel in 1860s and 1870s Victorian England was accompanied by widespread moral panic (documented in, e.g., Milne-Smith 2016). Newspapers regularly reported on how otherwise healthy individuals were driven to insanity by rail travel, or of how some passengers suddenly turned violent without any discernible reason. Doctors took to warning that the human body and mind were not *made* for the intense vibrations and unnatural speed of rail travel (21). This raises the question whether the championing of enhancement technology to upgrade our brains is actually—and paradoxically—somehow related to classic forms of panic at technological progress. Today we no longer fear unprecedented speed: once we got used to trains, the speeds of planes and rockets seemed to be smaller psychological steps. Now, instead, we fear the unprecedented advances in computing and artificial intelligence (see Bostrom 2014 for an overview).

In this way, I share with Buchanan and Powell (this volume) some of the worry that cognitive plasticity is being downplayed in this view of human success. However, I wish to draw a different conclusion, namely that the impact of the social environment on individual cognition means that the concept of individual autonomy is exaggerated and is an abstraction of how individuals actually conduct their reasoning and decision-making, especially regarding enhancement decisions.

As an illustration, consider one of the most primeval but psychologically powerful enhancements: enhancing the length of one's body. Sometimes children are short without any particular pathology causing the short stature (e.g., hormone deficiencies or insensitivities). This condition is known as "idiopathic short stature" (Argente 2016). Is it a disability? There are no physical health risks involved. Even so, short stature is sometimes viewed as a psychosocial disability by the parents of the child (see Allen 2017, 146). Children themselves may experience it as a psychological burden (Ranke 2013). Hence, as an obstacle to well-being, targeting short stature may seem like an instance of techno-libertarian success. In fact, idiopathic short stature is regularly addressed through administering human growth hormone (hGH or somatotropin) and is sanctioned by the U.S. Food and Drug Administration when the child is in the first percentile for height (or more precisely, when the child's height is 2.25 standard deviations shorter than the average height; Ranke 2013, 330).

But what is the underlying story? What explains why we experience short stature as a disability? If we shift from the medical-ethical domain to human evolution, the reason is straightforward: physical height (and, in general, physical formidability) is one of the most widespread indicators of dominance across animal species (together with strength and aggression; see Ellis 1994, 1995). Human status hierarchies are more complicated than those of, say, lizards, but height still plays a measurable role. Taller people are viewed as superior in leadership and intelligence, and taller males in particular are viewed as healthier and more dominant than shorter males (van Vugt and Tybur 2015; Blaker et al. 2013). Other studies show how height correlates with income, the likelihood of having a managerial position, and military rank (see Blaker et al. 2013 and references therein). While the effect of size should not be overestimated (i.e., many other factors predict income, such as education and socioeconomic class), sensitivity to physical height is ingrained in our inherited psychology, and this helps explain why parents and children may view short stature as a disability.

Techno-libertarian success dictates that hGH therapies should be made available to all, and that these would allow the parents and children to better pursue their well-being. The problem here for techno-libertarian success is that the meaning of "short" is largely comparative (average height has fluctuated significantly throughout human history; Steckel 1995) and that the value we place on height is—in a society where services are increasingly intellectual instead of physical—almost entirely due to the fact that others place value on height. Height enhancements thus raise the prospect of individuals feverishly enhancing their height in order to escape the first percentile. Each may be maximizing utility given the circumstances, but choosing height enhancement in such circumstances is clearly not an instance of a genuinely free choice. A libertarian could even deem that, in this circumstance, the sociocultural environment represents some kind of "tyranny," reducing the free choice of individuals: if parents did not have the sword of lifelong discrimination hanging over their heads, they would be in a better position to decide freely.

In itself, this consideration does not sink the techno-libertarian concept of success. One could respond that the sociocultural environment should be enhanced—for instance, by enhancing the moral psychology of the school bullies. In the context of enhancement ethics, social status has been recognized as important in how enhancing positional goods can create inequality (Mehlman and Botkin 1998) or perverse competitions (Sparrow 2019). However, the typical response is to point out how many competitions are beneficial for the community (e.g., Anomaly 2020, 11–13) and that the challenge lies in (technologically) promoting beneficial status competition and suppressing the perverse kind.

In the next section, I argue why this way of thinking about social status and individual decision-making does not work. However, first I will lay the ground for that by sketching just how crucial it is to take the sociocultural environment into consideration when evaluating decisions to enhance.

In general, it is difficult to overstate the importance of status for human psychology and for life outcomes. The desire for status has even been stated to be a "fundamental motive" across cultures, genders, ages, and personalities (Anderson, Hildreth, and Howland 2015). Social status is a currency that can be traded in for a whole host of goods. Higher status correlates with higher subjective well-being, higher self-esteem, and better mental and physical health (Anderson, Hildreth, and Howland 2015). Conversely, people with lower status, whether through lack of wealth or education, have higher levels of stress (Thoits 2010), less experience of having control over their lives (Ross and Wu 1995), and higher rates of mortality from all causes (Wilkinson 2001; Marmot 2005). Status is thus at the nexus of all sorts of other desirable goods. It could be compared to a kind of a "gatekeeper good": the gate of social status is narrow and competitive, but once one passes through, all sorts of benefits follow.

Enhancements are—per definition—interventions to increase human capacities, so if one considers why one would be motivated to enhance, it does not take many steps to suspect that status-related reasons may play a large role. Of course, they *need* not play a large role. In principle, individuals could be highly motivated to enhance their ear-wiggling capacity—just to pick out one trivial-seeming trait. This could presumably be a component of techno-libertarian success. However, absent Swiftian scenarios where those with a superior ear-wiggling capacity are admired and lauded, nobody currently cares much about this capacity.

Conversely, we do care a great deal about other types of bodily movement that determine the outcomes of athletic competitions. Athletes using performance-enhancing drugs are, at a superficial level, merely trying to win. But why do they want to win? Why risk one's health for an athletic competition? In one of the few studies on athletes' incentives to use doping (Kegelaers et al. 2018), athletes list a host of motivations related to improving social status: their image, respect from others, greater popularity among friends, obtaining what Kegelaers and colleagues call "hero status," and, finally, financial gain. It is doubtful there would be the same incentive to use performance-enhancing drugs if the status rewards of athletic success, both financial and in terms of respect and recognition, were not so great.

Cognitive enhancement is often viewed as intrinsically beneficial. Yet also here the motivations for cognitive enhancement seem closely related to status. Consider education, which is today still the most effective way of enhancing one's cognition (even if nontechnologically). Education is not merely undertaken for its intrinsic benefits; education credentials are perhaps the single most important means to gain access to socioeconomic status, since they allow entry into high-status professions (medicine, law, engineering, etc.). Students sometimes use technological cognitive enhancements (e.g., Adderall, Ritalin; see Ragan et al. 2013). Would they do this if their educational outcome did not determine their future in the way it does? According to a strict application of the techno-libertarian concept of success, permanent diminishments of cognitive capacities could also count as "success," as long as they increase well-being by satisfying preferences (Earp et al. 2014). However, whether such diminishment would occur with much frequency is doubtful, given the close link between cognitive capacity and status gains and human psychology's orientation toward status.

The role social status plays in decisions to enhance is documented in more detail in other work (Desmond 2021b). However, for purposes here, we can conclude that techno-libertarian success does not reflect how individuals are entangled with their social environments. The libertarian ideal of negative liberty (freedom from coercion) does not give direct guidance when the preferences underlying decision-making are themselves strongly influenced by status hierarchies, as seems to be the case with many (and perhaps all) decisions to enhance.

In this way, the response that techno-libertarian success offers to the problem of human success seems doubtful. The prospect of ever more powerful enhancement technologies to promote individual choice is consistent with perverse forms of status competition, and hence techno-libertarian success does not seem necessarily either desirable or sustainable. In such a regime of perverse status competition, increasing one's choice through technology would thus correspond to a de facto decrease in one's range of choice. For instance, actively promoting the ability to choose to enhance one's height yet further would also mean suppressing the ability to reject the importance of height. The libertarian would, of course, reject the latter cases as not genuine forms of techno-libertarian success. Then the question becomes how to distinguish genuine liberty from the merely apparent: how to distinguish genuinely autonomous choices from the apparent, or choices that deliver genuine well-being from choices we mistakenly believe will deliver well-being.

5. Technological Solutions to Status Competition?

Can the techno-libertarian conception of success not be rescued by targeting the sociocultural environment in some way? After all, one of the strengths of techno-libertarian success is its pluralism: as previously mentioned, *any* valued property P can be enhanced. In the ethics literature, this has allowed previous objections pointing to a decrease in "humility" or "appreciation of giftedness" (Sandel 2007) to be parried: "humility" and "appreciation of giftedness" are themselves experiences that could be the target of cognitive enhancement (see Roache and Savulescu 2016 for this argument). Thus, if techno-libertarian success is criticized as undesirable due to some property P, then the response could simply be "enhance P." If techno-libertarian success can lead to perverse status competitions, then why not simply enhance prosocial attitudes to avoid such status competitions?

Let us add some detail and plausibility to this objection. A unique dimension of human status hierarchies is that they are characterized by what has been termed "prestige" as opposed to "dominance" (Henrich and Gil-White 2001). Dominance indicates which individual would be the victor in a direct, physical confrontation, while prestige indicates some kind of competence or excellence. Since humans are biased toward learning from high-status individuals (Atkisson, O'Brien, and Mesoudi 2012), organizing status hierarchies according to prestige benefits social learning and cumulative culture—core elements of human eco-evolutionary success (see Demps and Richerson in this volume).

This distinction between dominance and prestige gives more detail as to what an "enhancement" of status competition would look like: it would enforce adherence to what some anthropologists call "service-for-prestige" norms (Price and Van Vugt 2014). High-status individuals are expected to act in the group's interests, and hence competition for prestige is more beneficial for the group over the long term than competition for dominance. Thus, the techno-libertarian promoting biomedical enhancement would target the moral cognition of high-status persons by increasing their prosocial tendencies to offer service to the group (in exchange for whatever status they may receive). In this way, the challenge to the sustainability and desirability of techno-libertarian success seems to be saved by yet further technological enhancement.

However, this attempt to save techno-libertarian success fails to consider a crucial question: *Who* should decide how such prosocial moral enhancements are administered? In reality, the distinction between prestige and dominance can be ambiguous. Consider the example of silencing someone in a public debate. This can be a service to the community when that someone is engaging in hate speech. However, it can also be a form of self-serving dominance, where the silencing mainly functions as a way to suppress challengers. The techno-libertarian response sketched above presupposes that some group of persons—"guardians," if you will—would have a deep understanding of social dynamics, and indeed of ethics itself. In this way, the techno-libertarian needs a further and yet more problematic assumption in order to address the problem of perverse status competition. Not only could it be doubted that such guardians are humanly possible, and not only does it raise the problem of infinite regress (who guards the guardians?), but the idea that there would be such arbiters deciding on which enhancements promote "true liberty" and which merely promote "apparent liberty" runs counter to the very concept of negative liberty at the core of libertarianism. In the effort of techno-libertarian success to engineer benign status competition, the core libertarian tenet that individuals should conduct their lives as they see fit is severely compromised—and even if we end up with benign status competition, it is no longer a form of *libertarian* success.

Hence the second and even more radical response to the problems facing techno-libertarian success: Could status competition itself be removed through technological progress? In this vision of the human future, humans would simply lead their lives and not be motivated by any type of status consideration. Perhaps technological progress would allow for abundant resources, removing the need to compete for social status, or perhaps the psychological tendency to be motivated by status would itself be pharmaceutically suppressed. Would not this benignly anarchical state be preferable?

Yet this response must be parried, because it ignores the basic function of social status hierarchies, which evolved in order to streamline group-level decision-making procedures concerning individual access to scarce resources such as mates, food, or shelter (see, e.g., van Vugt and Tybur 2015). In other words, without a status hierarchy, physical conflicts would determine who gets what, and such conflicts would leave the group as a whole worse off. Thus, a group of hens with a pecking order will be better off than one where each feeding session provokes conflict about who gets what. A group of humans making collective decisions about who gets nice houses, a good education, interesting jobs, and so on will do better than a group where these issues are decided through physical conflicts. In fact, the principle that decides status hierarchies would, in a different context, be referred to as a "principle of justice" (Rawls 1999). Status, at its most fundamental, determines which organism's needs are prioritized and prevents violent conflict over scarce resources.

For the sake of argument, one could grant that technological progress can alleviate most scarcity. Housing quality seems like something that could be "solved." Perhaps unrewarding lines of work could also be "solved" by advances in artificial intelligence so that computers and robots would take over all the drudgery. For the sake of argument, one could also grant that some future biomedical enhancement would suppress our unconscious obsession with social status. Yet the structural factors for which social status provides an adaptation cannot be enhanced away. Consider Arrow's theorem: in a group of at least three people deciding between three options, an impasse can be avoided only by some "dictatorship," where one individual's preferences weigh more on the collective decision. (This is a very rough formulation of Arrow's theorem; see Morreau 2019.) Similarly, unless one were to suppress human agency itself, it is inevitable that in a community, (1) the preferences of individuals do not coincide, (2) some collective decisions sometimes must be made, and (3) some principle is necessary to prioritize the preferences of some individuals over those of others. In other words, there is one scarce resource that technological progress cannot possibly alleviate: who gets to decide. From this perspective, it is not a coincidence that most institutions (whether corporations, governments, or charities) have "leaders," that is, individuals whose preferences are decisive for the collective, at least with regard to certain types of activity or subject matter. Competition for status (e.g., leadership positions) will not cease regardless of how much technological enhancement will occur. According to the line of reasoning presented here, the only way this could occur is if individuals ceased to be agents—that is, entities that act in order to realize their preferences. However, in that event there cannot be any techno-libertarian success.[9]

In sum, techno-libertarian success cannot resolve the problem of human success. It depends on concepts of autonomy and negative liberty that are implausible given how intertwined social environments and individual cognition are. We compete for social status, and this competition strongly influences our attempts to achieve well-being through enhancement, sometimes in a self-defeating way. Such challenges cannot be engineered away through further enhancement. Promoting benign status competition through further technological enhancement is not only inherently problematic (it presupposes guardians who can make prior distinctions between benign and perverse status) but goes against core libertarian tenets. Moreover, status competition cannot be engineered *away* either. It is here to stay, and human success concepts must take this into consideration.

6. Conclusion: Desiderata for Human Success

Individuals are embedded in their communities and are not only highly dependent on them for basic survival but also make core life choices in light of their sociocultural environment. This chapter focused especially on the role that status hierarchies and status competition—central components of any social-cultural environment—play in forming individual choices. Enhancement technologies merely alter or even intensify existing status competitions but cannot remove them. Because individuals actively adapt to sociocultural environments, the evolutionary rationale for techno-libertarian success (i.e., the deep flaws of inherited human genotypes and phenotypes) is likely overestimated. Moreover, because further promotion of human enhancement can involve promoting perverse status competition, the very desirability and sustainability of techno-libertarian success are undermined.

In such cases, "success" seems to lie in changing the social-cultural environment rather than simply individual capacities. In this way, the failure of techno-libertarian success suggests two desiderata for satisfactory concepts of future human success. The first is that a concept of human success should integrate community-level metrics. Eco-evolutionary success integrates only species-level metrics (population size, ecological dominance), and techno-libertarian success emphasizes individual-level metrics (i.e., individual choice). However, the desires of individuals are very often oriented toward the good of the community, and the community is organized so as to contribute to the development and well-being of its members (e.g., through social learning or division of labor). This means that a satisfactory concept of "human success" would need to refer to dimensions of "successful communities," for instance those with cultural environments where the flow of social learning is optimized. One specific metric could be the degree to which high status is accorded to individuals who have benefited the group the most, or who have the potential to benefit the group the most (via their competence or excellence).

The second desideratum concerns the content of such community-level metrics, which should acknowledge the importance of how status hierarchies are organized. Status hierarchies are a crucial feature of the structure of social environments; insofar as they allow multiple agents to coordinate in collective decision-making, they are perhaps *the* crucial feature. Status hierarchies can be organized in many different ways. Some reward actions that promote long-term interests, while others reward short-term ones. Some reward actions that involve overt self-sacrifice, while others celebrate individual status-seeking on the assumption that individuals chasing status will ultimately contribute the most to community-level metrics of human success. Yet others discourage individuals from pursuing status maximization for its own sake: ideals and virtues should be the primary values, with status and its various correlates (wealth, fame, recognition) mere afterthoughts.

In this way, ethics and politics cannot be excised from thinking about the future evolution of an intensely social species such as *Homo sapiens*. We compete and cooperate, and even at our most egoistic seek the approval of others. Concepts of future human success would need to identify successful forms of status competition, and in general would need to identify dimensions of what it means to be a "successful community."

Notes

1. For instance, in Genesis 22:17, Jahweh tells Abraham, "[I]n blessing I will bless you, and in multiplying I will multiply your descendants as the stars of the heaven, and as the sand which is upon the seashore."
2. Even Darwin (1871, 167), somewhat embarrassingly, spoke of how "the reckless, degraded, and often vicious members of society tend to increase at a quicker rate than the provident and generally virtuous members."
3. For instance, British eugenicists proposed reorganizing the House of Lords along eugenic principles in lieu of hereditary principles (reported in Kevles 1985, 73).
4. Liberal eugenics often employs utilitarian reasoning; however, the utilitarian logic is compatible with strong state intervention. Think of how the Benthamite line of thinking of "everybody to count for one, nobody for more than one" led to new charitable impulses, as well as reforms of public health and public education. This Benthamite line is not wholly absent in the literature on enhancement ethics (e.g., Savulescu 2001).
5. However, this need not be the case, given the primacy of individual autonomy. Techno-libertarian success can in principle entail sickly, short-lived, and miserable lives, if that is what the autonomous individual wants.

6. For a more detailed defense of this argument, see Persson and Savulescu (2012, chapter 7). For a challenge to this view, and a defense of the importance of policy and education in light of human cognitive plasticity, see Buchanan and Powell (2018) as well as their chapter in this volume.
7. For direct quotes, consider, for instance, Bostrom and Ord (2006, 665–666): "[O]ur current environment is in many respects very different from that of our evolutionary ancestors . . . [and] places very different demands on cognitive functioning than did an illiterate life on the savanna." Or, alternatively, Pugh, Kahane, and Savulescu (2016, 407): "[T]he relatively contingent and arbitrary features of human nature, selected as they were blind evolutionary processes."
8. In a recent response to criticisms by Buchanan and Powell (2018), Persson and Savulescu (2019) have stated that biomedical enhancement is merely one avenue to pursue moral progress, alongside social, legal, and institutional avenues. This seems like a dilution of techno-libertarian success, though in the response one can discern a similar structure or argument, where technological enhancement is necessary because "natural capacities for moral concern [are left] far behind" (818).
9. Harari (2017) seems to go down this path in speculating that, at some point in the human future, all decisions would be taken by artificial intelligence on the basis of large amounts of empirical data. However, that would also entail jettisoning techno-libertarian success altogether, along with the concepts of autonomy and liberalism.

References

Agar, N. 2005. *Liberal Eugenics: In Defence of Human Enhancement*. Malden, MA: Blackwell.

Allen, D. B. 2017. "Growth Promotion Ethics and the Challenge to Resist Cosmetic Endocrinology." *Hormone Research in Paediatrics* 87 (3): 145–152. https://www.karger.com/Article/FullText/458526.

Anderson, C., J. A. D. Hildreth, and L. Howland. 2015. "Is the Desire for Status a Fundamental Human Motive? A Review of the Empirical Literature." *Psychological Bulletin* 141 (3): 574–601. http://doi.apa.org/getdoi.cfm?doi=10.1037/a0038781.

Anomaly, J. 2020. *Creating Future People: The Ethics of Genetic Enhancement*. New York: Routledge.

Argente, J. 2016. "Challenges in the Management of Short Stature." *Hormone Research in Paediatrics* 85 (1): 2–10. https://www.karger.com/Article/FullText/442350.

Atkisson, C., M. J. O'Brien, and A. Mesoudi. 2012. "Adult Learners in a Novel Environment Use Prestige-Biased Social Learning." *Evolutionary Psychology* 10 (3): 147470491201000320. https://doi.org/10.1177/147470491201000309.

Barkow, J. H., L. Cosmides, and J. Tooby. 1992. *The Adapted Mind: Evolutionary Psychology and the Generation of Culture*. New York: Oxford University Press.

Barnosky, A. D., N. Matzke, S. Tomiya, G. O. U. Wogan, B. Swartz, T. B. Quental, C. Marshall, et al. 2011. "Has the Earth's Sixth Mass Extinction Already Arrived?" *Nature* 471 (7336): 51–57. http://www.nature.com/articles/nature09678.

Baron, R. S., J. A. Vandello, and B. Brunsman. 1996. "The Forgotten Variable in Conformity Research: Impact of Task Importance on Social Influence." *Journal of Personality and Social Psychology* 71 (5): 915–927.

Blaker, N. M., I. Rompa, I. H. Dessing, A. F. Vriend, C. Herschberg, and M. van Vugt. 2013. "The Height Leadership Advantage in Men and Women: Testing Evolutionary Psychology Predictions about the Perceptions of Tall Leaders." *Group Processes & Intergroup Relations* 16 (1): 17–27. http://journals.sagepub.com/doi/10.1177/1368430212437211.

Bostrom, N. 2014. *Superintelligence: Paths, Dangers, Strategies*. Oxford: Oxford University Press.

Bostrom, N., and T. Ord. 2006. "The Reversal Test: Eliminating Status Quo Bias in Applied Ethics." *Ethics* 116 (4): 656–679.

Boyd, R., and P. Richerson. 1985. *Culture and the Evolutionary Process*. Chicago: University of Chicago Press.

Buchanan, A., and R. Powell. 2018. *The Evolution of Moral Progress: A Biocultural Theory*. Oxford: Oxford University Press.

Buss, D. M. 2019. *Evolutionary Psychology: The New Science of the Mind*. 6th ed. New York: Routledge.

Caviola, L., A. Mannino, J. Savulescu, and N. Faulmüller. 2014. "Cognitive Biases Can Affect Moral Intuitions about Cognitive Enhancement." *Frontiers in Systems Neuroscience* 8: 1–5. https://www.frontiersin.org/articles/10.3389/fnsys.2014.00195/full.

Clark, A., and D. Chalmers. 1998. "The Extended Mind." *Analysis* 58 (1): 7–19.

Clarke, A. C. 1964. *Profiles of the Future: An Inquiry into the Limits of the Possible*. New York: Bantam Books.

Darwin, C. 1871. *The Descent of Man*. New York: D. Appleton.

Desmond, H. 2018. "Natural Selection, Plasticity, and the Rationale for Largest-Scale Trends." *Studies in History and Philosophy of Science Part C: Studies in History and Philosophy of Biological and Biomedical Sciences* 68–69 (April): 25–33. https://linkinghub.elsevier.com/retrieve/pii/S1369848617300961.

Desmond, H. 2021a. "The Selectionist Rationale for Evolutionary Progress." *Biology & Philosophy* 36 (3): 32. https://link.springer.com/10.1007/s10539-021-09806-1.

Desmond, H. 2021b. "In Service to Others: A New Evolutionary Perspective on Human Enhancement." *Hastings Center Report* 51 (6): 33–43. https://doi.org/10.1002/hast.1305.

Desmond, H. 2022. "Adapting to Environmental Heterogeneity: Selection and Radiation." *Biological Theory* 17: 80–93. http://link.springer.com/10.1007/s13752-021-00373-y.

Earp, B. D., A. Sandberg, G. Kahane, and J. Savulescu. 2014. "When Is Diminishment a Form of Enhancement? Rethinking the Enhancement Debate in Biomedical Ethics." *Frontiers in Systems Neuroscience* 8 (12): 1–8. https://www.ncbi.nlm.nih.gov/pmc/articles/PMC3912453/.

Earp, B. D. and J. Savulescu. 2020. *Love Is the Drug: The Chemical Future of Our Relationships*. Manchester: Manchester University Press.

Ehrlich, P. R. 1968. *The Population Bomb*. New York: Ballantine Books.

Ellis, L. 1994. "The High and the Mighty among Man and Beast: How Universal Is the Relationship between Height (or Body Size) and Social Status?" In *Social Stratification and Socioeconomic Inequality*. vol. 2, *Reproductive and Interpersonal Aspects of Dominance and Status*, edited by L. Ellis, 94–111. Westport, CT: Praeger.

Ellis, L. 1995. "Dominance and Reproductive Success among Nonhuman Animals: A Cross-Species Comparison." *Ethology and Sociobiology* 16 (4): 257–333. https://linkinghub.elsevier.com/retrieve/pii/016230959500050U.

Foley, R. 2005. "The Adaptive Legacy of Human Evolution: A Search for the Environment of Evolutionary Adaptedness." *Evolutionary Anthropology: Issues, News, and Reviews* 4 (6): 194–203. http://doi.wiley.com/10.1002/evan.1360040603.

Galton, F. 1869. *Hereditary Genius: An Inquiry into Its Laws and Consequences*. New York: Macmillan.

Galton, F. 1909. *Essays in Eugenics*. London: Eugenics Education Society.

Harari, Y. N. 2017. *Homo Deus: A Brief History of Tomorrow*. New York: Vintage.

Harris, J. 2007. *Enhancing Evolution: The Ethical Case for Making Better People*. Princeton, NJ: Princeton University Press.

Henrich, J. 2016. *The Secret of Our Success: How Culture Is Driving Human Evolution, Domesticating Our Species, and Making Us Smarter*. Princeton, NJ: Princeton University Press.

Henrich, J., and R. Boyd. 1998. "The Evolution of Conformist Transmission and the Emergence of Between-Group Differences." *Evolution and Human Behavior* 19 (4): 215–241. http://www.sciencedirect.com/science/article/pii/S109051389800018X.

Henrich, J., and F. J. Gil-White. 2001. "The Evolution of Prestige: Freely Conferred Deference as a Mechanism for Enhancing the Benefits of Cultural Transmission." *Evolution and Human Behavior* 22 (3): 165–196. http://www.sciencedirect.com/science/article/pii/S1090513800000714.

Hern, W. M. 1993. "Has the Human Species Become a Cancer on the Planet? A Theoretical View of Population Growth as a Sign of Pathology." *Current World Leaders* 36 (6): 1089–1124.

Hughes, J. D. 2014. *Environmental Problems of the Greeks and Romans: Ecology in the Ancient Mediterranean*. Baltimore, MD: Johns Hopkins University Press.

Kaebnick, G. E. 2014. *Humans in Nature: The World as We Find It and the World as We Create It*. New York: Oxford University Press.

Kahane, G., and J. Savulescu. 2015. "Normal Human Variation: Refocussing the Enhancement Debate." *Bioethics* 29 (2): 133–143. https://doi.org/10.1111/bioe.12045.

Kass, L. 2003. *Beyond Therapy: Biotechnology and the Pursuit of Human Improvement*. Washington, DC: President's Council on Bioethics.

Kegelaers, J., P. Wylleman, K. De Brandt, N. Van Rossem, and N. Rosier. 2018. "Incentives and Deterrents for Drug-Taking Behaviour in Elite Sports: A Holistic and Developmental Approach." *European Sport Management Quarterly* 18 (1): 112–132. https://www.tandfonline.com/doi/full/10.1080/16184742.2017.1384505.

Kevles, D. J. 1985. *In the Name of Eugenics: Genetics and the Uses of Human Heredity*. Cambridge, MA: Harvard University Press.

Krausmann, F., K.-H. Erb, S. Gingrich, H. Haberl, A. Bondeau, V. Gaube, C. Lauk, C. Plutzar, and T. D. Searchinger. 2013. "Global Human Appropriation of Net Primary Production Doubled in the 20th Century." *Proceedings of the National Academy of Sciences* 110 (25): 10324–10329. http://www.pnas.org/cgi/doi/10.1073/pnas.1211349110.

Marmot, M. 2005. *Status Syndrome: How Your Social Standing Directly Affects Your Health*. London: A&C Black.

Mehlman, M. J., and J. R. Botkin. 1998. *Access to the Genome: The Challenge to Equality*. Washington, DC: Georgetown University Press.

Mikkelson, G. M., A. Gonzalez, and G. D. Peterson. 2007. "Economic Inequality Predicts Biodiversity Loss." Edited by Jerome Chave. *PLoS ONE* 2 (516): e444. https://dx.plos.org/10.1371/journal.pone.0000444.

Milne-Smith, A. 2016. "Shattered Minds: Madmen on the Railways, 1860–80." *Journal of Victorian Culture* 21 (1): 21–39. https://academic.oup.com/jvc/article/21/1/21-39/4095587.

Morreau, M. 2019. "Arrow's Theorem." In *The Stanford Encyclopedia of Philosophy*, edited by E. N. Zalta (Winter). Metaphysics Research Lab, Stanford University. https://plato.stanford.edu/archives/win2019/entries/arrows-theorem/.

Ord, T. 2020. *The Precipice: Existential Risk and the Future of Humanity*. New York: Hachette.

Ordover, N. 2003. *American Eugenics: Race, Queer Anatomy, and the Science of Nationalism*. Minneapolis: University of Minnesota Press.

Parfit, D. 1984. *Reasons and Persons*. Oxford: Clarendon Press.

Persson, I., and J. Savulescu. 2012. *Unfit for the Future: The Need for Moral Enhancement*. Oxford: Oxford University Press.

Persson, I., and J. Savulescu. 2019. "The Evolution of Moral Progress and Biomedical Moral Enhancement." *Bioethics* 33 (7): 814–819. https://onlinelibrary.wiley.com/doi/abs/10.1111/bioe.12592.

Price, M. E., and M. Van Vugt. 2014. "The Evolution of Leader–Follower Reciprocity: The Theory of Service-for-Prestige." *Frontiers in Human Neuroscience* 8: 1–17. https://doi.org/10.3389/fnhum.2014.00363.

Pugh, J., G. Kahane, and J. Savulescu. 2016. "Bioconservatism, Partiality, and the Human-Nature Objection to Enhancement." *The Monist* 99 (4): 406–422. https://academic.oup.com/monist/article/99/4/406/2962101.

Ragan, C. I., I. Bard, I. Singh, and Independent Scientific Committee on Drugs. 2013. "What Should We Do about Student Use of Cognitive Enhancers? An Analysis of Current Evidence." *Neuropharmacology* 64 (January): 588–595.

Ranke, M. B. 2013. "Treatment of Children and Adolescents with Idiopathic Short Stature." *Nature Reviews Endocrinology* 9 (6): 325–334. http://www.nature.com/articles/nrendo.2013.71.

Rawls, J. 1999. *A Theory of Justice*. Revised ed. Cambridge, MA: Belknap Press of Harvard University Press.

Roache, R., and J. Savulescu. 2016. "Enhancing Conservatism." In *The Ethics of Human Enhancement: Understanding the Debate*, edited by S. Clarke, J. Savulescu, T. Coady, A. Giubilini, and S. Sanyal, 719–745. Wellcome Trust–Funded Monographs and Book Chapters. Oxford: Oxford University Press. http://www.ncbi.nlm.nih.gov/books/NBK401923/.

Ross, C. E., and C. Wu. 1995. "The Links between Education and Health." *American Sociological Review* 60 (5): 719–745.

Sandberg, A. 2014. "Cognition Enhancement: Upgrading the Brain." In *Enhancing Human Capacities*, edited by J. Savulescu, R. ter Meulen, and G. Kahane, 69–91. Oxford: Blackwell. http://doi.wiley.com/10.1002/9781444393552.ch5.

Sandel, M. J. 2007. *The Case against Perfection: Ethics in the Age of Genetic Engineering*. Cambridge, MA: Belknap Press of Harvard University Press.

Savulescu, J. 2001. "Procreative Beneficence: Why We Should Select the Best Children." *Bioethics* 15 (5–6): 413–426. http://doi.wiley.com/10.1111/1467-8519.00251.
Savulescu, J. 2005. "New Breeds of Humans: The Moral Obligation to Enhance." *Reproductive Biomedicine Online* 10 (Suppl 1): 36–39.
Savulescu, J., A. Sandberg, and G. Kahane. 2014. "Well-Being and Enhancement." In *Enhancing Human Capacities*, edited by J. Savulescu, R. ter Meulen, and G. Kahane, 1–18. Oxford: Blackwell. http://doi.wiley.com/10.1002/9781444393552.ch1.
Schaefer, G. O., G. Kahane, and J. Savulescu. 2014. "Autonomy and Enhancement." *Neuroethics* 7 (2): 123–136. https://doi.org/10.1007/s12152-013-9189-5.
Schultz, T. R. 2000. "In Search of Ant Ancestors." *Proceedings of the National Academy of Sciences of the United States of America* 97 (26): 14028–14029. https://www.ncbi.nlm.nih.gov/pmc/articles/PMC34089/.
Sparrow, R. 2011. "A Not-So-New EUGENICS: Harris and Savulescu on Human Enhancement." *Hastings Center Report* 41 (1): 32–42.
Sparrow, R. 2019. "Yesterday's Child: How Gene Editing for Enhancement Will Produce Obsolescence—and Why It Matters." *American Journal of Bioethics* 19 (7): 6–15. https://www.tandfonline.com/doi/full/10.1080/15265161.2019.1618943.
Steckel, R. H. 1995. "Stature and the Standard of Living." *Journal of Economic Literature* 33 (4): 1903–1940.
Street, S. 2006. "A Darwinian Dilemma for Realist Theories of Value." *Philosophical Studies* 127 (1): 109–166. http://link.springer.com/10.1007/s11098-005-1726-6.
Thoits, P. A. 2010. "Stress and Health: Major Findings and Policy Implications." *Journal of Health and Social Behavior* 51 (1): S41–S53. https://doi.org/10.1177/0022146510383499.
Tomasello, M. 2016. *A Natural History of Human Morality*. Cambridge, MA: Harvard University Press.
Trewavas, A. 2002. "Malthus Foiled Again and Again." *Nature* 418 (6898): 668–670. https://www.nature.com/articles/nature01013.
van Vugt, M., and J. M. Tybur. 2015. "The Evolutionary Foundations of Status Hierarchy." In *The Handbook of Evolutionary Psychology*, edited by D. M. Buss, 1–22. Hoboken, NJ: John Wiley & Sons. http://doi.wiley.com/10.1002/9781119125563.evpsych232.
Ward, P. S. 2006. "Ants." *Current Biology* 16 (5): R152–R155. https://www.cell.com/current-biology/abstract/S0960-9822(06)01204-8.
Wilkinson, R. G. 2001. *Mind the Gap: Hierarchies, Health and Human Evolution*. New Haven, CT: Yale University Press.
Wrangham, R. 2010. *Catching Fire: How Cooking Made Us Human*. London: Profile Books.

Index

For the benefit of digital users, indexed terms that span two pages (e.g., 52–53) may, on occasion, appear on only one of those pages.

Note: Tables and figures are indicated by *t* and *f* following the page number